EAST

AFRICAN MAMMALS

Jonathan Kingdon

EAST
AFRICAN MAMMALS

An Atlas of Evolution in Africa

Volume I

The University of Chicago Press

The University of Chicago Press, Chicago 60637
Academic Press, London

Library of Congress Cataloging in Publication Data

Kingdon, Jonathan
   East African Mammals
   Reprint. Originally Published: London, New York:
Academic Press 1971–1974.
   Bibliography, P.
   Includes Indexes.
   Contents: V. 1 (No title)—V. 2, Pt.A. Insectivores and Bats—
V. 2, Pt.B. Hares and Rodents.
   1. Mammals—Africa, East—Evolution. I. Title.
QL731.E27K56   1984   599.09676   83-24174
ISBN 0-226-43718-3 (V. I)
      0-226-43719-1 (V. IIA)
      0-226-43720-5 (V. IIB)

Printed in Great Britain by
BAS Printers Limited, Over Wallop, Hampshire

Preface

The aim of this book is to picture a wonderful variety of animals and at the same time to provide a long overdue inventory and atlas of the mammals of East Africa.

Both the variety of mammals and their distribution are manifestations of the evolutionary process and so it is evolution that is the central theme of this book.

It is in search of further information about the process that I have essayed into the behaviour, ecology and anatomy of species. It cannot be said that the inclusion of these topics will bring the volume any nearer to being comprehensive, but they may perhaps serve to increase awareness of the magnitude and magnificence of evolution.

The book is also intended to provide a broad background for the student of East African mammals, with information on local names, breeding, measurements, food and so on. As the animals have economic, medical and veterinary importance to the East African countries and a scientific value for the world at large, I have also included some data on these aspects.

Whether one is interested in their conservation, their exploitation or their control, a practical approach towards mammals in East Africa must be based on biological knowledge and I hope this work may be found useful by all those with an interest in this fauna.

The prime stimulus for the drawings, however, has been the contemplation of physical beauty in mammals; this is a reward in itself. Drawing is the discipline in which I am trained, and it has been a chosen form of note-taking and a useful adjunct in the study of mammals. The making of a drawing is not only a matter of technique for there is a constructive effort to "figure" the animal; looking at drawings can also be an active retracing of this figuring process and it is in this that I hope others will share the pleasure of looking at animals.

East Africa is not a natural geographic or faunal region, so the fauna discussed here really belongs to a very much wider area, and in many ways is broadly representative of the fauna of Africa as a whole.

Sub-saharan Africa is occupied by the fauna of two biotic extremes, moist forests and dry open savannas (see Maps 1 and 2), and East Africa has been an ancient theatre for the excursions of these habitats and their fauna. If maps of the forest and savanna faunal zones of contemporary Africa are superimposed, the result is a broad overlapping area in central and East Africa. At the present time forest mammals are confined in this region to numerous small islands of forest, but there have undoubtedly been several periods when the forest was very much more extensive, and other periods when arid conditions were widespread. Mammal populations have therefore been subjected to isolation and gradual but extensive climatic change. On a large continent with relatively few physical barriers to the movement of animal populations, climatic fluctuations leading to the isolation of populations over millions of years have been an important determinant in the evolution of species.

The "overlap" area in central and East Africa (see the third map, p. vi)

contains many endemic species, and these forms have received particular attention in this work.

The patterns of mammalian evolution seen today have been and are being continuously modified by man. Although greatly accelerated today, human interference is nothing new, and hominid fossils testify to the continuous presence of men and pre-men in this area over millions of years.

Our own emergence and survival as a species was within a rich community of mammals such as is found in East Africa today. The interaction of man and wild mammals dominated human culture for millennia, yet today the close and ancient connections with animals have long ceased to be a part of human culture. It is urgent that we gain some insight into this world that is so much a part of our inheritance and so much older than our civilization which is destroying it.

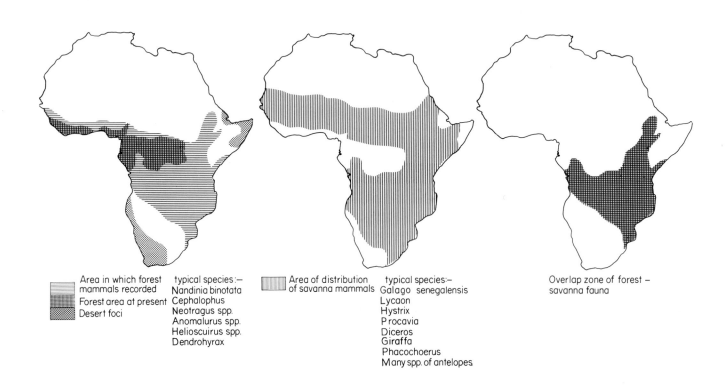

Area in which forest mammals recorded
Forest area at present
Desert foci

typical species:—
Nandinia binotata
Cephalophus
Neotragus spp.
Anomalurus spp.
Helioscuirus spp.
Dendrohyrax

Area of distribution of savanna mammals

typical species:—
Galago senegalensis
Lycaon
Hystrix
Procavia
Diceros
Giraffa
Phacochoerus
Many spp. of antelopes

Overlap zone of forest – savanna fauna

Acknowledgements

This volume has been some years in preparation and owes a great deal to the help offered to me by a very large number of organizations and individuals. I should first record my thanks to those who encouraged me to persevere in the early days of its preparation. Dr Peter Miller, Sir Julian Huxley, Mr Richard Hughes, Dr Julian Shelley and Mr Frederick C. Wood gave practical help and advice at this formative stage.

Since July 1967 The Wellcome Trust has supported field work and the preparation of this volume and also covered the cost of the colour plates. I am most grateful to the Trustees and Secretary of The Wellcome Trust who made it possible to achieve targets that seemed out of reach and relieved some of the anxieties of a project that had outgrown my resources.

For their greatly valued companionship in the field and for much practical help and data I owe a special debt to Robert Glenn, Eric Balson, Tony Archer, John Bindernagel, Angus McCrae, Akira Suzuki, Tim Synnott, Stephen Tompkins, Alan Walker, John and Richard White, Andy Williams and to members of my family.

The Chief Game Wardens of Kenya, Tanzania and Uganda have permitted me to collect and have answered innumerable queries, in particular Mr Ruweza, Chief Game Warden of Uganda and his predecessor Mr Tennant have patiently borne with my importunism and have kindly sent me a large variety of animal casualties to dissect. I would like to pay tribute to all the staff of the East African Game Departments, Wardens, Rangers and Scouts. I have been guided, befriended and helped by more Game Department men than it is possible to list here but I heartily thank them all.

I specially thank Mr Alex MacKay, mammalogist at the National Museum of Kenya, Nairobi for his practical interest, critical advice and for the facilities he has extended to me at the Museum.

The staff of the mammal section of the British Museum, Dr G. B. Corbett, Mr R. W. Hayman, Mr J. E. Hill and Miss J. Ingles have been unfailingly kind and helpful on my all too short visits to the Natural History Museum.

Various parts of the manuscript have been read and criticized by Mr A. MacKay, Dr P. J. Greenway, Professor B. Langlands, Mr D. Vesey-Fitz-Gerald, Dr A. Walker, Dr W. Bishop, Mr R. Hargreaves, Mr A. Hamilton, Dr R. Hannah and Dr T. Rowell. I am most grateful for the time and trouble they took. Also remembered with gratitude are the interest and criticisms of the late Mr R. E. Moreau, Mr C. J. P. Ionides and Dr M. Morrison.

I am most indebted to my colleague Tagelsir Ahmed who designed the format for the text and drawings and has patiently vetted the layout of the book. I thank Academic Press for their most friendly collaboration over this book.

In the collection of data and specimens, in advice, encouragement and all manner of practical assistance the following are gratefully remembered. M. J. Adams, K. Albrooke, P. Aldridge-Blake, J. Allen, P. Allen, J. G. Ament, W. F. H. Ansell, A. Archer, P. Arman.

E. W. Balson, W. Banage, E. Batema, W. Baumgartel, L. Beadle, F. Bell, J. Bindernagel, W. Bishop, N. Bolwig, P. Boston, M. Bowker, M. Brambell, A. Brooks, J. Brown, K. Brown, G. du Boulay.

N. Chalmers, G. S. Child, M. Coe, B. Cooper, G. B. Corbet, M. Croydon. A. Darwish, J. A. Davis, M. Delaney, T. Dormer, G. DeBeer.

D. P. Ebutt, R. D. Estes.

J. D. L. Fleetwood, F. Foltz, A. Forbes-Watson, B. Foster.

P. W. Gilbert, R. Glenn, L. Goodwin, P. J. Greenway.

A. Haddow, A. Hamilton, R. W. Hayman, H. Heim de Balzac, J. E. Hill, D. Hopcraft, R. Hughes, J. Huxley.

C. J. P. Ionides, M. Jellicoe, C. J. Jolly.

H. Kadoma, J. Kagoro, F. X. Katete, D. Kaya.

H. Lamprey, B. Langlands, R. M. Laws, J. N. Layne, L. S. B. Leakey, P. Leyhausen, H. Lock, J. Lock, B. Loka-Arga.

A. McCrae, T. McGinn, A. D. MacKay, H. S. Mahinda, G. Maloba, J. Manina, T. Mann, W. Marks, A. Martin, P. W. Martin, J. Maté, J. Matternes, S. Matti, J. Meester, A. J. Mence, P. Miller, G. Moller, E. M. Morrison, R. E. Moreau, D. Morris, Y. Mukasa, F. Mutere.

R. Ndabagoye, G. Ntenga, T. Nuti, J. Nzabonimpa, R. Neogy.

J. Oakley, J. Obondio Odur, D. Pye.

U. Rahm, L. Rogerson, T. Rowell, A. Ruweza.

J. Sale, G. Schaller, S. Shoen, C. Sekintu, K. Sempangi, J. Shelley, Y. Ssenkebugye, M. Stewart, J. Suckling, Y. Sugiyama, A. Suzuki, T. Synnott, K. L. Swarya.

M. Taylor, L. Tennant, C. Todd, S. Tomkins, C. Turnbull.

H. Van Lawick, J. Van Lawick, D. F. Vesey-FitzGerald.

J. Waite, B. Wakeman, G. Watts, A. Walker, J. Weatherby, W. A. Whimsatt, J. White, R. White, J. Whitehead, W. A. Wilkinson, A. Williams, E. H. Williams, J. G. Williams, F. C. Wood, J. Woodall, R. Woof, H. Ng. Weno.

Finally I would like to dedicate this book to the memory of Bill Moore-Gilbert, my first mentor in the world of mammals and a good friend.

Contents

Chapter I
An Introduction to Method

Of the worlds 900 genera of placental land mammals, nearly 20% occur in East Africa, a figure out of all proportion to East Africa's area compared with that of other lands.

Simpson (1945) classified 26 orders of placental mammals, ten of which have become extinct. Of the 14 orders of land mammals which remain 12 are represented in East Africa and they contain 47 families and over 360 species.

The mammals range from the four-ton elephant to mice, shrews and bats weighing a few grams. They eat practically every conceivable type of food and they are found in water, in the air, in the trees, among rocks and under the ground. For a list of E.A. mammal families see p. 401.

In the introductory chapters that follow, the geography, climate and botany of the region are discussed in terms of the physical habitat of mammals. There is a further chapter which discusses the ancestry and the past status of mammals as suggested by fossils and present distribution patterns, and there is an appendix of anatomy to aid the interpretation of the dissection drawings that illustrate this work.

It will be apparent to the reader that this book relies upon ample illustration; indeed the drawings and maps are a fundamental part of the work and a close correspondence between the visual image and the text has been attempted. It will be found therefore that there is an emphasis on form and pattern and that the interpretation of form is a major preoccupation throughout this work.

Beauty and precision of form are found everywhere and at many different levels of organization; there are striking patterns and colours, extraordinary horns and many types of fur, spines, bristles and scales, and there are impressively architectured bodies like those of the giraffe and the rhino.

The external appearance of a mammal is often dramatically different from its basic structural form. The latter is moulded by physical functions and conditions, but the appearance, achieved by the colouring and marking of the most superficial tips of hair, or sometimes the skin, has frequently evolved in direct response to the faculty of visual perception in other animals. Mammals cannot help being looked at, by their fellows or by other animals, whether neutral, predatory or dependent, and although sight is not the dominant sense in the majority of mammals, the external form of many species is a compromise between being easily seen in certain circumstances and being decidedly cryptic in others. External patterns are sometimes bizarre and often serve to concentrate attention on a particular part of the body— face, genitals or mammary glands. There are often dramatic visual differences between closely related forms that are achieved by very small changes in tone, colouring, length of fur and other superficial alterations.

Because of the pictorial nature of this book, it may be appropriate to discuss the drawings and their function before considering the text and some of the problems involved in compiling a checklist and atlas.

| Rodents 27·5 |
| Bats 27 |
| Artiodactyls 13·5 |
| Carnivores 11 |
| Insectivores 10 |
| Primates 6 |
| other Orders 5 |

Mammal orders in E. Africa expressed as percentages of the total number of species.

The Drawings

The large-scale drawings in this work are intended to display some of the detail of an animal's external appearance, and the direction of hair and the tonal values of the animal's coat are indicated wherever this is possible. However, the limitations of a single representative animal drawn from a single viewpoint should be remembered, for animals vary and they change with growth, with sex and with posture.

Mammals are only static when they are stuffed in a museum so that the sheets of quick sketches are a reminder of their plasticity, complexity and ceaseless activity. Small movements can bring about a revolution in an animal's appearance. Such changes may have evolved as part of the animal's communicative system and must, in the majority of cases, confer some benefit on the animal. Thus visual effects may act as a warning, as in the zorilla, or a confusion to predators, as in the bobbing hindquarters of a gazelle, or they may act as flags or signals, stimulating action between individuals of a species.

The visual effects are often the outcome of independent development in different parts of the body, combining coloured areas of skin, fur or horn in overall patterns. The analysis of these patterns and their correlation with behaviour is one of the most stimulating aspects of modern ethology.

In the sheet sketches those designs, structures, actions and postures that seem most characteristic of each species have been shown. In the text, these characteristics have been related whenever possible to the behaviour towards which they appear to be directed. These patterns have favoured the species' survival, so a visual bias in exploring distinctness of appearance in mammals needs no apology.

The drawings of mammal dissections and the exposure of their skeletal structures are made in order to underline the fundamental form of an animal. Also, the dissected animal may be more readily compared with others, for the gross musculature is fairly easily recognized and the changing position and emphasis of muscles on body frames that are adapted for differing modes of life can be appreciated.

I have adopted the naturalist's and artist's procedure of viewing the whole. This is admittedly superficial in more senses than one, for I have not attempted analytical division and description. Anatomy is a specialist field, but the subject matter of anatomy should not be regarded as closed to the naturalist or layman for want of a disciplined methodology. I know that my own curiosity about the "form" of mammals has been one of the incentives for preparing this work and I believe that this curiosity is widely shared and not easily satisfied.

It is perhaps trite to remark that when we look at the distinctive form of a mammal we are seeing the outcome of millions of years of natural selection, for we are no nearer comprehension of the process with the knowledge. However, we do know that even slight dissimilarities in appearance between species can usually be related to functional differences in the way of life to which the respective species are adapted. In considering these formal dissimilarities, drawing seems to me to be, in its own way, as appropriate an expression of thought as mathematical formulae or tables, even if the only common ground shared by the scientist's mode of thought and the draughtsman's is the effort of the mind to extract meaningful relationships from "forms".

It is possible to view the gross form of an organism as the manifestation of forces and stresses, and this form has been organized to cope with certain environmental conditions. For instance, the form of a dolphin or a dugong has been moulded by certain properties of the environment, water, in which these animals live and the submarine's resemblance to aquatic animals is no accident. Land animals are dominated by gravity, and the body framework is the means by which the animal is supported against gravity. A mammal's limbs and tissues demonstrate principles of efficient construction, using the organic materials of the body in a variety of ways but always with the utmost economy.

As any athlete knows, bone and muscle develop in a growing animal according to use, and although the disposition of bone and muscle is hereditary, muscular atrophy will set in if, for any reason, the part concerned is not used. A similar reduction of unused or useless parts is found in phylogeny although arising for different reasons; the trend towards the loss of hindlegs and pelvis in aquatic mammals is an example, and less extreme cases could be cited.

The materials of the body, particularly bone and muscle, are, like any other material, susceptible to modification by stresses and strains; these stresses derive from an interaction of mechanical forces within the body itself, with influences coming from the external environment.

The physical form of an individual wild animal is the result of an interplay of hereditary and environmental forces; proper development depends on both elements and the two are normally inextricable. But there are instances where the effect of depriving an animal of normal activity can be immediately seen. A wild lion, which frequently seizes struggling prey, tears tough tissues and carries heavy, dead weights in its jaws, has a lean, long skull with taut fully stressed bones. A zoo lion on the other hand, fails to exercise certain muscles which probably results in uneven stress on the skull with consequent alteration of proportions (see Hollister, 1918).

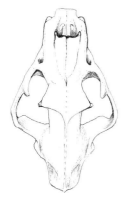 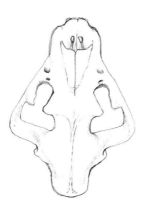 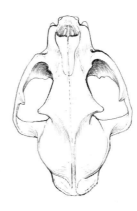

Skull of wild lion—zoo lion—golden cat (*Felis aurata*).

Wild Zoo *Felis aurata*

In nature different proportions are often correlated with size and weight differences. This may occur in related species, within a species, or within an evolutionary sequence of "chronospecies". A phylogenetic increase or decrease in a dimension sets off a series of alterations and does not simply involve shrinking all over or enlarging all over. An example of this type of allometric growth can be illustrated by comparing a wild lion skull with that of the smaller golden cat, *Felis aurata*. In the example of the two lions the influence of the environment on structure is evidenced by comparing two phenotypes with the same genotype; in the second comparison the difference is between genotypes and the differences are not *directly* attributable to the environment. Modifications in structure can be observed, measured and drawn. Furthermore, they can be subjected to comparison, feature by feature in exactly the same way for both examples. This sort of comparison is often avoided for fear of error, but if the distinctions between phenotype and genotype are appreciated, the comparison will serve to show the inextricable link between physical environment and adaptive structure at every level. In other words, changes in an individual animal's form resulting from a direct interaction with the environment have their parallels in phylogenetic adaptations evolved in response to the demands of the environment.

The original observation of such differences in proportion is an initial step in taxonomic practice. The outcome can be a synthesis suggesting answers to at least some of the questions raised when differences are observed between one animal and another, whether the animals are phenotypes or genotypes.

There is, I find, a relationship between the mental processes involved in this sort of biological thinking and those linked with the creative looking which is involved in drawing. It is hardly possible to compare animals without asking questions, and drawing is an exercise in comparisons, comparing the proportions of parts with parts, parts with wholes and comparing one form with another. Drawing also seeks to parallel in lines on paper those subtle and unique combinations of interacting forces that characterize a particular organism.

Convention has assigned to drawing two limited functions. The first is that of depicting an idea, the executor of the illustration and the originator of the idea being usually two individuals. The second function, being expression of emotion, is generally closed to the scientist. The fact that drawing is a language in its own right has been ignored or forgotten. The comparison of forms, as I have already remarked, raises questions, and drawing can be employed as a wordless questioning of form; the pencil seeks to extract from the complex whole some limited coherent pattern that our eyes and minds can grasp. The probing pencil is like the dissecting scalpel, seeking to expose relevant structures that may not be immediately obvious and are certainly hidden from the shadowy world of the camera lens. At the very least I hope the drawings will leave the reader with a feeling of more intimate knowledge of the animals discussed in this work, and the sketches, which are a form of "field note", should be of some help in pinpointing those unique characters that allow a field naturalist to recognize an animal instantly in the field, an ability that distinguishes him from the cataloguer of descriptions. Two-dimensional representations suffer from similar limitations to a distant view;

4

there is more implied in them than the explicit silhouette or lines that meet the eye. Coward, the ornithologist, asked an old fisherman how he could name his birds when they were a great distance away; "by the *jizz* of them" was the answer. It is one of the intentions of this book and particularly of the drawings to cultivate *jizz*, which is both an unconscious summing up of form, stance and movement, often called intuitive "feeling" for an animal and, more plainly, an accurate assessment of that balance of forces which is manifested in "shape". Many of the drawings here are the outcome of an attempt to formulate *jizz*.

The Text

The first and most fundamental task in a work of this sort is to make an inventory of species. The principal reference work in Africa has for many years been G. M. Allen's *Checklist of African Mammals* (1939). Other important references are *East African Mammals in the U.S. Natural History Museum* by Hollister (1918) and a *Checklist of Land Mammals of Tanganyika* by G. H. Swynnerton and R. W. Hayman (1951). More recently the Smithsonian Institution's *Preliminary Identification Manual for African Mammals* has lightened the task.

The nomenclature of the larger groups, orders, families and genera no longer presents a serious problem, and this work will follow the classification of G. G. Simpson (1945). At the specific level there are still numerous problems as to the exact status of mammal populations, but the Smithsonian Manual will be generally followed for the more detailed nomenclature.

Mayr (1958) has shown the fundamental importance of the separation of populations in speciation. He has pointed out that large animal populations vary more on the periphery of their range and that minor barriers, both physical and ecological, may help to reinforce clinal differences. Interruption of geneflow between separate populations allows more important differences to appear.

In East Africa the many levels of difference amongst related forms are particularly interesting. Populations of mammals can be found that are intermediate between two different species; others even bridge subgeneric distinctions; while numerous fragmented populations have been separated into a bewildering number of subspecies.

It must be recognized that our subjective categories of likeness or dissimilarity and our finite terms are ultimately inadequate to describe populations of animals that exist at different levels of distinctness. Naming must therefore, in many instances, be regarded as provisional as it suffers from many limitations of knowledge, and this work will pay special attention to the definition of distinctness in mammal populations and the mapping of their distributions.

The maps follow these conventions:
1. Symbols or dots for museum and sight records.
2. The same, augmented with probable distribution or total range in stipple or texture.
3. Continuous shading of total range, normally used to show pan-African distribution.

4. Past and present distribution distinguished by different textures or symbols.

For the distribution patterns to acquire any meaning they must be correlated with the physical environment, and there are brief outlines of East African geography, climate and vegetation in Chapters II and III, where the role of mountains, lakes, rivers and valleys as barriers or agents of faunal dispersal are discussed.

A principal agent in the isolation and dispersal of mammal populations has been climatic fluctuation in the past; this is discussed in Chapter IV.

In spite of many interesting topics of general zoological interest that are raised, these preliminary chapters are intended simply to serve as a background to the East African region and to assist the reader in interpreting distribution patterns. A general discussion is beyond the scope of these chapters, but specific situations are discussed in the profiles of species.

The profiles have been written as a continuous narrative, for the focus of interest tends to change from species to species. For instance, a fruit bat or common rat offers unrivalled opportunities for the study of population dynamics or breeding cycles, while a primate may be of greater interest for its social behaviour. The contemporary importance of large ungulates may be economic, while certain rodents, primates and carnivores may be of particular importance in the epidemiology of disease in stock or man. The profiles reflect these changing foci of attention.

It has been tempting to use headings, as there are clearly advantages for the reader in being able to refer directly to a paragraph on say "range" or "population density", but with this system there would be a great many headings with no data below them and a gross disproportion in the treatment of certain subjects. Nonetheless, the presentation generally follows the sequence of my preliminary data sheet which is shown in Appendix III (p. 400).

As this work is in three parts there will also be a perceptible change in emphasis in each volume. The present volume, apart from providing general background information, is largely concerned with primates and the peculiar interest that their habits excite. The small mammals of the second volume are the least known and the focus of attention is upon their distribution, taxonomic relationships, ecology and something of their habits and behaviour. The large mammals of the third volume are the best known and the most studied. The great importance of large ungulates as meat is the only long term justification for their continued occupation of many thousands of square kilometres of fertile, well-watered land, while much of our rapidly increasing population lives at subsistence level on inadequate protein. The greatly superior productivity of indigenous ungulates over the exotic cattle, sheep and goats can only be realized by the scientific management of large-scale ecological units, including the national parks, for the maximum yield of meat for the benefit of man. A special effort is made in Volume III to bring together some of the biological information relevant to the economic exploitation of wild ungulates.

I am acquainted with the great majority of mammals treated in this work, I have made drawings in the field and the laboratory and zoos and have had the good fortune to watch numerous species in captivity and in the wild.

It is only fair to sound a warning note about the visual bias which is apparent in a preoccupation with designs, structures and postures. This may well lead to some arbitrary or unnatural emphasis on particular traits, since the drawing of living animals has often been limited by time and by being made in difficult, uncomfortable or unnatural conditions, either for the animal or for the observer.

Many mammal species live in a variety of habitats in which their behaviour is radically modified to suit conditions, and nowhere is this more true than in that special "habitat", confinement, where so many observations have been made. Recognizing how adaptable wild mammals are, field behaviourists are becoming increasingly scrupulous in specifying the temporal and spatial limitations of their study and its subjects. From their patiently accumulated data of field observations, patterns will emerge on which predictions may be based, but there are as yet few species of mammals in East Africa whose behaviour can be said to have been thoroughly studied.

It is not possible, nor is it intended that this work should be definitive, but rather it aims to provide a broad background and an agreeable introduction to the mammals of East Africa, and its inadequacies will, I hope, stimulate others to the further studies that are urgently needed.

Chapter II
East African Environment

The Topography of East Africa

This book concerns itself with the fauna of three countries, Kenya, Tanzania and Uganda. As a political and not a geographical unit East Africa does not represent a natural faunal area; nonetheless much of the area is defined by formidable physical boundaries.

The coast of the Indian Ocean is the eastern boundary. On the west there are four major rift valley lakes, Lake Malawi, Lake Tanganyika, Lake Edward, Lake Albert and also the Ruwenzori Mountains and the Nile. These natural features constitute the greater part of the western boundary, but corridors connect East Africa with the Congo basin and, further south, with the Zambian plateau. This chain of lakes and mountains serves as a political frontier with the Congo Republic, but north of Lake Tanganyika it also marks a major change in vegetation and fauna, separating the lowland forests of the Congo basin from the higher altitudes and more open habitats of East Africa. To the south the Rovuma River is the political boundary but Mozambique is ecologically and geographically inseparable from southern Tanzania and a more natural boundary would run further south.

Northern Uganda is contiguous with the Sudanic Savannas that cover a vast area further west and to the north, but the Didinga and Imatong mountain blocks provide some definition to the northern frontiers of Uganda. Further east, in Kenya, there is a broad semi-desert zone beyond which are the highlands of Ethiopia and arid Somalia. Both these regions have faunal affinities with East Africa, but also display important differences.

As was stressed in the Preface, East Africa has a special zoological interest as the principal theatre of an ancient interaction between forest and savanna and, although the area is not itself strictly a faunal unit, it is representative of a rather more extensive region of "overlap" between forest and savanna faunas and one in which there are a number of endemic mammal species (see margin chart).

East Africa can then be usefully regarded as a limited faunal unit, if the close relationship of Mozambique, Zambia and Angola with the southern parts of Tanzania is borne in mind. For the rest, the fauna is broadly representative of both the savannas and the forests of Africa.

The physical relief of the country has played a part in assisting or limiting the dispersal of mammal species and the role of Plateaus, Rifts, Mountains and Lakes will be discussed in the following pages.

A relief map (see facing page) shows that the region is made up of a coastal plain and a broad plateau, which includes the Lake Victoria depression. Cutting into this plateau there is a double line of rift valleys, in which or along the flanks of which are waters and many isolated mountain masses. The total area is 1,768,450 sq km of which 100,750 sq km are lake waters.

East Africa is predominantly a region of plateaus in which the rock material is largely Precambrian sedimentary rock. The ancient crystalline rocks are usually overlaid by deep soils and plains, which are sometimes up-

SOME EAST AFRICAN ENDEMICS

Species of restricted distribution

Rhynchocyon chrysopygus
Myosorex (Surdisorex)
Galago inustus
Paraxerus lucifer
Paraxerus vexillarius
Paraxerus byatti
Paraxerus flavivittis
Delanymys brooksi
Lophuromys woosnami
Thamnomys venustus
Bdeogale jacksoni
Dendrohyrax validus
Equus grevyi
Cephalophus adersi
Damaliscus hunteri

Species more widely distributed in East Africa "overlap" region

Galago zanzibaricus
Galago crassicaudatus
Paraxerus ochraceus
Tachyoryctes splendens
Lophuromys flavopunctatus
Beamys
Pelomys fallax
Zelotomys hildegardae
Otomys denti
Bdeogale crassicauda
Rhynchogale melleri
Cephalophus spadix
Neotragus moschatus

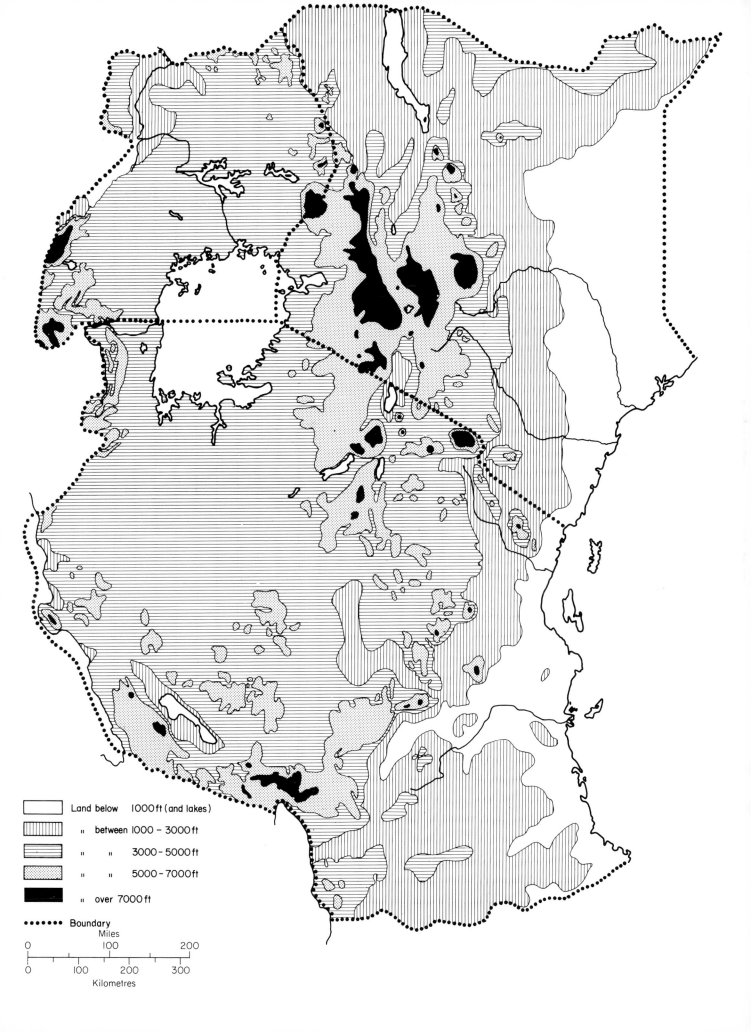

Land below   1000 ft (and lakes)

"   between 1000 – 3000 ft

"   "   "   3000 – 5000 ft

"   "   "   5000 – 7000 ft

"   "   over 7000 ft

•••• Boundary

Miles

0   100   200

0   100   200   300

Kilometres

warped and are generally high (between 900 and 1,500 m). The faulting that has created the great rift valleys may have started very early, but the most dramatic features, the volcanoes and the rift valley escarpments are mostly Tertiary and Quaternary. Some faulting in the eastern rift may have occurred as recently as the Middle and Upper Pleistocene. These rifts and mountains occupy well-defined and limited areas, and their abrupt definition is a sign of their relatively recent origin.

The Lake Victoria depression, which is formed by a downwarping between the eastern and western rifts, occupies a very large area. The lake is 1,130 m above sea level and covers 68,800 sq km and may be of mid-Pleistocene origin, in which case the climatic influence of the lake on vegetation belts would also be relatively recent (although earlier lakes are thought to have existed in the region). The lake causes a wetter climate along its northern shores, where there was a belt of continuous forest in the recent past. The fauna to the north of the lake is quite distinct from that of the country to the south, where a drier regime prevails. To the west of the lake the boundary between these faunas can be regarded as being the Kagera River. To the north and east there is a series of geographic and ecological boundaries with which the distribution limits of various mammal species coincide. These natural barriers include the Victoria Nile and the eastern rift wall, and there are interesting zones where the faunas of the north and south overlap (discussed in some detail in Chapter IV). To the north of Lake Victoria, downwarping has formed the Kyoga, and to the south the Wembere and Malagarasi basins.

The ancient rocks bordering the south margins of the main East African plateau have been tilted and faulted, forming a broken range of hills and mountains, the "African Ghats" of early travellers. These are the Nguru, Rubeho and Uzungwa mountains and they have been most important in the dispersal of fauna for during wet periods they linked the East African coastal forests with the central African forests and, up to the present, this chain of mountains retains relic forests in which a variety of important "indicator species" still survive. The Uzungwa Mountains form an almost unbroken extent of highlands with the Livingstone Mountains, which are a series of fault blocks that mark the steep edge of the Nyasa rift; here the mountains drop abruptly into the waters of Lake Malawi. To the north of Lake Malawi the eastern and western rifts meet and the area is the site of a volcanic range, the Poroto Mountains, and a volcano, Mount Rungwe.

The rifts are important faunal barriers, usually because of ecological differences associated with altitude, but the coincidence of a geographic feature with vegetation differences makes the role of a physical barrier difficult to assess. For instance, the dry-country spring hare, *Pedetes*, is not found beyond the western wall of the Kenya rift, while another dry-country species, the aardwolf, *Proteles*, occurs as far west as Karamoja.

In central Kenya the rift floor around Nakuru is elevated to 2,130 m above sea level, while the Aberdare Mountains on the eastern flank and the Mau on the northern flank are well over 3,000 m; in this region there are faunal differences in the forests of the two mountain blocks. The rift discourages effective mixing between related subspecies and species, for instance, *Cephalophus callipygus* occurs on the Mau and *Cephalophus harveyi* on the Aberdare

Mountains. The size of an established population also seems critical here. For instance, some mixing is apparent wherever different *Cercopithecus mitis* races come into contact along the eastern rift, but neither population has been decisively superior. Racial stability has generally been achieved by each population being sufficiently large to absorb genetic material crossing the geographical barrier of the rift. The geography of this monkeys distribution and the role of the rift are discussed in the profile.

Many volcanic mountains are associated with rifting. Those in eastern Uganda and western Kenya are Miocene, the largest of which is Mount Elgon. In Tanzania and west Uganda there are many volcanoes that were active in the Pleistocene (Oldonyo-Lengai is still active). Volcanic soils and wetter climates are the base for the fertility of many areas, notably the Kenya Highlands, the Kilimanjaro-Meru area, Rungwe and Kigezi. Pockets of forest fauna are found in the forest reserves of these areas. There are often important differences between the forest faunas of the older mountain masses and those of some more recently colonized volcanoes.

Lakes of varying extent and depth have formed in the rift depressions and important lacustrine deposits cover areas where there are now no lakes, indicating former expanses of water, which may have acted as barriers to the dispersal of mammals in the past.

The coastal plain, which narrows to about 18 km opposite Tanga, opens out to the north and south forming two very different hinterland plains in North Kenya and southern Tanzania. Southeastern Tanzania is a moderately humid area, while northeastern Kenya is arid. The faunas of these areas are related to the vegetation and climatic conditions.

A coastal strip, sometimes no more than a few kilometres in breadth is distinguished by a fairly high rainfall. A species of squirrel, *Paraxerus palliatus*, is largely limited to the coastal strip, being found from South Africa to Somalia, and there are other populations of mammals restricted to relic forests on the coast.

The larger islands off the East African coast have interesting faunal differences (see Moreau and Packenham, 1941). Pemba has 19 mammal species and a 7% endemic vertebrate fauna to Zanzibars 39 mammal species and 5% of peculiar vertebrates. Squirrels are inexplicably absent from Pemba. The implication of these differences is that both islands have been physically isolated from the mainland for some time but that Pemba has been isolated longer and probably more thoroughly (an 800 m deep channel surrounds it).

An odd feature of the islands Pemba and Mafia is the presence of *Pteropus* an Oriental bat genus which is quite absent from continental Africa. *P. voeltzkowi* is endemic to Pemba island alone while the Mafia bat *P. comorensis* also occurs on the Comoro islands.

Oriental mammals introduced by man onto the East African coastal islands are the rat *Bandicota bengalensis* on Manda, the Indian civet *Vivericula indica* on Pemba and Zanzibar, the Oriental mongoose *Herpestes javanicus* on Mafia and the shrew *Suncus murinus* on Zanzibar.

Both Zanzibar and Pemba have a high annual rainfall (between 150 and 230 cm) and before the advent of cultivation were largely forested. These islands and other moist coastal localities probably escaped the climatic vicissitudes that much of East Africa suffered in the past and so provided the stable conditions on which forest fauna is dependent.

A number of forest species have survived on these islands, some of which have become differentiated from mainland populations. There are also several mammals which may be survivors of ancient stocks, which through isolation have avoided the competition of later forms. The duiker, *Cephalophus adersi*, occurs only on Zanzibar and in the forests north of Kilifi, the tree hyrax, *Dendrohyrax validus*, is found only on Zanzibar, Pemba and the upper reaches of some Tanzania mountains. Neither of these species occurs on the coast opposite Zanzibar. The arid conditions found just inland of the Kenya coast seem to have isolated the forests there as effectively as the sea around Zanzibar; hence the peculiar distribution of *Cephalophus adersi*. Another very isolated relic species, the mangabey, *Cercocebus galeritus*, is restricted to the forests bordering the Tana River and some localities in the Congo basin. Other peculiarities of these coastal forests are discussed in later chapters. The occurrence of the relic *Dendrohyrax validus* on the oceanic islands and on the "ecological islands" of mountain forest is also discussed in Chapter IV.

The physical features of East Africa cannot fail to impress. However, their influence on fauna is often indirect, as the all-important climatic and ecological environment is superimposed upon topography. Environmental influences have moulded the distribution patterns of mammals within the area and on the continent as a whole, and these are discussed in Chapters III and IV.

Climate

The direct influence of climate on mammals is most obvious in those areas where seasonal contrasts are violent, when water becomes scarce, where bush-fires scorch hundreds of square kilometres and where the whole country appears to be without the green growth that directly or indirectly sustains all mammals. Food habits change and animals move to new pastures, herds assemble or disperse and in many species breeding rhythms are clearly accommodated to the discipline of the seasons. Sometimes epidemics are a reminder that the frequency of disease and the general vulnerability of species alter with the seasons.

Climate determines vegetation, but as the vegetation types are dealt with in a separate chapter it is only necessary here to observe that natural vegetation has a correlation with the duration of moist or arid periods that can be expected in the course of a year (see pages 9 and 10).

The most regular climatic contrast is that between day and night, which at equatorial positions gives an even length to each condition and a rapid transfer from one to the other. Further away from the equator the length of day varies seasonally. Particularly in the more arid parts of East Africa the contrast in temperature between a hot day created by the cloudlessness of the skies and the rapid loss of heat at night create a daily alteration of climatic conditions to which animals respond.

Rainfall is the single most important element in annual climatic cycles. On the equator, vertical solar rays tend to be associated with rain; the over-

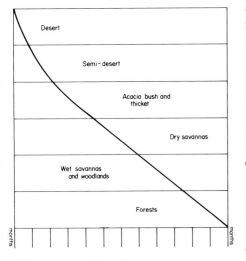

Diagram correlating the duration of humidity (in months) with vegetation zones. (NOTE: For E.A. a monthly rainfall exceeding 5 cm is indicative of a humid month.)

12

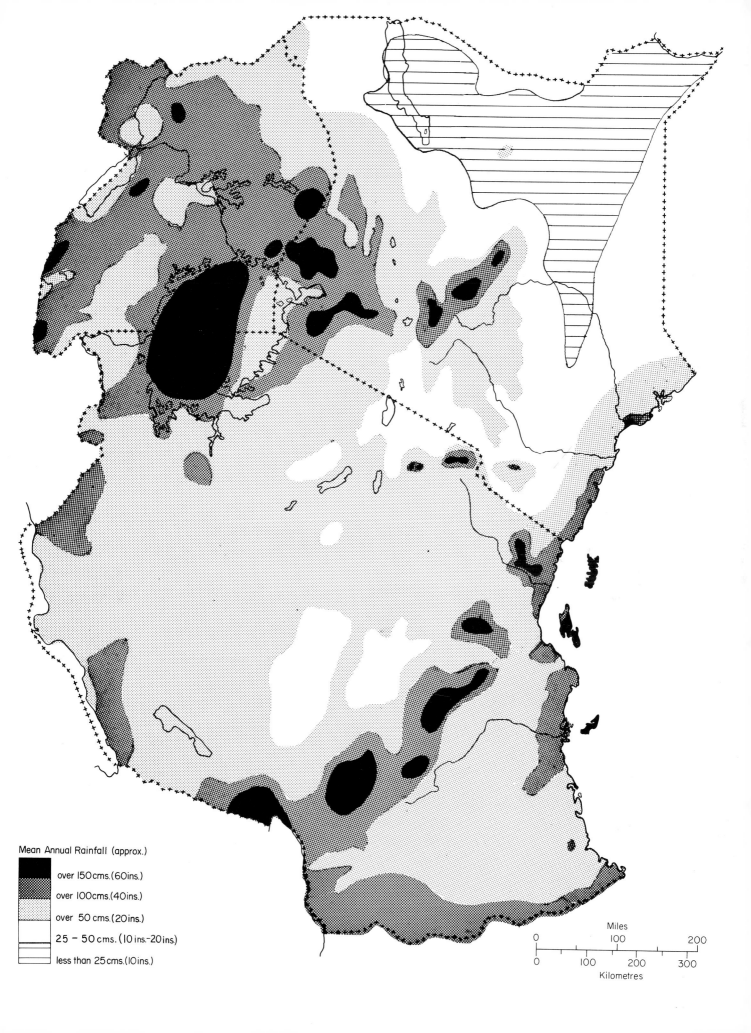

Mean Annual Rainfall (approx.)

over 150 cms. (60 ins.)

over 100 cms. (40 ins.)

over 50 cms. (20 ins.)

25 – 50 cms. (10 ins.–20 ins.)

less than 25 cms. (10 ins.)

Miles

0   100   200

0   100   200   300

Kilometres

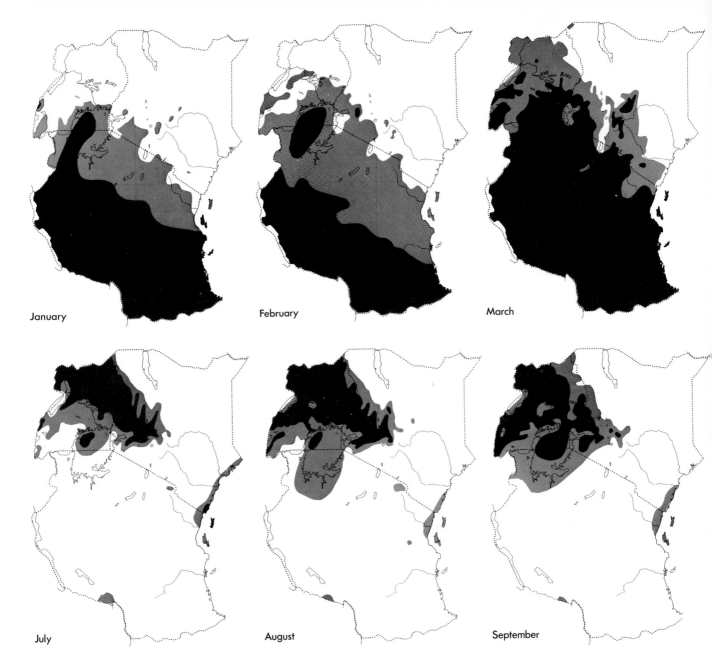

January

February

March

July

August

September

Mean monthly rainfall showing approx. 5 cm approx. 10 cm isohyets.

head sun creates a zone of low pressure which brings about a convergence of air streams which causes uplift, cloud and rain. This zone of low pressure is known as the Equatorial Trough. In East Africa there is generally a time lapse of about a month between the movement of the sun over the equator in September and March and the occurrence of maximum rain in April and October. East Africa lies between 5° N and 12° S and this pattern of two rain seasons is only applicable to the equatorial latitude. As one goes south the two rain seasons tend to merge, so that the dry season of the southern solstice in December tends to become shorter, while the dry season of the northern solstice becomes much longer, as the sun is at a low angle in the south from May to August. The climatic pattern in southern Tanzania is therefore very different from that in Karamoja and northern Kenya.

The general climatic pattern is modified by altitude and temperatures decrease (with increasing altitude) on the East African plateaus and highlands.

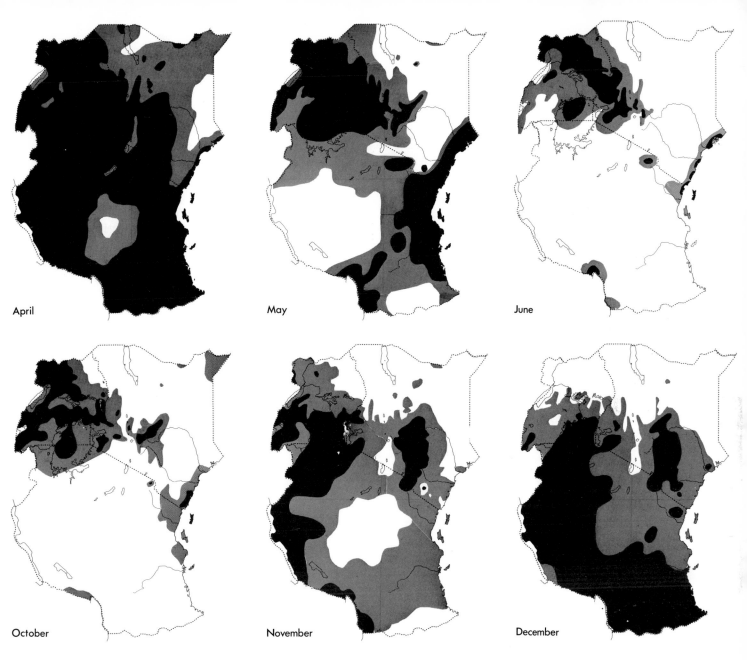

April

May

June

October

November

December

East Africa is therefore in general cooler than the low-lying Congo basin and West Africa. The region is also drier than equatorial West Africa, which benefits from moist Atlantic winds. The drier climate of East Africa is a meteorological anomaly not fully understood. Rain is an atmospheric feature associated with pressure and wind movement, and the trade winds may be partly responsible for the relative aridity; being stable winds they only produce rain with organized disturbances at the equatorial trough. The high plateau and the alignment of the coast are also thought to inhibit rainfall. The dominant southeast trade winds blow in from the Indian Ocean from April to October while the northeast monsoon blows from November to March. The accompanying maps show the mean monthly rainfall for East Africa and it will be seen that the only universally wet month is April, after which there is a fairly general drying out, so that September is dry in most areas. In November there is generally heavy rain everywhere except over central Tanzania and Turkana. In December, Uganda and North Kenya are dry but the equatorial low pressure trough moves south, bringing rain to the whole of Tanzania, where it is wet until May.

The mean monthly averages do not, however, give any idea of the uncertainty of timing or the variability in the amount of rain falling from year to year. East African weather is notoriously unpredictable and the maps are offered as a guide to average conditions which may not agree very closely with the actual climate experienced in any one year; nonetheless these average conditions probably determine vegetational complexes and many aspects of mammalian distribution.

Turning to the more direct effects of climate on animal life, the day-night fluctuations in temperature mentioned earlier are of great importance for many mammals. Many of the smaller species that are active at night die rapidly if exposed to the sun's radiation during the day. During the dry season many large mammals such as the elephant, the zebra and the buffalo drink regularly, while others like the eland, *Taurotragus*, the reedbuck, *Redunca*, the topi, *Damaliscus*, must seek shade for the hottest part of the day. Many fossorial mammals and hyrax seek the cool protection of burrows and rock shelters. On the other hand, gazelles and oryx are adapted to withstand considerable heat without needing shade or drink, and consequently can remain on the open plains in the dry season when other animals are more restricted.

The competition between mammals for the available food and moisture during the dry season ensures that these differing responses have a considerable influence upon both their local and overall distribution. The competition for a single resource may be avoided by a nocturnal-diurnal specialization. For instance, on the Ruwenzoris, where night temperatures may drop below freezing in the upper reaches of the montane forest and daytime radiation may be very severe, a large number of a mouse, *Hylomyscus denniae*, are caught at night in traps which, with identical bait and in identical sites, catch another species, the harsh-furred mouse, *Lophuromys flavopunctatus*, in the day. The activity periods of these species can be predicted to within a minute or two of nightfall and sunrise. Some mammals are only active in the nocturnal "climate", and as such have specialized eyes, i.e. the bushbaby, *Galago*, and the spring hare, *Pedetes*.

A large number of mammal species have annual breeding seasons and often bear their young at the end of the dry season or at the beginning of the rains. However, the seasonal rhythms vary from species to species; furthermore within a species it is possible for patterns to vary from population to population and for quite different reasons. Environmental, climatic, nutritional and social factors determining breeding patterns are often interdependent and may be difficult to separate. Breeding patterns cannot be inferred for a population on the basis of data from another latitude or climatic belt, nor can patterns seen in captive groups be extrapolated for wild populations, even when the captive groups are kept in the same locality and under "ideal" conditions. The breeding patterns subsequently discussed in this work are usually suggested on the basis of scattered and rather inadequate data and should in many instances be regarded as tentative suggestions, for long-term quantitative data have only been collected for a few species.

In temperate regions one stimulus triggering sexual activity in mammals is increasing daylength; the seasonally breeding hare, *Lepus europaeus*, is an example. In East Africa, a closely related species, *Lepus capensis*, appears to breed continuously throughout the year, but the rain season does have an effect for the litters increase from one young to two, while the parents grow fat (Flux, 1969).

The problem of seasonal breeding is highlighted in species bearing young twice a year. The bushbaby, *Galago senegalensis*, seems to have a biannual birth season in Uganda where there is an equatorial pattern of two wet seasons. In southern Tanzania, with a single long rain season, births occur only once a year towards the end of the dry season. A greater preponderance of twins in the annual breeders and single births in biannual breeders has been observed (see Butler, 1967). The mating peaks for the biannual breeders coincide with periods when the animals have been well-fed for some months. The mating peak for the annual breeders coincides with a nutritional optimum following the long rains. In both instances the young are weaned during a period of optimum feeding conditions.

Whether a species is nocturnal or diurnal in habit, it seems unlikely that solar radiation has any direct influence on breeding patterns. Rather does natural selection find an accommodation between the nutritional condition of the breeding pairs and, perhaps more important, the needs of the young for suitable food at critical periods of their growth. Likewise, in some species the vulnerability of young animals may exert a selective pressure in favour of the young being born while there is adequate natural cover from enemies. The dik-dik's biannual breeding in central Tanzania (Kellas, 1955) may be influenced by this factor and according to Estes (1966) the reproductive cycle of the gnu, *Connochaetes*, is largely shaped by predation.

Thompson's gazelle, *Gazella thompsoni*, is another biannual breeder, on Serengeti most young tend to be born at the beginning of the dry season in June and another birth peak occurs in the short dry spell between the rains in January and February. Calving peaks show significant local variation and further north in Kenya most fawns are born in the rains during April and November (Percival, 1928). The variations show no correlation with gross rainfall patterns, but Brooks (1961) found a local correlation between births and fresh grazing. This species is dependent on short grass and dry ground which must be found by seasonal migration, and the need for movement must subject the animal to pressures that do not operate on species that are more local in their habits. Whatever the ultimate factors in this breeding pattern, its rhythms presumably depend upon the females and young, as the males are continuously sexed. By contrast, male topi, *Damaliscus korrigum*, have rutting peaks which must largely determine the two birth peaks found at Serengeti, where the largest number of calves are born in August—September and rather fewer in January—February. The same species has single calving peaks in the Rukwa Valley (September) and in West Uganda (January—February). In all cases young topi are born when it is relatively dry and the grass is short. They grow very fast so that optimum feeding conditions for the growing calves may be an important factor in determining the timing.

There are also variations in populations living under exactly the same climatic conditions. For instance, Laws and Parker (1968) found differences in the breeding pattern of two elephant populations separated by the width of the River Nile in the Murchison Falls National Park. The southern population, which is very dense and virtually confined by the park boundaries, has a seasonal peak in conceptions which is five months later (i.e. in the second half of the rains) than the population north of the river, where the conception peak is associated with the first half of the rains which is considered the norm for the region. Higher densities and possibly nutritional deficiencies in the southern population are thought to be responsible for the discrepancy.

Other evidence for local differences has been presented by Sale (1969) for colonies of the hyrax, *Procavia abyssinica*, living ten km apart on the floor and the walls of the rift valley. He correlates the contrasts with local differences in rainfall. Differences in diet between related species may also be critical, *Heterohyrax brucei* living in the same latitude near Nairobi and under a similar climatic regime to that of *Procavia* breeds before the rains instead of after as *Procavia* does. Sale suggests that the different feeding habits of these hyraxes influence their breeding.

For many small mammals the dry season is a harsh period and there is minimal breeding. In a detailed examination of the breeding seasons of seven rodent species from one small locality in the savanna of West Uganda, Delaney and Neal (1969) confirmed that this was the only feature common to all the species studied. While all species showed peaks and some had definite seasons, the breeding was not synchronized and followed a somewhat different pattern in each case. Two forest rodents, *Praomys morio* and *Lophuromys flavopunctatus*, were found to have breeding peaks at the end of the rains (Delaney, 1964).

Interesting contrasts can be found in bats. The molossid bat, *Tadarida pumila*, breeds continuously throughout the year at Jinja (Marshall and Corbett, 1959). However, the larger *Tadarida condylura*, which often lives together with the former species, breeds twice a year in the same area with births in February and July (Mutere, 1969). Another common little bat, *Hipposideros caffer*, which is also found in the same area, has a single birth season in February and March.

In Gaboon, also on the equator, different colonies of this species have been found to have incompatible breeding rhythms in the same area, some colonies giving birth in March, others in October. Brosset (1968) suggests that this may be due to the colonies having different geographic origins as *H. caffer* from north of the equator tend to follow a "boreal" cycle with the young born in March, while southern populations follow an "austral" cycle in which the young are born in October. All Uganda populations of *H. caffer* recorded to date seem to follow the "boreal" cycle. It would be interesting to know if colonies with an "austral" cycle also occur in the East African tropics.

The fruit bat, *Roussettus aegyptiacus*, breeds twice a year in Uganda, the young being born in March and September, both periods immediately preceding a rainfall peak (Mutere, 1968). A slightly larger species of fruit bat, *Eidolon helvum*, has a single birth season in February—March in Uganda. Mutere (1965) investigating the biology of this species found a three months delayed implantation. The consequent delay in the timing of the birth season allows the young to be weaned and grow to adult size during the height of the rains (and presumably during optimum feeding conditions). This period

(April to June) coincides with the adult mating period and is followed by a splitting up and dispersal of bats from the huge communal roost. Implantation coincides with a rainfall peak in October—November, but it is perhaps more significant that it coincides with peak numbers of bats returning to the main roost. Furthermore, it is by no means certain that delayed implantation is the rule for all populations of this species which is very widely distributed in Africa, so that the device could even be a peculiarly local adaptation.

The work of Laws and Parker, of Sale and of Mutere show particularly clearly the local and specific nature of an animal's adaptation to tropical seasons and conditions. The various and contrasting patterns instanced above show that there is little room for generalization.

Strictly seasonal breeding patterns or the numbers of young may also be upset by irregular rainfall or drought and, in view of the variability of the East African climate, it is perhaps not surprising that really well-defined seasons such as those experienced in temperate regions are rare or unstable in East Africa.

In the semi-desert of northern Kenya where rainfall is very unreliable, a mole rat, *Heterocephalus glaber*, lives in colonies which greatly extend their burrow system during the rains in order to harvest food collectively and store it at the social nest sites. The colony's fortunes can therefore be expected to follow fluctuations of the climate, as it is probable that no breeding takes place in very dry years (Jarvis, 1969).

Seasonal changes alter the choice of foods in all habitats and lead some ungulates to migrate relatively long distances. The gnu, *Connochaetes*, may travel as much as 18 km in a day and at least 620 km in a year. These movements are largely determined by local rains bringing on fresh flushes of grass. At the end of the dry season this encourages great concentrations of animals at a time which coincides with the calving season. The concentration of *Connochaetes* when there are many calves about may lead to their young being trampled, and Talbot and Talbot (1963) have shown how the relative success or failure of the rains becomes an important factor in regulating population in this species. A generally wet year leads to more dispersal and less movement, with negligible mortality from trampling. A dry year causes great concentrations and much movement, so that many calves are lost through trampling and leg breaking.

Very few mammals do not change their feeding habits with the seasons, and it will not be necessary here to give more than a few well-documented and interesting examples. The arrival of the rains in the drier areas of East Africa brings a reawakening of life itself, dusty, grey bushes burst into leaf and insects and animals that had remained hidden during the drought suddenly reappear. The choice of fodder for ungulates is greatly increased and for many ungulate species the rains are a time for getting fat. An interesting contrast in seasonal foods was evidenced by kudu, *Strepsiceros strepsiceros*, at Shinyanga (H. Harrison, 1936). Nearly half the rainy season diet consisted of succulents, *Markhamia acuminata*, *Hymenodictyon parvifolium* and *Abrus schimperi* were other favourite plants. In the dry season, *Combretum* spp., *Cassia* spp., *Thylachium africanum* and *Cadaba adenotricha* accounted for over half the diet, while succulents were not eaten at all; *Markhamia acumi-*

nata was eaten in both seasons but in general there was a completely different regimen.

It has been frequently observed that outbreaks of man-eating in lions are seasonal. In the southern region of Tanzania there have been many such outbreaks, Ionides (1965) found that lions along the coastal belt fed on bushpigs, warthogs and such other small game as survived the expansion of human population. During the rains, the dispersal of these animals made them difficult to find and the long grass hampered a quiet approach. To the hungry lions, particularly young animals just learning to fend for themselves, the vulnerable and numerous humans became a tempting source of meat.

Seasonal changes in diet are most obvious in arid habitats, but even in the forest there are periodic swarms of caterpillars and many other insects or fruiting seasons that alter the choice of food for forest species (see section on forest).

The weather is also an important factor in disease. Rinderpest, for example, tends to be associated with the first rains after a long dry season. The incidence of some diseases increases with the wetness of the habitat, and the lowered resistance of animals after a long dry season may assist outbreaks of disease at the beginning of the rains. Wilting by certain grass species also causes poisoning at this time as toxins may be generated by the wilt.

Several species appear to be less active during the dry season; this may or may not deserve the term aestivation. The aardvark, *Orycteropus*, the hedgehog, *Erinaceus*, the fat mouse, *Steatomys*, and the dormouse, *Graphiurus*, may all store up fat for the lean periods of the dry season and the case of *Erinaceus* is particularly interesting, as hedgehogs hibernate in temperate countries where their reduced activity is induced by lower temperatures. Hedgehogs from North Africa taken to Berlin hibernated when the temperature fell to 19°—23° C (Herter, 1963). However, factors other than temperature must operate if East African hedgehogs do in fact go through a phase of reduced activity.

Catastrophic droughts or floods affect mammals in rather obvious ways, the East African drought of 1949 killed over a thousand hippo in Lake Rukwa alone. Fires take an annual toll of many species and are a major limitation on surface-living rodents in the savanna.

An adaptation to climatic conditions, that can be demonstrated in many wide-ranging groups of mammals, is the presence of darker coats in populations from humid areas; and lighter ones in populations from arid habitats. The darkening of colouring with high humidity is a trait that can be discerned in local populations living on mountains and in very humid forests. The blue monkey, *Cercopithecus mitis stuhlmanni*, today lives in several dry forests in Karamoja where their colouring is not materially different from that of blue monkeys in the moist forests of western Uganda, although populations in the very humid eastern Congo are certainly the darkest. *C. m. stuhlmanni* probably evolved under very humid conditions, but the dark colouring seems to be a feature that is not immediately responsive to changed ecological conditions as it is in many other species. This may be partly due to the recent expansion of *Cercopithecus mitis stuhlmanni* or to a lack of selective pressure, but it is indicative of the way in which characters evolved in relation to a climatic condition may become "fixed" and so be found in a different situation to that in which they evolved.

Many differences between populations can be attributed to the discipline of climate and habitat. These differences have become to variable degrees inherited and will be discussed in the profiles of species, as are other facets of mammals and their relationship with their physical environment.

Chapter III
Vegetation

In describing the variety of environments or habitats that mammals occupy there is a difficult choice to make of those features or plants which seem to influence mammalian life most. Usually one can do little more than point out landmarks, for the truth is that the interconnections in biological systems are a web that has not begun to be untangled particularly in this tropical climax of evolutionary activity. In any case, the scale required to map or describe the individual mammal's environment is quite beyond the scope of this book, and it should be borne in mind that both the map and the description of vegetation types can only outline some of the more obvious features of major communities. For more detailed accounts of vegetation the bibliography should be consulted.

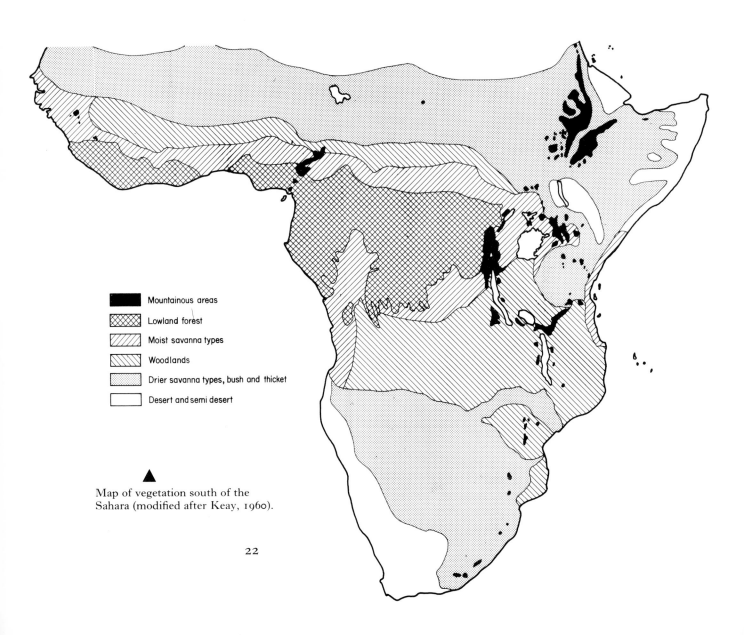

Mountainous areas
Lowland forest
Moist savanna types
Woodlands
Drier savanna types, bush and thicket
Desert and semi desert

Map of vegetation south of the Sahara (modified after Keay, 1960).

22

In describing East African vegetation I have consulted P. J. Greenway's *Report on Vegetation Classification* and C. G. Trapnell and I. Langdale Brown's *The Natural Vegetation of East Africa* and have adapted the map illustrating the latter. I have followed the latter authors in using the term "savanna" which has found general acceptance instead of "wooded grassland" (Greenway, 1943) which is a more exact designation.

The general pattern for the African continent is an equatorial belt of forest with belts to the north and south of increasingly arid type vegetation. Between the West African forests and the Sahara these belts form a simple strip-like pattern, but the eastern part of the continent is marked by high relief and a more complicated vegetation pattern is apparent. The *Vegetation of Africa* map published by L'Association pour l'Etude de la Flore d'Afrique Tropicale (A E T F A T) (scale 1:10,000,000) has explanatory notes on pan-African plant communities (Keay, 1957). This publication is very useful for a continental perspective.

In East Africa the graduated series of vegetation zones have a complicated pattern with numerous opportunities for ecological isolation. The evolutionary significance of this pattern for mammals is discussed in a later chapter.

Superimposed upon the latitudinal zonation of tropical Africa there is an altitudinal zonation on the East African mountains with which the description of vegetation types may conveniently start.

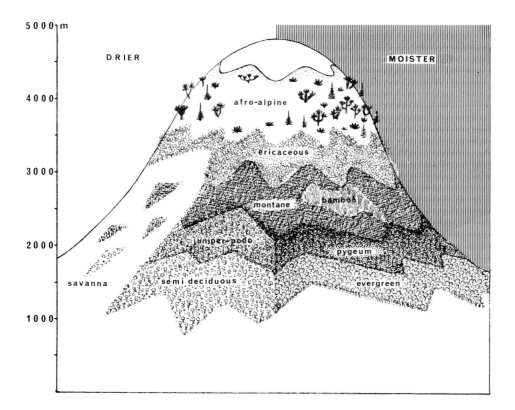

Montane and forest vegetation in relation to altitude and moisture (modified from Langdale Brown *et al.*, 1964).

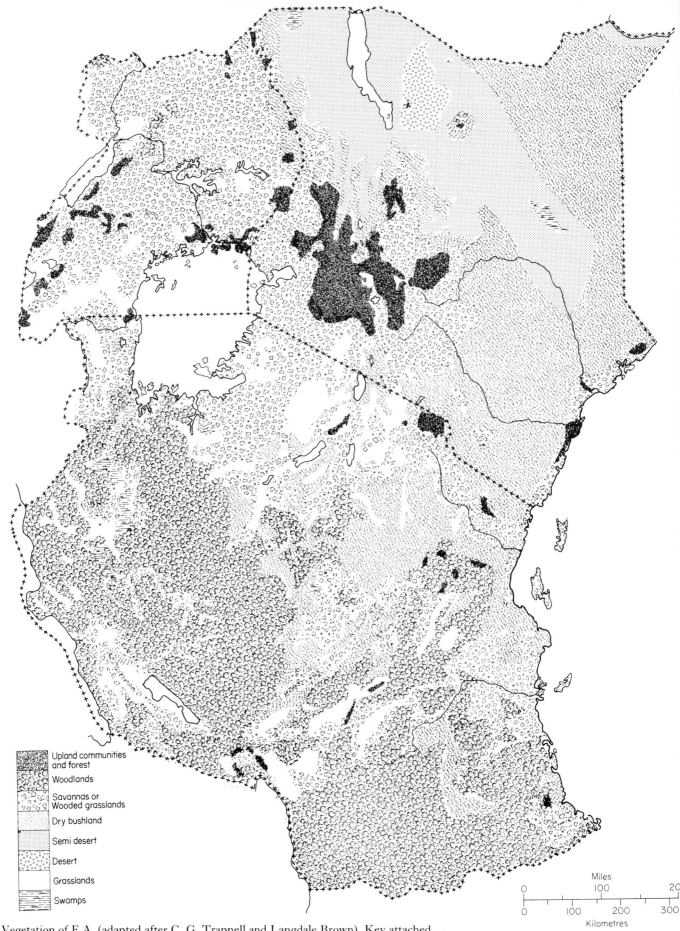

Upland communities and forest

Woodlands

Savannas or Wooded grasslands

Dry bushland

Semi desert

Desert

Grasslands

Swamps

Miles
0   100   20
0   100   200   300
Kilometres

Vegetation of E.A. (adapted after C. G. Trapnell and Langdale Brown). Key attached.

Paramos or Afro-Alpine

The mountain snows occasionally yield the frozen corpse of some wandering leopard, monkey or elephant, but no species has become adapted to this habitat. Mammalian life begins among the mosses and alpine grass and among the giant lobelias and groundsel. The afro-alpine belt grows at altitudes of more than 4,000 m under very peculiar climatic conditions of great cold at night and intense insolation during the day. Hedberg (1964) has divided the afro-alpine belt into five principal plant communities: *Dendrosenecio* "woodland", *Helichrysum* scrub, *Alchemilla* scrub, tussock grassland and *Carex* bogs.

4m

Alpine zone, Lobelia,Carex,Senecio.

On the steep slopes of Ruwenzori there are dense and extensive "woodlands" of *Senecio adnivalis* and *Lobelia wollastoni*. On the other mountains there are different species, generally growing more sparsely or in small groves. The leaves of these giant plants are eaten by hyrax.

Helichrysum and *Alchemilla* scrub is very extensive on all the moister mountains but often gives way to tussock grassland where there is burning. Tussock grassland is therefore characteristic of the drier mountains and of areas repeatedly fired by pastoralists and hunters. *Carex* bogs are common in waterlogged valleys and depressions.

The afro-alpine habitats occupy a relatively narrow belt on a number of isolated mountains and their isolation has some interesting reflections in the mammalian fauna. Many mountains are of relatively recent origin and the species colonizing their higher reaches have been drawn from different lowland stocks. Rockdwellers, grazers large and small, grass-seed eaters and subterranean insectivores have opportunities as long as they can cope with the cold. Cold selects for certain characters; according to Allen's rule low temperatures favour shorter limbs, ears and tail. Bergman observed that body sizes are larger in colder environments, hair grows longer and thicker. In East Africa the relative youth of some of the mountains has meant that colonizing mammals have tended to be from stocks that are highly versatile or that already possess characteristics that adapt harmoniously.

The alpine rocks are colonized on all major East African mountains by hyrax, of four species and three genera. On Mt Kenya, the alpine *Procavia* is larger, heavier and thicker-coated than the lowland race, as is the Ruwenzori *Dendrohyrax*.

25

Earthworms and other cryptic fauna flourish on Ruwenzori and this food supply is exploited by the golden mole, *Chlorotalpa stuhlmanni*. Primitive shrews, *Myosorex*, are found on some East African mountains and the sub-genus *Surdisorex* has acquired a decidedly mole-like character on Mt Kenya and the Aberdares (where *Chlorotalpa* is absent).

The alpine meadows are visited by several species of antelopes where they have access (as on Kilimanjaro) but the typical antelope of this belt is the duiker, *Cephalophus nigrifrons*, on Ruwenzori, Bufumbira, Mt Kenya and Mt Elgon. On the latter mountain, *Sylvicapra grimmia lobeliarum*, has also colonized moorland; both these duikers have longer, thicker coats than low-land forms. Buffalo are common grazers of the tussock grass.

The principal alpine rodents are *Otomys* and *Dasymys* which live under the mosses and grass, *Dendromus mesomelas* is a seed-eater also common in alpine swamps and grasses.

Ericaceous

The ericaceous belt is dominated by tree heaths, *Erica arborea* and *Philippia*. On Ruwenzori, which is a very wet mountain, these form an extensive forest belt between 3,000 and 3,500 m with tussocky bogs of *Carex runssoroensis* in the valleys. On the other drier East African mountains the *Erica* is more patchy. Heaths burn easily and on the mountains of the north *Erica* has been replaced by grassland. *Stoebe* is an important tree on Elgon

Subalpine or Ericaceous zone, *Erica aborea*.

and is increasing due to its greater recovery after fire. The fauna of this zone is generally identical to that of the afro-alpine, although species characteristic of the higher reaches of the montane forest such as *Lophuromys* species and *Hylomyscus denniae* are also common on Ruwenzori. Leopard, *Felis pardus*, and golden cat, *Felis auratus*, occur where hyrax and duiker are sufficiently numerous.

Bamboo

Bamboo, *Arundinaria*, grows on most of the wetter East African mountains in a zone between 2,500 and 3,000 m sometimes mixed with *Hagenia*, *Dombeya* and other trees (*Hagenia* can form a pure stand and is a favourite haunt of the gorilla on the Virunga volcanoes). Other locally important members of this zone may be *Rapanea* and tree *Hypericum*. Associated with bamboo are *Mimulopsis* amd *Fleurya* species; both *Mimulopsis* and bamboo flower

26

Bamboo, Arundinaria.

and die gregariously at irregular intervals. Simon (1962) reported that *Mimulopsis* is toxic at some stage of its growth and that numbers of forest hog, *Hylochoerus*, and bongo, *Boocerus*, die feeding on it at this time—a curious case of "population control". *Thamnomys rutilans* is characteristic of the bamboo forests of West Uganda, and in Kigezi sedgy marshes associated with bamboo form the only known habitat of *Delanymys brooksi*.

Montane Forest

Montane forest differs from mountain to mountain existing within a rainfall range of 635—2,000 mm and between 1,700—2,700 m above sea level. The wetter montane forests may have many lowland forest elements but lack the buttressed species, they may be dominated by *Pygeum, Podocarpus, Aningeria, Ocotea, Dombeya, Croton, Macaranga* or *Olea*. The lower margins of these montane forests end in an abrupt line, usually a forest department boundary at about 1,850 or 2,200 m against which grassland and cultivation press. North Kigezi is one of the very few areas where mountain forest merges with lowland forest in an unbroken succession; here *Chrysophyllum* and *Parinari* form large stands between 1,500 and 2,200 m. The area is of great scientific interest and is also the last permanent refuge of the gorilla in East Africa. Other interesting areas exhibiting a montane-lowland forest succession are Kalinzu-Maramagambo and the Usambara Mountains.

The drier montane forests suffer from frequent fires and in most of the northern mountains forests are confined to valley relics. Stands of juniper,

Montane Forest, Podocarpus, Cyathea, Ocotea, Aningeria.

Juniperus, occur on the drier sites and dense forests of *Podocarpus* grow in more favoured localities; exposed trees are generally covered with species of *Usnea* lichen. Stunted forests dominated by *Olea, Cassipourea* and *Euclea* are found on the northern Kenya massifs where the unique crested rat, *Lophiomys ibeanus*, is found. These drier forests grade into thicket and scrub. Linked in

27

the past with western forests, their subsequent isolation and changes in floral composition have exerted pressure on isolated animal populations to adapt to drier conditions. The long-term evolutionary significance of this type of forest may therefore be greater than its present very limited distribution suggests.

Lowland Rain Forest

In spite of its relatively small area in East Africa this is quite the richest habitat in numbers of mammal species. One of the reasons for this is the great age of this life form; it is thought that the ecology of the early Angiosperm communities resembled the present rain forest, and the survival of the chevrotain, *Haemoschus aquaticus*, is some indication of the stability of conditions in the forest. In West Kenya, fossil *Dorcatherium* closely resembling the chevro-

Diagram of forest profile, showing changes in ground cover.

tain have been found in Miocene sites, in the same sites are numerous *Celtis* fruits which are indistinguishable from present day species.

There are over 400 species of trees in East African forests and an extraordinary degree of diversity is found in insects, birds, mammals and other forms of life.

In addition to the opportunities offered by the floor, the trees of the forest create a three-dimensional habitat for primates, rodents, hyrax, bats, viverrids and pangolins. High, relatively constant temperatures (min. 19°—max. 29°C), an evenly spread rainfall and an abundance of foods and niches reduce competition, so that at ground level the forest supports many species of small mammals, but the density of individuals tends to be markedly lower than in ecologically less diverse habitats. For instance, many more individual rodents but fewer species are caught in montane habitats than in mixed lowland forest. Predation of small mammals particularly by snakes is very considerable in lowland forest and it is probable that these forest predators may have specialized to some degree in their favourite prey species. The forest branches by contrast with the floor, are a uniform and exacting habitat. Monkeys in particular have colonized the tree tops very successfully and the densities of individuals may be remarkably high.

The forest has a number of layers or stories, these are often difficult to distinguish in practice and seldom make the neat picture presented in books, especially in East Africa where almost all forests have been exploited or disturbed. The lower strata have characteristic tree species as well as young individuals of the tall emergents and these trees may provide the preferred food of monkeys at different times of the year, so that the allocation of species

28

to stories may be misleading. The following simplified table indicates strata by strata some typical tree species of mixed lowland forest, correlated with fruiting times and the fruit preferences of monkey species giving some idea of the levels at which a species might be found at different times of the year. While still immature many of the top-storey trees are deciduous, conse-

60 m

▲

Diagram of forest storeys.

Fruiting periods of some forest species in Bunyoro, Uganda, forests (from Karani (1968), Suzuki (personal communication) and personal observation).
NOTE: The fruiting periods listed above are based on limited observations and consist of months in which fruiting has been noted. Some trees may have irregular fruiting times or fluctuating seasons may alter the pattern.

TOP STOREY	FRUITING MONTHS
Piptadeniastrum africana	January—February
Cynometra alexandri	peak January and June
Antiaris toxicaria	February—March
Cola gigantea	February—June
Aningeria altissima	February—June
Maesospis emini	peak April—September
Canarium schweinfurthii	April—May
Desplatzia lutea	May—June
Entandophragma spp.	June
Treculia africana	August
Ficus mucuso	November

MIDDLE STOREYS	
Morus lactea	January—February
Erythrophleum guineense	January—February
Monodora myristica	February
Cordia millenii	peak April—July
Pseudospondias microcarpa	May—June
Celtis spp.	May—October
Pycnanthus angolensis	all year

LOWER STOREY	
Myrianthus arboreus	February—June
Alchornea cordifolia	April—May
Cordia abyssinica	peak April—July

GROUND LEVEL	
Calamus	June—August
Aframomum sanguineum	December—January
Marantochloa spp.	December—January

quently colobus species, which prefer to eat young leaves or buds, may be seen at any level. They may be in the top canopy feeding on *Piptadeniastrum* at one time of the year, or down in the bottom storey on *Coffea canephora* at another.

Many species of trees and shrubs are adapted to fruit dispersal by mammals and birds. In natural forests, fruit, foliage and flowers are well-distributed throughout the year (see calendar) ensuring a food supply for arboreal species and a banquet of fallen fruits for antelope, pig and rodents. The practice of poisoning uneconomic trees is however drastically changing the ecology of forests. In unexploited or uncontrolled forests there are many large mammals: antelope, buffalo, elephant and pig may be numerous. Elephant maintain favourite browsing areas to which they return time after time, these spots are known as "elephant tangles". Flowers support the nectar-feeding bat, *Megaloglossus* (pollination of some tree species is assisted by bats and prosimians) and flowers are eaten by several mammal species. Insects flourish at all levels of the forest and the temperate conditions of the forest floor support amphibians, snails, myriapods and other cryptic fauna and thus many invertebrate-eating mammals, from shrews, pangolins and galagos to squirrels and other rodents.

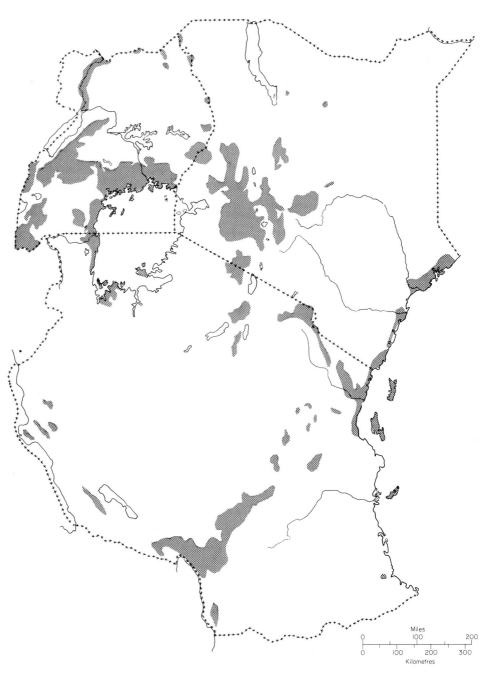

Map of forest areas together with communities thought to have derived from forest.

Lowland forest has stages of succession. The first recognizable forest type is characterized by colonizing trees such as *Symphonia* or *Maesopsis eminii* in Uganda and northwest Tanzania, the fruit of these trees are sought for and distributed by many species of birds and mammals. In grassland or in savanna near the forest's edge, colonizing trees and shrubs may be found focused around ant-hills, established there through the agency of birds and mammals. The forest edges are visited by many animals and provide a refuge and a food supply for savanna species in the dry season. While the trees are fruiting, mammals and birds of the forest are attracted to the margins where a concentration of animal species may at times be very noticeable. Distinctions are usually sharp between savanna and forest fauna, so that these ambiguous conditions are interesting. There are, for instance, some animals like the potto, *Perodicticus potto*, and the red-flanked duiker, *Cephalophus rufilatus*, which prefer the forest edge and a detailed study of this zone should be rewarding.

Colonizing trees do not regenerate under shade and after one generation mixed forest appears with many more species of trees. This is the richest forest type and contains the largest number of animal and plant species. Plant species found in Uganda are listed in the *Vegetation of Uganda* (1964), floral lists for other regions may be found in other papers listed in the bibliography.

Some Uganda forests are dominated by *Celtis*, which may pioneer for *Cynometra alexandri* (thought to be a climax on poor soils and in the excessive shade of old mixed forest), it is common in western Uganda below 1,240 m. Another important dominant in western Uganda is *Parinari*, found above 1,380 m; its seed is distributed by bats, *Eidolon*. Due to the need for higher yields and a quicker turnover of valuable timbers like mahogany, many of the natural forests are now subject to management, so that the spectrum of tree species is being reduced by poisoning, leaving those with an economic value.

Extensive areas that are now grassland, "pseudo-savanna" or cultivation were derived from forest. Grazing by stock and wild herbivores and frequent burning maintain these derived types. The map opposite lumps the present forest areas, both lowland and montane, together with the much more extensive grass and savanna communities that are derived from forest. The tiny relic patches of present forest may be separated by very long distances and these distances may obscure the groupings to which these forests belong. Representing the forests of the immediate past, the map may help to explain odd occurrences of forest mammals in what is now savanna or grassland and to illustrate that recent forests have covered a not inconsiderable area. The forests can be seen to form several major islands with varying degrees of interconnection, these islands have corresponding affinities and differences in their mammal fauna which are discussed in Chapter IV. The map is also useful with reference to rainfall and the distribution of mammals.

Swamp Forest

Swamps and waterlogged areas are common and have a rather specialized flora and fauna. Dominant trees at lower altitudes are *Mitragyna stipulosa*, *Pseudospondias microcarpa* and *Uapaca guineensis*. There are many palms,

Phoenix reclinata, *Raphia*, climbing rattans, *Calamus*, and *Elaeis guineensis* in some western Uganda forests. There is often a dense undergrowth of wild ginger, *Aframomum*, and wild arrowroot, *Maranthocloa*, this is a favourite habitat of the hero shrew *Scutisorex*, and the rats *Colomys* and *Malacomys*. These swamps can be very unpleasant to get into and several monkey species commonly found there are in secure refuges, particularly *Cercocebus* species, *Cercopithecus neglectus* and *Colobus badius*. In the Sango Bay area on the Uganda—Tanzania border, there is a unique seasonal swamp forest showing

30m

<u>Swamp forest</u>, Phoenix, Marantocloa, Pseudospondias, Elaeis, Mitragyna, Calamus

many apparently montane elements. The Uganda forest department have found that the *Podocarpus* is not regenerating under present conditions, so that it is possible that local edaphic conditions suiting *Podocarpus* and other typically montane species may have preserved in this locality a vegetation type dating from a time when temperatures as a whole were lower. This forest has faunal peculiarities that are discussed in Chapter IV.

Swamps at higher altitudes are often dominated by *Syzygium*.

Woodlands

Woodlands in East Africa are faunally ambiguous, it would seem that very few mammal species rely exclusively on this habitat, yet woodlands cover about a third of East Africa and a very large area of southern Africa.

Uganda woodlands of *Terminalia* are in some places mixed with elements of colonizing forest (i.e. *Sapium ellipticum* and *Phyllanthus*). As an extension of a long line of forests along the edge of the western rift these woodlands probably represent a succession towards forest, yet this area together with Geita and Lindi in Tanzania are the only places where there is evidence of such a succession. Extensive areas of woodland in the Murchison Falls National Park have been totally destroyed in recent years by elephant. This suggests that too many elephant, men or other mammals, or even a few protracted dry seasons reinforced by fires can very easily and very quickly tip the balance towards grassland; the fragility or ecological instability of this habitat is evident.

Isoberlinia doka woodlands in the northwestern corner of Uganda are the most easterly extension of the typical West African woodlands. These are the habitat of the giant eland, *Taurotragus derbianus*, which browses on this tree and in East Africa is found only in this area.

Most East African woodlands occur in Tanzania; the dominant species are *Brachystegia*, *Isoberlinia* and *Julbenardia*, trees which are adapted to a long dry season. The commonest grasses are *Andropogon* and *Hyparrhenia*.

Vesey-FitzGerald (personal communication) considers these woodlands to be a fire-degraded primary formation developed after an ancient recession of forest.

These *Brachystegia* woodlands, locally known as "miombo", grow under a rainfall of 800—1,200 mm but this falls between November and May. In the waterless months an uneven pattern of fires lit by the honey-gatherers and hunters creates a mosaic of different stages of growth. Areas burnt early carry a flush of young grass and leaf growth, but on later-burnt or higher, drier ground, regeneration may not begin again until the rains. The dispersal of ungulates tends to follow the pattern of burning, as fresh growth is encouraged by the earlier fires and the local edaphic conditions found along drainage lines. Elephants and shifting cultivation are important in breaking up the woodland canopy: *Isoberlinia* and *Brachystegia* provide fodder for elephants which strip these trees in the dry season for their bark and break down many more, thereby encouraging grasses and woody secondary growth. These disturbed areas provide shelter and food for smaller species which would otherwise be unable to survive under the mature canopy.

13m

<u>Woodland</u>, Brachystegia, Terminalia spp.

The tse-tse fly which is immensely successful in miombo woodland excludes domestic stock and is the main discouragement to human settlement. Like other forms of life, insects must adapt to the violent annual change from cool, shady conditions to a sun-seared desolation. Termites and with them ant-eating mammals may be locally common, i.e. *Orycteropus*; the ground pangolin, *Manis temmincki*, and elephant shrews, Macroscelididae, but, on the whole, insectivorous animals are relatively scarce. Bees flourish and their abundance influences the conspicuous success of the honey badger, *Mellivora capensis*, in miombo. Larger animals that appear to be broadly confined to the woodland vegetation belt are the hartebeeste, *Alcelaphus lichtensteini*, and the sable, *Hippotragus niger*. It is interesting that the sable is depicted in rock paintings in the Sahara, far to the north of the present limits of *Brachystegia* woodland. Bush rats, *Aethomys*, gerbils, *Tatera*, mole rats, *Heliophobius*, and zebra, *Equus burchelli*, are strikingly numerous in *Brachystegia* woodland. Throughout the miombo there are fringes of *Combretum* and *Commiphora* savanna or "hardpans" around wide grassy "mbugas" of seasonally flooded black cotton soil where the dominant grass is often *Echinochloa pyrimidalis* which provides an important dry-season pasture for many woodland ungulates.

Savanna or Wooded Grassland

Savanna occurs in woodland as a catenary zone, but in Kenya and

33

northern Tanzania it is a major vegetation type and savanna types cover most of Uganda.

The classification of types of wooded grassland is a difficult problem and a variety of systems and terminologies is in current use. It should be appreciated that different factors or combinations of factors control this ecosystem, some types of savanna being fire-degraded woodland or even forest while others are primarily edaphic and hence catenary.

Tall grasses and herbs with scattered groups of bush and trees characterize savanna, the dominant trees are very often gnarled, broad-leaved *Combretum* species that are tolerant of the fires that sweep through the grass each year (even bi-annually in parts of Uganda). Moister types of *Combretum* have larger, and drier types have smaller leaves, they may grow in association with *Terminalia*, *Acacia*, *Albizzia* or in Uganda, *Butyrospermum*.

Dominant grasses vary locally, they may include *Hyparrhenia*, *Themeda*, *Cymbopogon*, *Rhynchalitrum*, *Andropogon*, *Setaria*, *Eragrostis* and *Trichopteryx*, on poor soils *Loudetia* and *Heteropogon* may be dominant. In flat sandy areas there may be a scattering of *Borassus* palms or extensive groves standing in *Hyparrhenia*; elephants are very fond of borassus nuts and as the nuts germinate more readily in their dung the distribution of elephant and *Borassus* have been linked. Euphorbia and baobabs grow in drier areas, the latter being restricted to lower-lying areas of Kenya and Tanzania.

Savanna provides both browse and grazing, supporting most species of the typical "game" animals, sometimes at very high densities.

Talbot (1963) concluded that savanna in East Africa could support a year-long biomass of wild ungulates 2—15 times higher than that of domestic livestock, and that in terms of digestive efficiency, water requirements, growth rates, age of maturity, disease relationships and carrying capacity wild ungulates make more efficient use of the savanna rangelands than domestic livestock.

ANIMALS	APPROX. YEARLONG BIOMASS (lb/sq mile)
Wild ungulates	70,000—100,000
Wild ungulates with domestic stock	30,000 +
Cattle ranched on managed savanna	21,300—32,000
Domestic livestock on tribal grazing	11,000—16,000

Yearlong ungulate biomass data from East Africa Savannas (from Talbot, 1963).

Although savannas are not generally of very high agricultural productivity, intense cultivation also occurs within this zone, the soils are generally well-drained and fertile and crops like maize, cassava, etc. do well. Many small

Savanna, Combretum, Acacia, Borassus.

mammals flourish in the mosaic of cultivation and fallow found over so much of this habitat.

Parts of Uganda are scattered with pockets of forest and the vegetation has been called "Forest Savanna Mosaic". In these areas a well-distributed rainfall eliminates severe drought and, small mammals apparently dependent on moisture or green fodder, like *Lophuromys* and *Oenomys*, live in or near cultivation; similar moist, derived savannas occur on the Coast.

Acacia Savanna

Acacias are an interesting group of trees, growing on uplands, in swamps, along river courses and in semi-deserts. Most species are associated with special edaphic conditions or erratic seasonal rainfall. "An *Acacia* cycle is only now being recognized and is as yet little understood, acacia formations (and similar types) are probably cyclic in nature, rather than climax or successional" (Vesey-FitzGerald, personal communication).

Savanna dominated by *Acacia* and *Albizzia* species may follow persistent cultivation; strips of *Acacia rovumae* fringe "mbugas" in Tanzania, growing on particular soils which are less water-logged in the wet season than the "mbugas" and less deeply dehydrated in the dry season than the higher ground. The flat-topped *Acacia tortilis spirocarpa* also grows round flood

Acacia Savanna, Acacia spp.

plains on hard-pan areas. Their fruits are shed in the dry season, after burning has left the earth ashy and bare, so that there are no opportunities for flood waters or rivers to distribute their pods. The pods are highly nutritious and their seeds are covered by a sweet pulp that attracts impala, giraffe, eland and other antelopes; indeed, while grazing is scarce the animals may eat little else and large numbers of the seeds are passed out in the dung. The rate of germination of some species of acacia seeds eaten by antelopes has been shown to be higher than that of those left uneaten, while depredations of a borer that

35

attacks pods are not found in the dung-protected seeds. The dung is often shed along paths or in the resting places of the animals and so provides an opportunity for the young plant to establish itself without the competition of heavy grass growth. *Acacia tortilis* tend to be found in broad swathes round open plains in Masailand, Wembere, Usangu and Rukwa, all areas that have or have had concentrations of game animals.

In common with some other antelope, the eland is very dependent on shade during the dry season although it can do without water if moist foods, dew and shade from the sun are available; in the case of *Acacia tortilis* and the eland a mutual advantage is apparent.

Some *Acacia* savanna in western Uganda (dominant grasses *Cymbopogon*, *Themeda*, and *Setaria*) is thought to be derived from thicket by heavy grazing and burning; these formerly rich wildlife lands in Ankole supported topi, roan, eland, impala, reedbuck, buffalo and many other species.

Many *Acacia* species grow in grey alluvial clays and in cotton soils along drainage lines, on flood plains and in valleys that are water-logged for the rain season and dry for the rest of the year. Valleys in Tanzania with open *Acacia* bush, *Commiphora*, *Lannea*, and *Grewia* are known as hard-pans, a term which only applies in the dry season. They are a quagmire in the rains

Hard Pan, Commiphora, Combretum, Baobab, Lannea, Acacia formicaria.

when most animals avoid them for the higher ground. During the dry season, however, giraffe, impala, kudu, dik-dik, zebra and rhino congregate there and the hard-pans become a distinctive habitat for mammals. The combined pruning of the bushes by fire and browsing maintains fresh, green shoots during the dry months and gives some of the bushes a characteristic shape.

Thicket and Bushland

Bush and thicket cover about half of Kenya and large areas of northern and central Tanzania. The driest bush which merges with the semi-desert is dominated by *Acacia* species. Important fruiting trees are *Balanites*, *Grewia*, *Croton* and *Sclerocarya*. *Lannea humilis* is a dominant and widely distributed species, which is an important food plant for elephant and rhino, and *Commiphora*, a small gnarled tree, dominates large areas of bush; its roots and bark are eaten by warthog, elephant and rhino. Both species of kudu seem particularly well adapted to this habitat.

There are many types of bush depending on a variety of soil conditions and many of these communities are unstable and have been induced by man, stock and wild animals. Bush is encroaching on savanna in many areas as a

36

result of overgrazing by stock. Fires cannot spread so that the less fire-resistant shrubs and thicket bushes flourish, these in turn repress the further

Thicket, Commiphora, Combretum, Acacia, Teclea, Maba.

growth of grass. It is thought that elephant may reverse this process by opening up the thickets. Thickets with many succulent species reflect a low rainfall or a rapid runoff of water (sometimes induced by overgrazing and erosion), they succeed bush and form climax communities. The Itigi thicket (about 6,000 sq km) in central Tanzania is the largest and best known climax thicket. It is dominated by *Pseudopropsis fischeri*, *Baphia* and *Grewia* which coppice and interlace, impassable to anything but the smallest and largest mammals. Elephant and rhino find refuge here, the pliant branches springing back after the animal's passage. Termites are particularly common on the thicket floor feeding elephant shrews (Macroscelididae) and hedgehogs (*Erinaceus*); dik-dik (*Rhynchotragus*), are also numerous. Klipspringer (*Oreotragus*), hyrax (*Hyracoidea*) and rock hares (*Pronolagus*) are successful and typical mammals of the climax thickets which are found on rocky outcrops, on hillsides and in ravines.

On the northern mountains, dry montane forests merge with evergreen thicket dominated by *Acokanthera schimperi*, *Euclea*, *Rhus* and *Carissa*.

The shore and islands of Lake Victoria are often fringed by thickets of *Alchornea cordifolia*, a favourite browse for sitatunga, *Tragelaphus spekei*. Coastal thickets with *Manilkara* and *Rhus* grow up to the ocean shore sheltering a wide range of mammals, notably suni, *Neotragus moschatus*.

Semi-desert

Semi-desert suffers over six months without rain and an annual total of less than 250 mm of rain falling during the months of April, May and November. Dominant shrubs are *Acacia* spp., *Balanites orbicularis*, *Euphorbia* spp., *Jatropha* and *Disperma* species. Some grass dominants are *Aristida*, *Chrysopogon*, *Chloris* and *Eragrostis*. Specialized insectivores, rodents, carnivores and ungulates survive in this habitat; many of them never drink, conserving and synthesizing their moisture requirements from their food, but some of the larger or more mobile animals rely on periodic visits to springs and waterholes. Rhino and elephant also chew quantities of *Salvadora persica*, *Euphorbia*, *Sanseviera* and succulents. Roots, stems and tubers supply moisture to

Semi Desert, Aloe, Calotropis, Sansevieria, Commiphora, Balanites, Euphorbia.

burrowing rodents (*Heterocephalus*, *Gerbillus* and *Xerus*). Grasshoppers, locusts, termites, beetles, other insects and their larvae support elephant shrews, *Elephantulus*, hedgehogs, *Erinaceus*, and pigmy mongoose, *Helogale*. The carnivores live off the ungulates, rodents, birds and reptiles.

Desert

Sand and stones cover most of the desert. In the northern frontier district of Kenya a few stunted shrubs may grow in sheltered spots and a little green grass may also appear when there is rain. At such times, oryx, gerenuk, Grant's gazelle and Grevy's zebra may enter the desert area but for most of the year no mammals are apparent. The area is little studied.

Grasslands or Rangelands

The majority of the grassland areas shown on p. 24 are derived by a combination of grazing, fires and soil factors. While there may be some highland and valley grasses that are maintained by peculiar climatic and edaphic conditions, the majority of grasses would be rapidly succeeded by woody growth in the absence of ungulates or fire.

The dominant grass species alter very rapidly if the intensity of grazing changes or if there is a burn early or late in the dry season. The drainage or chemical composition of the soil also shows an immediate reflection in the species of grass present, as does the amount of shade from trees. Many grassland types are defined by reference to the woody growth with which they are associated and some authorities would have much of the semi-desert and bushland of northern and eastern Kenya defined as grassland.

There are extensive areas of grassland occurring within almost all the vegetation types mapped, which cannot be shown on a small scale and which would tend to obscure still further the already complicated pattern of East African vegetation belts. Some of the dominant grasses associated with trees in woodland and savanna have already been listed and the grassy "mbugas" were mentioned as the lower part of the catena in "miombo" woodland. Under natural conditions grass communities are complex and numerous and are subject to continuous change. Today they are being altered by the uses to which man decides to put the land.

The influence of human activity can be most readily seen along the sides of any road where disturbed soils encourage *Hyparrhenia* and *Loudetia*, grasses that often make the banks and verges of roads a well-marked vegetation "zone". Similarly late burning by pastoralists tends to favour *Themeda* at the expense of other species. On the other hand, felling and fires in formerly forested areas above 2,100 m leads to a very uniform mixture of grass and herb species, all of which seem to have been "borrowed" from other habitats (Vesey-FitzGerald, 1963). Chemical factors control the presence of *Sporobolus* and *Diplachne* which grow on alkaline pans.

These examples show the numerous interacting forces that determine the condition of rangeland and the species of grass found there; however it is pastoralists and their stock or wild ungulates that have been responsible for the broad pattern and the maintenance of many East African grasslands and this aspect will be of special interest to the student of mammals.

Some common grasses in East Africa:

a. *Cymbopogon validus* e. *Aristida adscensionis* i. *Hyparrhenia filipendula*
b. *Heteropogon contortus* f. *Eragrostis superba* j. *Setaria sphacelata*
c. *Themeda triandra* g. *Andropogon chrysostachyus*
d. *Loudetia kagarensis* h. *Echinochloa pyrimidalis*

39

Large ungulates have an immediate and drastic effect on pasture and in few other habitats is the relationship between fauna and flora as obvious as in the rangelands. Both plant and animal species may benefit or be discouraged by heavy trampling and pruning, but clearly those species that can benefit will be at an advantage, indeed some mammals may well come to depend on the pasture being "processed" for them to a condition fit for their own metabolism.

The action of animals on the environment should in these circumstances be considered in the same way as other physical factors, sun, rainfall, fire and drainage conditions. Any animal and any plant living in the habitat must turn the peculiar conditions to its own advantage, or mitigate the disadvantageous effects by some means or other.

Mammals living in grassland cope with annual fluctuations in the availability of water, day-night differences in temperature and fluctuations in the height and condition of the grasses on which they are all ultimately dependent. Large numbers of animals help to maintain the range, and gregariousness is certainly more common on the plains and on a greater scale than in other habitats.

Migration for water, for grasses that are moist or in another stage of growth allows the plain ungulates to exploit large areas and rotate the pasture. It seems that the plain ungulates have generally emancipated themselves from being tied to a small locality, and the evolution of greater size and mobility are linked to allow the effective exploitation of this habitat by a wide range of large animals. Several species are non-drinkers, so long as they have access to green grass and shade.

Differentiation in ungulate feeding habits and their movement maintain the particular pattern of grasses that we see on the great "game plains" in East Africa. The ungulates are of a wide range of morphological types, zebras, gnus, gazelles, buffalo, kongoni, reedbuck and so on, and each species tends to exploit a particular condition of the habitat.

Pioneer in elucidating the relationship of wild ungulates to grassland has been Vesey-FitzGerald; his papers and others listed in the bibliography should be consulted as no more than a brief outline can be given here.

The dry season is a period of concentration for most herbivores, which congregate in well-watered or shady localities. With the arrival of the rains most of the larger animals disperse as grazing conditions are almost universally favourable. In the Rukwa Valley in southern Tanzania (see Vesey-FitzGerald, 1960, 1965), elephant, buffalo and hippo start at the end of the rains to feed on the marsh grass, *Vossia*, that has grown tall and rank in the flood waters. Feeding and trampling by very large animals have a physical impact on vegetation that is hard to imagine; three-metre-high canebrakes are chewed, stripped, pulped, excreted and trodden flat, then when requirements for bulk food can no longer be met and surface water is becoming scarce these large animals move out, the flattened grass puts out fresh shoots which now attract eland, zebra and hartebeeste. Herds of topi, *Damaliscus korrigum*, also come down from the higher ground where they had concentrated while the valley pastures were flooded. The grass is kept short and green by constant grazing and, before it dries up completely, showers falling towards the end of the dry season bring on a flush shortly after the birth of the topi calves.

The dependence of the topi on this cycle was highlighted by the unprecedented rains of 1962—63, when there was extensive flooding around Lake Rukwa and the largest animals left the area. When the topi moved down to the flood plains they stood around in the reeds and water and died in numbers. Furthermore on moving back into long grass animals would panic and the long sharp seeds of the tall *Hyparrhenia* grass are reported to have blinded many individuals. Later the topi learned to trample the grass themselves but for two years their cycle was disrupted and they failed to breed (E. Balson, personal communication). Other ungulates also show some dependence on one another: zebra, *Equus burchelli*, prefer green grass in flower, while the gnu, *Connochaetes*, is reported to prefer the sprouting perennial grass left in the zebra's wake.

The grazing succession in the Rukwa Valley may be contrasted with that in northern Tanzania and parts of Kenya. Here, the gazelle (which are able to do without water) disperse over the empty plains during the dry season and feed on stubble or hay and browse the scrub, while the larger animals are concentrated in the valleys. With the arrival of the rains the large mammals disperse but the gazelles move into the over-grazed dry-season concentration areas to feed on the short fresh cushion grass.

Changes, whether they are temporary seasonal ones or long-term changes induced by man, elephant or climate set off immediate chain reactions over a much wider range of associations than is first apparent, altering the pattern of fauna and flora. There are many varieties of ungulates, there are hundreds of grasses and many situations or conditions in which they are found, but they are all part of a range of biological associations with ramifications the complexity of which we are only beginning to grasp. So improbable are the links found in ecological chain reactions that one is reminded of Gibbon's observation that "the orders of a Mongol Khan who reigned on the borders of China lowered the price of herrings in the English market". For instance, the demand for ivory in 1910 may have a connection with the expansion of lesser kudu in parts of Karamoja today: the ivory was exchanged for cattle, elephant and their ecological role in maintaining rangeland disappeared, the cattle have since multiplied and overgrazed and dense thicket and bush became established which are ideal for lesser kudu.

Ungulates are of course not the only mammals to be found in the rangelands and the presence of a large biomass of meat inevitably invites predators which have to be fast and socially well-organized to hunt in the open. Lion, cheetah, hyaena, hunting dogs and black-backed jackal are well adapted and are numerous wherever there are large concentrations of herbivores.

Fires, trampling and predators are hazards for small mammals, and rodents are rare on the open plains. Those that succeed best are fast nocturnal jumpers like gerbils and spring hares which shelter by day in burrows. Vesey-FitzGerald (personal communication) has found that rodents frequently increase in exclusion plots, even under metre-square frames. Natural "exclusion plots" are territaries and other pockets of sheltered vegetation where rodents, mongoose and other small mammals may be common. Mole rats have a localized distribution but are common in some grasslands, particularly *Tachyoryctes* in highland areas.

Swamps

Permanently waterlogged swamps are often dominated by *Papyrus*, bulrushes, *Typha*, or reeds, *Phragmites*. A common succession along the shores of Lake Victoria starts with aquatic vegetation, water lilies, *Nymphaea*, and Nile lettuce, *Pistia*, followed by a floating mat or "sudd" on which ferns and sedges are common; *Papyrus*, *Typha* and *Phragmites* follow, with *Miscanthidium* on the seasonally flooded land. Then there may be a narrow belt of *Phoenix reclinata* and swamp forest before higher ground is reached. In drier or seasonally flooded areas, there may be a floating sudd of *Leersia*, *Vossia*, *Echinochloa stagnina* and *Pennisetum glaucocladum* during the wet season. *Thalia geniculata* and *Oryza perennis* are a common association in water one to two metres deep. Swamps are the home of sitatunga, *Tragelaphus spekei*, the marsh mongoose, *Atilax paludinosus*; hippo, *Hippopotamus*, may also be common where there is some firm standing.

Swamp, Pistia, Nymphaea, Phragmites, Papyrus; Miscanthidium, Phoenix.

Cultivation and Induced Vegetation

Areas under cultivation in East Africa are becoming very considerable, but their mapping on the vegetation map would obscure many zoogeographic points and the "original" vegetation type has been mapped instead. Reference to the map of human distribution (p. 109) will give a fair indication of the relative density of cultivation. Cultivation covers a very wide range of human agricultural activities, from temporary millet fields in partially cleared woodland to permanent plantations of banana, sugar cane, sisal, coffee, tea or coconuts.

Many of the most intensively farmed areas were once forest, as in the highlands of Kenya, Tanzania and Kigezi. The plantations of sisal, coconut, mango, cashew, etc. covering large areas of coastal plain are derived from forest, woodland and savanna. The cereal and cotton growing steppes of Usukuma and Unyamwezi were formerly woodland savanna and seasonal swamp grassland. The banana and coffee gardens of Uganda have mostly replaced lowland forest and papyrus swamp.

Mammal species surviving in cultivation vary from place to place and may or may not reflect the original fauna. The rat *Arvicanthis* is very common in subsistence cultivation and becomes rarer away from settlement. The arboreal mice *Dendromus* and *Thamnomys* are common in banana gardens and in suburbia. Hedgehogs, *Erinaceus*, flourish in several over-grazed and

42

settled areas and also in Nairobi gardens. Mole rats, *Tachyoryctes*, flourish in the highland sheep pastures of Kenya.

Small harmless scavengers such as the white-tailed mongoose, *Ichneumia albicauda*, and the civet, *Civettictis civetta* may be common around villages and towns, and many species of bats and rodents are familiar in buildings. Fruit bats, *Eidolon*, like weaver birds, *Ploceus*, sometimes roost by preference in towns where they are presumably safer from predators.

Termitaries

Termitaries are an important focus for animal life wherever they are found, which is practically in all vegetation zones below 3,000 m.

The mounds vary enormously in size, shape and composition, the largest and most obtrusive are built by *Macrotermes* using a clay-saliva cement, *Microtermes* and *Odontotermes* also build in this way but they are among the many termites which make their hives below ground. All these species make fungus gardens in the centre of the nest, filling chambers with spongy masses of chewed wood which becomes coated with fungi on which the termites feed. During the rains several termite species take fungus up above the nest

▲ Cross section of termite mound
showing vents, chimneys and fungus
gardens, also form of various types
of termitaries.

and spread it on the ground where, after a day or two, a dense growth of edible mushrooms appear (these attract several mammal species). The spore-laden humus is then collected by the termites to replenish the fungus combs in the termitary. Another type, *Cubitermes*, feed on humus and make small hard-capped umbrella-shaped mounds. The third major group, *Amitermes*, feed on wood and make their nest from their own excrement.

Mounds are in effect piles of subsoil, their form is determined by the termite species, by the soil and by the climate. In poorly drained valley soils the termitaries are often well-drained and become vegetation centres, mounds even emerge from two metres of water in swamps where they may be capped by *Phoenix reclinata*.

Many mounds have either ventilation chimneys (whence warm air can be felt rising in active termitaries), or galleries leading out at the base; inside there are mazes of interconnecting flues, corridors and fungus gardens. These termitaries shelter a wide variety of vertebrates and invertebrates. Depending on the habitat, elephant shrews, *Elephantulus*, bats, Microchiroptera, various rodents, carnivores and warthog, *Phacochoerus*, may breed or shelter there. Ants may invade parts of the termitary and orycteropus and pangolin in opening up the mound often make themselves a shelter in the centre.

Vervet monkeys use mounds as perches and with birds assist the spread of vegetation by depositing seeds in their dung. Fruit bats often use the termitary thickets for diurnal shelter and, as the principal agents in the pollination and seed dispersal of *Chlorophora excelsa*, they may thereby deposit seed in an ideal bed, protected from fires, floods, sun and weeds. *Maesopsis*, *Sapium* and *Vitex* species are other trees established in this way and it is likely that many other typical thicket and forest colonizing plants depend on a similar chain of circumstances involving mammals and birds. *Harungana*, *Lannea*, *Cassia singueana*, *Grewia*, *Balanites*, *Strychnos*, *Teclea*, *Rhus*, *Ziziphus*, *Tamarindus* and *Ficus* all have fruit eaten by mammals and birds and all tend to grow round termitaries, attracting mammals large and small to these centres.

Sometimes old mounds are used as mineral licks by ungulates. Rutting antelopes may use them as a territorial marker, and zebra, rhino and buffalo use them as rubbing posts.

Orycteropus holes, both in and out of termitaries are shelters for many species of mammals, including many of those mentioned above; warthog, porcupine, ratel, pangolin and hyaena are particularly common tenants.

Appendix I
Bwamba Forest

Bwamba county lies to the west of the Ruwenzori massif and although still within the Nile watershed, the faunal affinities of this northeastern corner of the Semliki Valley are with the Congo basin from which it is separated by low hills. It is fortunate that included in this area are the 218 sq km of the Semliki Forest Reserve (lying below 750 m) and part of the Ruwenzori Forest Reserve ranging from 1,830—3,350 m. These reserves contain 10 species which are unknown anywhere else in East Africa, 10 species which are known only from one or two other localities and 5 or 6 other species which either formerly inhabited the area or may yet turn up.

MAMMALS CONFINED TO BWAMBA IN THE E.A. PART OF THEIR RANGE	MAMMALS NEARLY CONFINED TO BWAMBA IN THE E.A. PART OF THEIR RANGE
Colobus badius ellioti	*Colobus polycomos ruwenzori*
Cercopithecus mona denti	*Cercopithecus l'hoesti*
Myonycteris wroughtoni	*Cercopithecus neglectus*
Nycteris argae	*Megaloglossus woermanni*
Haemoschus aquaticus	*Scutisorex somereni*
Syncerus nanus	*Potamogale velox*
Cephalophus dorsalis	*Neotragus batesi*
Anomalurus beecrofti	*Genetta victoriae*
Idiurus zenkeri	*Crossarchus alexandri*
Aethomys longicaudatus	*Colomys goslingi*

Okapia johnstoni is no longer found in the area but wandering individuals have been seen and one was hunted and wounded in Bwamba in recent years. The species is known to occur a few kilometres across the border. There is some evidence that the following species may also be found in the area:

Manis longicaudata	*Cephalophus leucogaster*
Micropotamogale ruwenzori	*Galago inustus*
Cercopithecus talapoin pilettei	*Funisciurus anerythrus*

Fifteen species of primates can be found within six or seven kilometres of the hot springs and detailed study of the ecological distribution of these species is therefore possible.

45

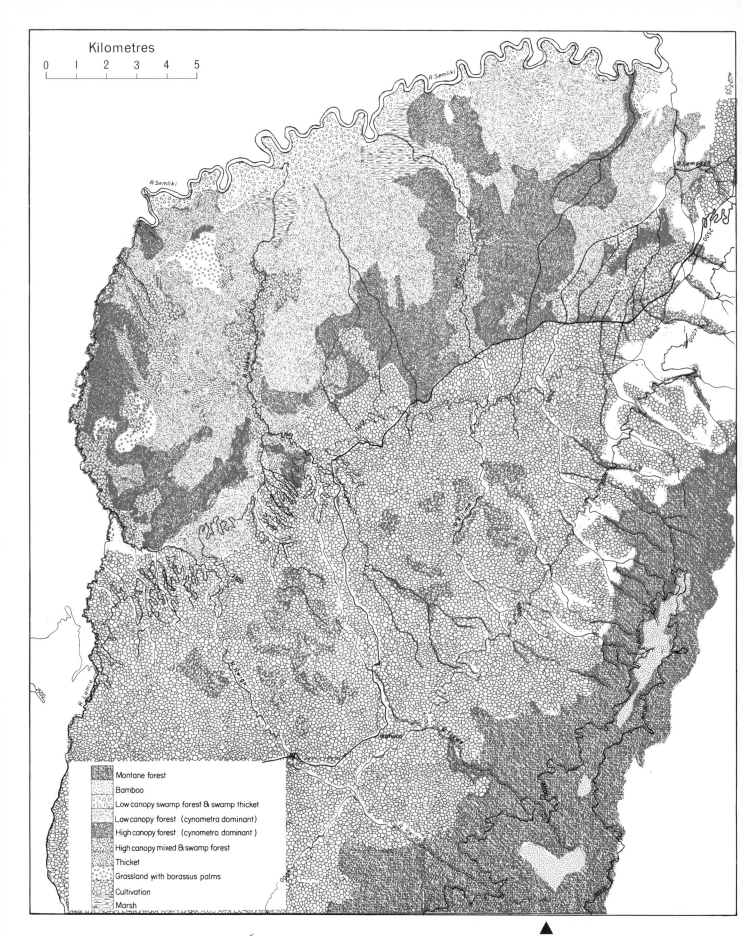

Kilometres

0 1 2 3 4 5

Montane forest
Bamboo
Low canopy swamp forest & swamp thicket
Low canopy forest (cynometra dominant)
High canopy forest (cynometra dominant)
High canopy mixed & swamp forest
Thicket
Grassland with borassus palms
Cultivation
Marsh

R.Semliki

R.Semliki

R.Lamia

R.Lamia

Vegetation of Bwamba.

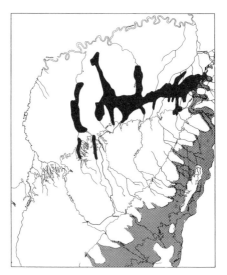
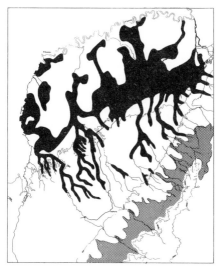
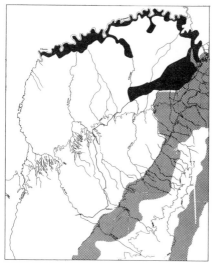

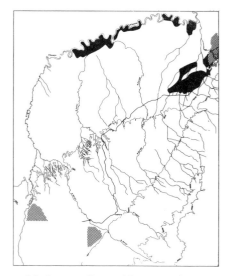
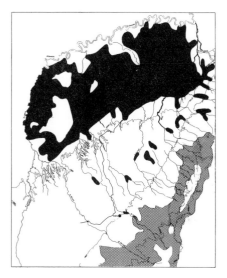

1. black area: Cercopithecus mona
shaded: Cercopithecus mitis

2. black area: Cercopithecus ascanius
shaded: Cercopithecus l'hoesti

3. black area: Cercocebus albigena
shaded: Papio

4. black area: Cercopithecus neglectus
shaded: Cercopithecus aethiops

5. black area: Colobus abyssinicus
shaded: Colobus polykomos

Distribution of Cercopithecoid monkeys in Bwamba.

NOTE ON BWAMBA

As from 1971 much of the vegetation mapped on p. 46 will undergo considerable change due to adoption of the "Taunguya farming system" of forest management. The forest will be cleared by farmers who can harvest crops until this is made impossible by the growth of replanted trees that are of economic value.

All vegetation west of the Kirimia River will be felled and replaced with more valuable timbers, particularly the West African *Terminalia ivorensis* and *Terminalia superba* together with the colonising tree, *Maesopsis emini*.

Fortunately the area east of the Kirimia is designated as a Nature Sanctuary working cycle and this section will be left undisturbed.

Bwamba illustrates in microcosm the increasing problems that are faced by authorities administering refuges of crucial value to fauna and flora.

In this case the under-staffed and under-financed Uganda Forest Department is making a commendable experiment in harmonising three powerful pressures. These are severe and increasing population pressure, the need for more adequate financial returns from timber, and scientific considerations of conservation, some of which have been outlined in this appendix.

Further details on Bwamba may be found in Uganda Forest Department working plans and Fort Portal forest office mimeograph ref. No. W.R./90. This area is also discussed by Moreau (1966), pp. 164–288.

PRIMATES RECORDED FROM BWAMBA

Perodicticus potto
Galago demidovii
Galago senegalensis
Pan troglodytes
Colobus polycomos ruwenzori
Colobus abyssinicus uellensis
Colobus badius ellioti
Papio cynocephalus
Cercocebus albigena
Cercopithecus ascanius
Cercopithecus mona denti
Cercopithecus mitis stuhlmanni
Cercopithecus neglectus
Cercopithecus l'hoesti
Cercopithecus aethiops

The local Bambuti pigmies are knowledgeable guides and naturalists and the area is of relatively easy access and cannot be paralleled in East Africa, or perhaps anywhere else in Africa, for the comparative study of primates. As the fauna of Bwamba is of unique zoological interest I have included a vegetation map to assist future workers and have tentatively mapped the local distribution of some of the species found there. The map is based on:
a) Personal field-work in Bwamba over several years.
b) Uganda Forest Department boundary plan No. 1036 (scale 1:50,000).
c) Uganda Land and Surveys Department sheet No. 56/1 (scale 1:50,000).
d) TO/41 Bwamba (A. J. Haddow, 1944) (scale 1:75,000).

The distinctions drawn between vegetation types on this map have been determined by obvious physical and floral differences that appear to be significant in the distribution of some mammal species in this area. For instance, some monkeys are not found in low canopy forest although the floral composition may not be very different. On the other hand, mixed types have been lumped although important differences in botanical composition may appear to have been glossed over thereby. It should be appreciated that the ecological situation is constantly changing, the most obvious example being the clearing of forest and the development of new agricultural practices, which outside the reserve cause rapid and dramatic changes in the vegetation.

Less conspicuous is the fact that large areas of the central and western parts of the reserve were once cultivated, this is betrayed by the presence of glades of open grassland and secondary scrub in the low canopy forest, much of which is also secondary growth. In the eastern area of the reserve a more stable situation is apparent and there is no evidence of past or present human interference with the forest. Deep flashfloods during the rains, poorly drained saline soils and perhaps the presence of diseases have ensured that the low-lying eastern parts of the reserve never suffered serious human disturbance in the past and this eastern half the Semliki Forest has been designated by the Uganda Forest Department as a Nature Reserve and Animal Sanctuary. It will be immediately obvious from the map that cultivation stops sharply on the forest reserve boundary and that only a strictly enforced legislation is at the present time capable of preserving the forest and its fauna for the future.

The whole of the Semliki Forest Reserve is on poorly drained clay soils on the flats of the Semliki Valley (an area of low agricultural productivity). The Reserve is mostly below 750 m. The montane Ruwenzori Forest Reserve grows on very steep slopes and has a lower boundary at about 1,830 m. These two forest areas were probably linked at one time (as they still are further south in the Congo). Today, the foothills and lower slopes of the Ruwenzori mountains, which have deep loamy soils that are highly fertile and very productive agriculturally, are heavily populated. It is therefore extraordinary that a rich fauna should survive in spite of interspecific competition, in spite of extensive seasonal flooding and in spite of heavy and continuous hunting and snaring. This reservoir of numerous species exists under conditions that appear to be far from optimum in a rather unpromising habitat; it is a reminder that the immediate role of the environment and of competition can sometimes be oversimplified or overstressed.

East of this low-lying valley at the heart of the Central Forest Refuge the chances increase of forests having been subjected in the past to extreme climatic vicissitudes, isolation or even destruction. The dwindling lists of forest mammal species as one travels east is most marked and it seems that the botanical composition of forests or their area influence mammal faunas less than their distance from the Central African Refuge. The number of species from a forest area can therefore be regarded as some indication of the forest's age and its ecological stability.

Notwithstanding these remarks a simple breakdown of forest types in Bwamba is worthwhile and they have been grouped under the following categories (Derived from Uganda Forest Dept "Notes on Semliki C.F.R. Forest type map T/044").

High Canopy Forest (*Cynometra* dominant)
Low Canopy Forest (*Cynometra* dominant)
High Canopy Mixed and Swamp Forest
Low Canopy Swamp Forest and Swamp Thicket
Thicket
Marsh
Grassland (with *Borassus* palms or Forest Galleries)
Cultivation Mosaic
Montane Forest
Bamboo

High Canopy Forest

(24—30 m) *Cynometra* dominant, sometimes in pure stands but may have local admixtures of *Morus*, *Holoptelea*, *Alstonia*, *Albizzia*, *Ficus* and *Celtis*, generally light undergrowth.

Low Canopy Forest

(9—18 m) Dominated by *Cynometra*. In some central and western areas thickets of *Chaetacme* and *Clausena* may be secondary growth on old settlements or on sandy soils. In the eastern areas there are many *Euphorbia* and the soils are seasonally flooded and poorly drained.

High Canopy Mixed and Swamp Forest

Cynometra, Chlorophora, Desplatzia, Croton, Mildraediodendron, Alstonia, Holoptelea, Cola, Antiaris and *Celtis*. May be frequently waterlogged and, particularly in the east, may include many swamp forest elements, i.e. *Mitragyna* (locally in pure stands) *Elaeis guineensis, Ficus vogeliana, Kigelia moosa, Markhamia* and frequently an undergrowth of *Pandanus, Marantochloa* and Zingiberaceae, and numerous epiphytes.

Low Canopy Swamp Forest and Swamp Thicket

Elaeis, Markhamia, Kigelia moosa, Schrebera with impenetrable undergrowth of *Phoenix, Pandanus,* Zingiberaceae, *Maranthocloa* and Cycads. (There are a few local stands of *Acacia milbraedii* around the hot springs and near the banks of the Semliki which are mapped in this category.)

Thicket

Dense growth of *Dombeya, Chaetacme* and *Clausena, Cussonia* on hilltops and old settlements.

Marsh

Open, seasonally flooded area with swamp grasses.

Grassland

Pennisetum and other grasses with
    a) *Borassus* palms on the flats and
    b) forest galleries and savanna trees on the hillsides.

Cultivation Mosaic

Principal crops: coffee, bananas, cocoa, cassava, sweet potatoes, groundnuts, colocasia, beans, pineapples, cotton, maize, millet, sugar and rice. This category includes fallow dominated by elephant grass (*Pennisetum purpureum*). Relict forests (often *Cynometra*) form galleries along the banks of rivers and along the steeper ravines and hillsides.

Montane Forest

On steep slopes above 1,800 m *Pygeum, Aningeria, Albizzia, Olea, Dracaena, Cyathea, Ilex, Rapanea, Musa, Lobelia* and a dense undergrowth of herbs, shrubs and climbers.

Bamboo

Arundinaria with some *Hagenia* and sometimes dense undergrowth of *Mimulopsis* and other herbs and climbers.

Chapter IV
Time Perspectives in
Mammalian Evolution

No aspect of animal life is without roots in time. Our own observations of living phenomena are a witness of moments in a process, for we are seeing patterns on the surface of ancient substrata and the animals or communities we see are but fragments of a vast organic system driven and activated by events and happenings of a universe of space and time. These surface patterns led Darwin to infer the hidden workings of the evolutionary process. He observed the structure, distribution and ecology of living animals and he also saw resemblances between the fossils and the living forms he found in South America. His deductions were based on the detailed comparisons of the patterns he saw. Since Darwin, the ramifications to the study of evolution have become immense; physiology, genetics, embryology, behaviour, palaeontology, biochemistry and ecology have all enlarged and complicated our understanding. Yet there remains the central inspiring spectacle of countless life forms that have suited themselves, by natural selection, to the succession of opportunities that are offered to them by physical changes occurring in time and space. His enthusiasm for the spectacle led Darwin to say: "From so simple a beginning, endless forms, most beautiful, most wonderful have been and are being evolved".

In East Africa we have glimpses of the course of evolution in a very wide range of mammals at every level of adaptive organization and phyletic age, and the life histories of such a wide range of mammals involve us in very different time scales. A diagrammatic tree of the mammal orders represented in Africa (see p. 52) shows that the fundamental radiation of orders took place in the Mesozoic, about 70 million years ago. Families may be in the region of 25 million years old, while Species and Subspecies vary greatly but are mostly relatively recent in formation.

It may be useful here to single out three major "time perspectives": the present, the more recent past, the course of which has determined the present distribution and occurrence of mammal forms, and the Tertiary, which saw the radiation of mammals and established the major orders, families and genera of mammals—a period of 70 million years.

We also have to deal with three rather different sets of data. The present offers us animals living in what must usually be taken to be "natural" conditions. The recent past must be deduced by the study of distribution patterns and specific differences with some help from geographers, palaeontologists and botanists. Beyond are tempting vistas for the imagination but data become limited to scattered fossil remains, which are the only material examples we have of the operation of evolution in the most distant past.

Fossils have acquired greater significance by reference to living animals and there are numerous organisms evolved a long time ago that have survived with little change up to the present; the coelacanth, the ginko tree and the

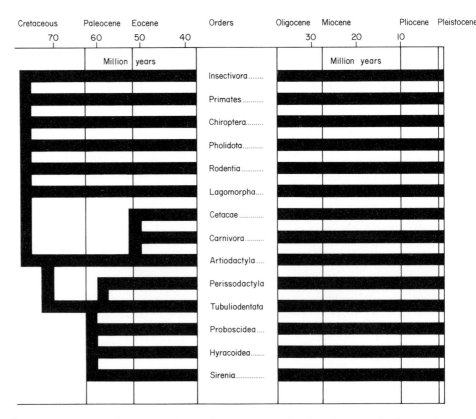

		Million years			Orders		Million years			
Cretaceous	Paleocene	Eocene				Oligocene	Miocene		Pliocene	Pleistocene
70	60	50	40			30	20		10	

Insectivora........
Primates ....
Chiroptera........
Pholidota..........
Rodentia...........
Lagomorpha....
Cetacae ....
Carnivora.........
Artiodactyla ....
Perissodactyla
Tubuliodentata
Proboscidea ....
Hyracoidea.......
Sirenia.............

Probable affinities of mammal orders represented in Africa (modified after J. Z. Young and A. S. Romer).

crocodile are popular instances. Of African mammals, the chevrotain (scarcely changed since the Miocene) and the okapi (Pliocene) are favourite examples of "living fossils", but many less dramatic types have also survived, particularly in refuge areas. For instance, where forests have not suffered from past climatic changes, the peculiarly stable conditions of that very rich habitat put little pressure on well-adapted animals to change or to become extinct. Where an archaic form has competition, perhaps from linear descendants, the potential competitors have often with time become more specialized, or the ancestral species becomes limited to those areas or ecological niches where the descendant is inferior.

It is often not appreciated that the capacity to survive is widespread and occurs at every level of time. The list of African mammal species is so long, partly because of relic species, whose survival has been assisted by that mosaic of ecological conditions, particularly marked in East Africa, which allows ancient types to maintain themselves in pockets of suitable habitat. No apology is therefore necessary for including in this account of living mammals an outline of fossil history in Africa. Any attempt at understanding or interpreting mammals leads us down long corridors of time. However, it may be simplest to start by defining the distinctness of African fauna in the most general way, by comparing it with that of other continents and particularly with that of its neighbouring continent, Eurasia.

Continental Zoogeography

African fauna is very distinctive today, but in considering the groups exclusive to the continent, we must recognize two categories: those that have

52

been found as fossils outside Africa but are now limited to this continent, and those for which no such outside fossil evidence has been found and which may therefore have evolved in Africa.

Listed below are groups exclusive to the continent and their earliest known appearance.

MAMMAL GROUP	FOSSILS
Potamogalidae	not known
Chrysochloridae	Miocene
Macroscelididae	Oligocene
Anomaluridae	Miocene
Pedetidae	Miocene
Thryonomyidae	Oligocene ancestor
Petromyidae	Oligocene ancestor
Bathyergidae	Miocene + Oligocene ancestor
Cephalophinae	Late Miocene
Neotragini	Lower Pleistocene
Tragelaphini	Lower Pleistocene
Galaginae	Miocene
Cercocebus	Lower Pleistocene
Cercopithecus	Lower Pleistocene

Hippopotami, orycteropi, giraffes, alcelaphines and the reduncines are known from outside Africa only as fossils, mostly from India and southern Europe.

In both fossils and living forms Africa's closest faunal relationship is with Eurasia and in living fauna more specifically with the tropical Oriental region. Africa shares the lorises, baboon-macaques, apes, colobines, pangolins, elephants, rhinoceroses and chevrotains with the Oriental region.

Hyraxes, gerboas and horses are found in common with the Palaearctic region, while hedgehogs, porcupines, civets, hyaenas and pigs are common to the Old World.

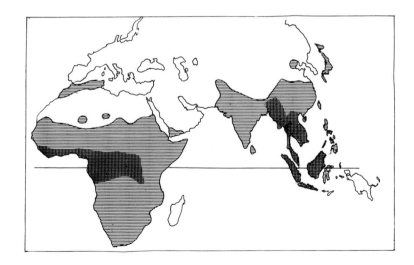

World distribution of *Cercopithecidae* (horizontal lines) and apes (vertical lines).

Not surprisingly the more distant Americas have fewer common elements but there are such widely distributed types as shrews, bats, hares, squirrels, cricetid and murid mice, dogs, cats, mustelids and some bovids. Fauna in common between continents is some measure of their past connection, for animals do not share a common ancestry without ultimately having a common geographic origin. On faunal grounds alone, it is quite clear that Africa and Eurasia have had several connections over a period of many millions of years. The nature and age of these connections and the evolutionary implications of changes and faunal assemblages have begun to leave the realm of pure speculation as new fossils are found and improved dating techniques are developed. Cooke (1968) has discussed in some detail the history of Africa's fossil fauna and its relationship with that of Eurasia, and his paper should be read by those wishing to pursue the subject further. The brief outline presented here will provide some background to further discussion in the subsequent profiles of species or larger groups.

Mammal fossils are known from as early as the Jurassic up to the present, but the present fossil history of African mammals only starts with the Eocene and Oligocene.

The Eocene—Oligocene

Africa is very poor in early mammal fossils. The earliest deposits known at present are the Eocene—Oligocene beds at Fayum in Egypt, estimated to be between 30—50 million years old. After this deposit there is a gap of about 6—8 million years before the Miocene beds of East Africa.

Africa (Arabia was still part of the continent) was isolated from Eurasia by the sea of Tethys, which had cut the continent off for much of the Mesozoic and Tertiary. Nonetheless, the continents clearly already shared a basic mammalian stock. "From a Palaeocene ferungulate stock already possessing early anthracotheres and hyaenodonts, the proboscideans, hyracoids and sirenia developed as new elements in Africa. Insectivores and primates must have been part of this early fauna" (Cooke, 1968).

Fayum was an outlying corner of the continent, but it is the only fossil indicator we have at present of the evolutionary activity that must have been taking place in the great equatorial regions to the south, where mammals presumably flourished then in the greatest abundance and variety, as they do now.

The fauna may not be broadly representative of Eocene—Oligocene Africa as a whole, but the groups unique to Fayum are highly significant. Anthropoid primates, ancestral elephants, hyracoids and sirenians had almost certainly evolved in Africa. The absence of these groups from Eurasian deposits of the same period is a strong indication of African isolation, while their relatively advanced evolution is suggestive of the hinterland of time needed to develop their distinctive characteristics.

Some contrasts between Eurasian and Fayum fossil fauna are summarized on p. 55 (from Simons, 1968).

As the Oligocene beds contain numerous fossils, some idea of the relative importance of orders is possible. The diagram opposite based on Cooke (1968) compares the composition of the Fayum fauna with that of succeeding African deposits.

FAYUM	EURASIA
MACROSCELID (elephant shrew) Metoldobotes	
EARLY APES AND MONKEYS *Parapithecus* *Apidium* *Aegyptopithecus* *Propliopithecus* *Aeolopithecus*	*SPECIALIZED ADVANCED PROSIMIANS* Adapids and early Tarsids
PROBOSCIDS AND SIRENIA *Palaeomastodon* *Phiomia* *Moeritherium* *Eotheroides*	*PERISSODACTYLS* *Palaeotherium* *Protapirus* *Baluchitherium*
HYRACOIDS 6 Genera	
EMBRITHOPODA *Arsinotherium* (unique monospecific order)	

Hyracoids, proboscids and anthracotheres appear to be the dominant herbivores, while hyaenodonts were the dominant carnivorous animals. As very small mammal bones tend to break up or be overlooked, one gets a limited idea of the relative importance of smaller groups from fossil deposits; nevertheless, the successive displacement of the primitive hyracoids and anthracotheres by bovids and that of the hyaenodonts by true carnivores is very striking.

Little is known of physical conditions in Africa during this period but the continent was, in general, flatter and with less relief. The climate is thought to have been more equable than today.

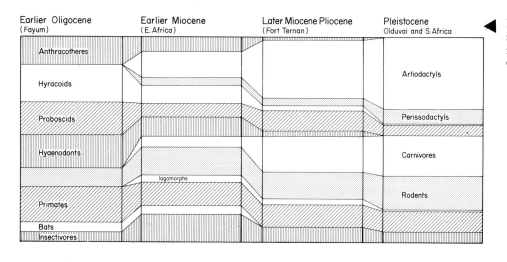

Diagrams showing relative importance of orders in African fossil sites (modified from H. B. S. Cooke, 1968).

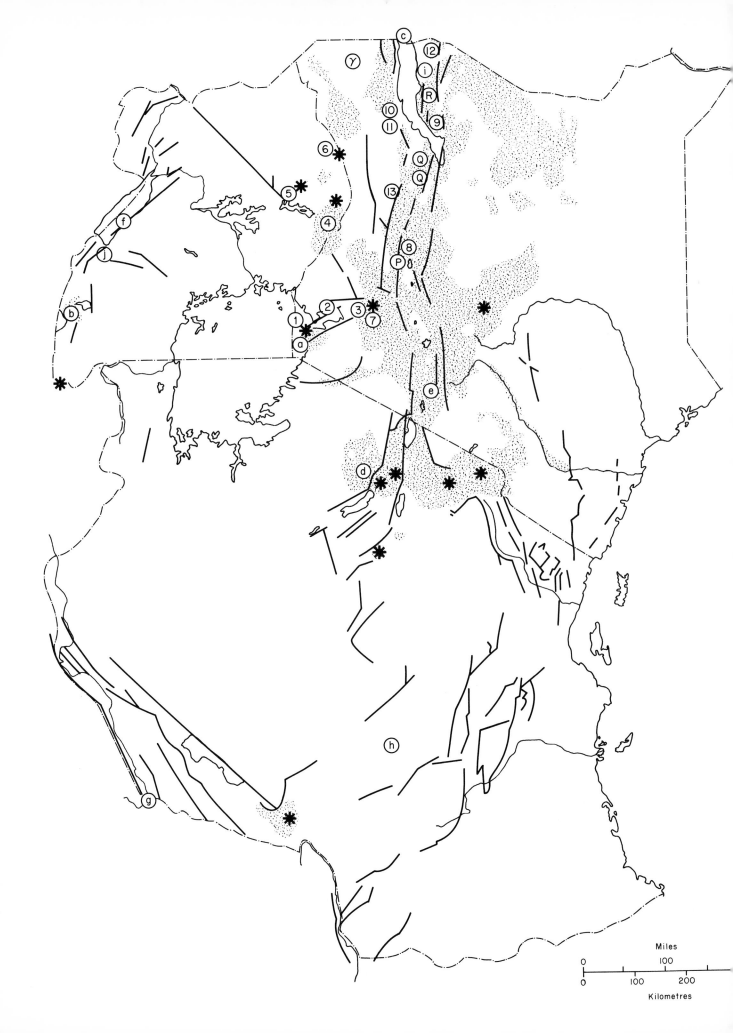

The Miocene

It is with the Miocene that our first knowledge of the past mammals of East Africa begins. Deposits are numerous in eastern Uganda and western Kenya, mostly occurring on the sides of extinct volcanoes that were intermittently active during the period. The facing map shows some of the best known sites and indicates the extent of the Tertiary volcanics to which we owe the preservation of so many Miocene and Pleistocene fossils. The first major faulting of the eastern rift valley is also late Miocene.

The vegetation was probably not very dissimilar in type to that found on East African mountains today. For instance, the commonest fossil fruits in many of the Miocene deposits are those of *Celtis*, a forest tree that is still widespread in Uganda.

With the Miocene there are marked changes in the composition of fauna both in Africa and in Eurasia, and some faunal interchange had taken place at the end of the Oligocene or early Miocene. Many elements can be recognized as having developed from an indigenous Oligocene stock; hyracoids and anthracotheres were still common but artiodactyls, principally represented by tragulids and palaeomerycids (early giraffes) were present for the first time. These are among several groups that are thought to have come in from Eurasia. Others are shrews, hedgehogs, rhinos, chalicotheres and a few carnivores (see diagram on following page). Lavocat (1959) has suggested that cricetid mice also entered Africa at the end of the Oligocene and that the dendromurines evolved from them during the African Miocene.

The presence of proboscids and apes in the Miocene deposits of Eurasia suggests that these groups had migrated out of Africa at the end of the Oligocene.

In spite of great morphological similarities the African apes and the Asiatic orang-utan have almost certainly been separate since the early Miocene at the latest (Washburn, 1963), which emphasizes that a close resemblance between members of Oriental and Ethiopian families can give a misleading idea of their temporal relationship. The modern chevrotains resemble the Lower Miocene *Dorcatherium* and Verheyen (1961) in a comparative study of the crania of African colobus monkeys and Oriental langurs concluded that "their genetic patrimony has remained virtually unchanged". While the history of the smaller forest groups common to the Oriental and Ethiopian regions, i.e. cercopithecoid monkeys, lorises and chevrotains remains uncertain for lack of fossils, it is not impossible that ancestral populations of these groups also became established on both continents during the late Oligocene or Miocene.

◀ Map of Tertiary—Recent volcanics with volcanic centres and some important fossil localities.
Key-texture: Tertiary Recent volcanics:
——— = major fault lines.
* = Tertiary volcanics centres.
0 = Some important fossil localities.
Miocene or pre-Miocene site: γ. Lokitaung.
Miocene sites: 1. Rusinga and Mfwanganu islands, 2. Ombo and Maboko, 3. Songhor and Koru, 4. Bukwa, 5. Napak, 6. Moroto, 7. Fort Ternan, 8. Baringo (Ngorora), 9. Loiengalani (Kajong), 10. Lothidok, 11. Moruarot, 12. Buluk, 13. Loperot.
Pliocene sites: P. Baringo (Chemeron, Lukeino Kaperyon), Q. S. Turkana (Kanapoi, Lothagam Ekora), R. E. Turkana (Kubi Algi).
Pleistocene sites: a. Kanam, b. Kazinga, c. Omo, d. Olduvai and Laetolil, e. Olorgesailie, f. Kaiso, g. Kalambo, h. Isimila, i. E. Turkana (Koobi Fora, Ileret), j. Kisegi Wasa.

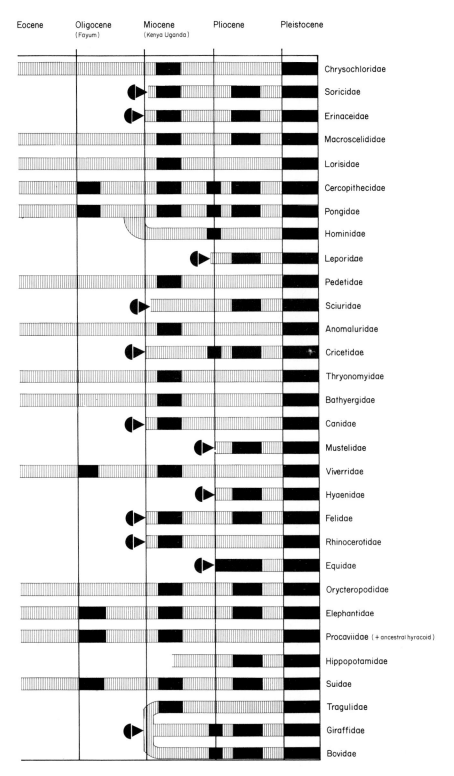

▲ A diagrammatic reconstruction of the geological history of contemporary Families represented as fossils before the Pleistocene.
Black lines = known fossil occurrences of members of Family or ancestral form.
Broken line = inferred existence in Africa.
◖▶ = group thought to have come into Africa from Eurasia. (Adapted after Cooke, 1968 and Simons, 1968.) *NOTE :* Diagram does not represent a time scale.

At the very end of the Miocene there is a change in the composition of fossil fauna. The evidence for this comes from a single deposit at Fort Ternan in West Kenya, which is about 12—14 million years old, placing it on the boundary between the Miocene and Pliocene. In this deposit an important hominid fossil, *Ramapithecus wickeri*, appears and there are for the first time antelopes, mastodonts and giraffes.

There was probably a good land connection between Africa and Eurasia during the late Miocene and the sudden appearance of true giraffes and antelopes in the fossil record has lent weight to the long-held theory that these groups, together with many others, invaded Africa from Asia at this time. This view is no longer so firmly or so widely held. Cooke, 1968, has suggested that these groups are probably indigenous. "It is most unlikely that the *Bovidae* are directly related to the *Antilocapridae* of the new world, which probably arose independently from a similar basic pecoran stock. This would imply that giraffids, and bovids are essentially of African origin and that the cervids and antilocaprids are parallel developments outside Africa from a common stock of late Oligocene age". All that can be said at present is that the time and place for the evolution of bovids is obscure. Mayr (1954) has pointed out that too much reliance can be placed on isolated fossil sites, "many palaeontologists have postulated various kinds of typostrophic "saltations" in order to explain the absence of crucial steps from the fossil record. If these changes have taken place in small peripherally isolated populations it would explain why they are not found by palaeontologists".

The Pliocene

After Fort Ternan there is very little fossil evidence until the Pleistocene, but Africa is thought to have been isolated from Eurasia again for much of the Pliocene (Cooke, 1968). There are two minor Pliocene deposits in Egypt and another in Tunisia, and new Pliocene sites recently discovered north of Lake Baringo may help to fill some of the gaps in the knowledge of Pliocene mammals in eastern Africa, for there remains a great deal to be learnt about the fauna of this important period.

The temporal and spatial relationship of Africa with Eurasia is of particular interest, as late Pliocene fauna from a large number of widely distributed sites in Eurasia shows great consistency and has some similarities with the Pleistocene fauna of Africa (Romer, 1945).

The Red Sea opened into the Indian Ocean during the Pliocene having been initially invaded in the Miocene by the Mediterranean—Tethys Sea (Cooke, 1968). The faunal stock in Arabia would have been an Ethiopian one and the firm attachment of Arabia with Asia must have effectively transferred many Ethiopian faunal elements before the region was faunally impoverished by becoming almost totally arid. The climatic stability of the earlier Tertiary is thought to have given place in the Pliocene to a more complex pattern with greater seasonal and zonal variation. In the Pliocene and Pleistocene, eastern Africa was subject to great tectonic disturbances with rift faulting and widespread uplift.

New elements appearing in the African Pliocene deposits are mustelids, hyaenas, hipparionids and hippos, while mastodonts give way to elephantids.

Two important sites with similar fossils Omo—north end of Lake Rudolf—and Kaiso—Lake Albert—were formerly thought to be early Pleistocene. The former site has on isotopic dating been shown to be in the region of four million years old, which brings into question the definition of the end of the Pliocene in Africa. "The most unsatisfactory aspect of late Tertiary and Quaternary geology is the inadequate definition of the base of the Pleistocene—the Plio-Pleistocene boundary—and the lack of agreement concerning the criteria for determining the integral stages of the Pleistocene period" (MacCall, Baker and Walsh, 1967).

The Pleistocene

Years	Olduvai	Other E.A. sites	Congo Basin	Sahara and Chad	Earth Movement	Fauna and Climate
3000				Sahara hot / Lake Chad present size		
8,850				MegaChad extensive		
10,000			Semi-arid in west	Sahara temperate		
15,000		Ruwenzori glaciers begin to shrink				
18,000						Forest belt from Congo to West Kenya
22,000		E.A. Mountains colder and drier		MegaChad full (after long dry period)		
54,000			Kalahari sands extensive			
200,000					4th Faulting episode (minor, local)	
400,000	Bed 5					Olduvai fauna as today
500,000	Bed 4	Kanjera beds				Many extinct mammals at Olduvai
	Bed 3	Rawi beds				
	Bed 2	Lake Victoria formed				Giant extinct herbivores
1,700,000	Bed 1				3rd Major Faulting	
250,000 to 4,000,000		Omo Laetolil, Kaiso				

Pleistocene events : A table indicating the type of events and correlations that may assist the reconstruction of mammalian evolution. (In many cases the dating given above should be regarded as provisional and tentative.) (Data from Moreau, 1966; Leakey, 1966; Livingstone, 1962 and Bishop, 1967.)

Unlike the Pliocene, Pleistocene beds are very rich in fossils representing many species of extinct herbivores, carnivores and primates that lived in open habitats. The climate was subject to great fluctuations and the boundaries of climatically controlled vegetation zones would have expanded and contracted very considerably during this period. The tectonic activity that marked the later Pliocene continued in the early Pleistocene and there was a short recurrence in the late Pleistocene. Attempts have been made to correlate Pleistocene climate in Africa with the northern glaciations. The dating of the

"Pluvials" that were thought to coincide with the glaciations has not stood up to geological evidence, and the belief that the series of sediments at Olduvai and other Pleistocene sites represents a sequence of "Pluvials" is now untenable. "It is apparent . . . that the breaks in the succession are in most cases due to episodes of tectonic activity or volcanism" (MacCall, Baker and Walsh, 1967).

Dating in the Pleistocene is therefore under revision at present and few data are certain. However, a chart of Pleistocene sequences (see facing page) with some tentative correlations within the continent and even more tentative datings is offered as an indication of Pleistocene "events" that are relevant to the history and evolution of contemporary fauna.

The principal site in East Africa is the famous Olduvai gorge where Dr L. S. B. Leakey continues to excavate an astounding and increasing number of extinct species of mammals, as well as the widely publicized hominids. The intimidating list of species shows antelopes and pigs to have been particularly numerous. Living forms are present in the early Pleistocene, but only begin to be numerous in the middle Pleistocene while the fauna of the uppermost bed is basically modern.

It is interesting to consider the contemporary status of some of these species represented at Olduvai. The blackbacked jackal, *Canis mesomelas*, the lesser galago, *Galago senegalensis*, the giraffe, *Giraffa camelopardis*, the hyaena, *Crocuta crocuta*, the caracal, *Felis caracal*, and the black rhino, *Diceros bicornis*, are found throughout the Pleistocene deposits. These species are exceptionally successful, stable and well-adapted types with a very wide distribution in Africa today. Other species known from early Pleistocene deposits have now a rather limited distribution. The tree mouse, *Thamnomys dolichurus*, the striped hyaena, *Hyaena hyaena*, and the white rhino, *Ceratotherium simum*, are examples. The Otomyinae are amongst the most abundant of rodents in all African Pleistocene deposits and were probably widely distributed and successful animals. Today they are found in scattered local refuges where conditions have remained fairly stable and where competition from ecologically equivalent species may not be so keen. One species, *Otomys denti*, that is found at all levels at Olduvai, now occurs as a relic species on a few widely separated mountain massifs and in parts of Uganda and the western Congo.

The shaggy rat, *Dasymys incomptus*, the grass mouse, *Rhabdomys*, the pouched mouse, *Saccostomus*, the mole rat, *Tachyoryctes splendens*, and the brown hyaena, *Hyaena brunnea*, are all rather locally distributed today but their presence in the Olduvai deposits implies a former abundance and suggests that these species also may be in decline.

In the deposits at Omo, north of Lake Rudolf, a bongo, *Boocerus nakuae*, has been found associated with such dry-country forms as gazelles, oryx and kudu. The contemporary bongo is a forest species with a very scattered and local distribution in both lowland and montane areas. In this there may be some resemblance with *Otomys denti* and with the okapi, *Okapia*, for bongos also may have previously occupied a wider geographic and ecological range but have now retreated into forest. Entry into the forest could be a late development and as for the okapi may be as much a product of competition (and possibly predation) as a special ecological adaptation. It must be admitted however that the area sampled by the Proto-Omo River might have covered a range of ecological habitats.

The contemporary fauna of the horn of Africa shows perceptible affinities with some of the fauna from Olduvai. Naked mole rats, *Heterocephalus*, an early dik-dik, *Praemadoqua*, and numerous gazelles suggest that the climatic fluctuations of the Pleistocene brought Olduvai more than once within the Somali arid or semi-arid focus. During wet periods lakes formed even in Somalia, yet the area remained largely semi-arid and in Olduvai too the wet periods never established forest. Throughout the rich succession of beds at Olduvai no trace of a forest mammal has been found and the evidence for wet periods has been inferred from aquatic fauna, lake levels and dark soil deposits. Olduvai lies today within a rain shadow, but the improvement in both total and seasonal rainfall would have had to be immense to allow the establishment of forest, so that the Olduvai fossils may represent a relatively small part of the fauna of the period, and this is incomparably the richest Pleistocene site in Africa.

The contemporary fauna of East Africa is a remnant of a great Pleistocene radiation of mammals in Africa. Apart from fossils our only guide to the history of mammals is provided by an examination of the distribution of contemporary mammal communities, species and races. The particularly complicated distribution patterns in East Africa are largely due to change and long-term oscillations of climate that have isolated some species and led to the expansion of others. A discussion of some interesting distribution patterns follows.

Climatic Change

Discussion of climatic change in Africa has tended to centre on evidence indicating climatic extremes, with a corresponding interest in relics of forests or deserts. This interest has given a special value to many tiny patches of forest in East Africa and to their fauna.

Extreme climatic changes took place in the northern hemisphere in the Pleistocene, and there is evidence that Africa was considerably influenced by the lowering of temperatures and by changes in atmosphere circulation, particularly during the last glaciation. The tying up of vast bodies of water in ice to the north lowered the ocean level by some 100 m and the consequent extensions of land surface may have some bearing on the fauna of Zanzibar and other oceanic islands.

Although forests may have been extensive during wet periods, it would be wrong to blanket Africa with forest or desert at any time in the past. East Africa in particular has a topography and climate of unusual complexity and it is not likely that a detailed reconstruction of past climates or vegetation distribution will ever be possible, although broad patterns are apparent.

The influence of environment on mammals is immediately obvious in their uneven distribution. Often, the occurrence of mammal species in any particular locality seems to be capricious, and this is due to the extraordinary complexity of East Africa's topography and vegetation, the considerable climatic changes that have taken place in the past and the very varied responses of mammals to their environment. These factors and the parts played by disease and competition are hidden. Apart from the well-known and widespread extermination of larger mammals in this century, almost all the decisive events leading to present distribution patterns have occurred in a past that is without written record.

Where there is a consistent pattern in the distribution of mammal communities inhabiting a specialized habitat, and where there is some supporting evidence from the natural history of these species, it is possible to suggest how (but scarcely ever when), mammals got where they are today.

The listing of species within an ecologic or geographic unit and the comparisons of lists with other units, extracting endemics and species in common, is a useful and objective way of measuring the degree of connection or affinity. By these means, "indicator species" exclusive to certain special habitats or areas can be used to demonstrate connections. However, we must be aware of differentiation within a vegetation zone which may thereby offer refuges for animals from a different habitat as occurs for example in valleys or hill tops. Also, we must be reasonably certain of the limits of the animal's adaptability.

Deserts of the Past

In addition to the Sahara, which need not concern us here, there are in northeastern and southwestern Africa animal and plant communities that are adapted to dry conditions and, prominent in the fauna, are species with a broken distribution that are common to both areas. These species have been invoked as evidence of a past arid corridor connecting what have been called the desert foci of Africa. Both these areas had a comparatively dry climate throughout the Pleistocene and also probably during the Pliocene, providing a stable focus for animals and plants adapted to dry country. It is believed that some forms could not have survived a high rainfall or even a mild amelioration of drought conditions, but it is unlikely that this is true for any species common to the two areas.

There are a number of endemic species of mammals, birds and plants confined to each area, clearly indicative of an ancient ecological separation. The species common to the two areas have a much wider distribution in East and in South Africa; these are animals of the fringing communities which can adapt themselves to various stages in the succession towards (or away from) moister conditions. Populations of spring hare, *Pedetes*, caracal, *Felis caracal*, aardwolf, *Proteles*, bat-eared fox, *Otocyon*, dik-dik, *Rhynchotragus*, and oryx, *Oryx*, are separated by as much as 1,000 km, but it is misleading to regard them as desert forms. Their discontinuous distribution only suggests that areas of dry acacia bush and savanna, rather than really arid communities, were sufficiently widespread in the past to connect the horn of Africa with Southwest Africa.

The Forests of the Past

The more adaptable the animal, the more unsatisfactory it is as an indicator. It is for this reason that forest animals have received most attention as they are generally fairly rigidly tied to their habitat. Even so, forest mammals

▲
Past and present distribution of *Oryx gazella* to illustrate discontinuous distribution of arid and semi-arid adapted species.

63

may disperse into woodlands adjacent to forest areas, particularly during the wet season. Thus it is possible, if dry seasons were not severe over a long period, that more adaptable forest mammals might be able to pass through large tracts of woodland that never achieved a full succession to rain forest, and could thereby have colonized isolated forest relics. Notwithstanding this theoretical qualification, the distribution of forest fauna in East Africa is the most impressive evidence for very much wetter climates in the past.

Naturalists and scientists have from the earliest times remarked on the striking cleavage of life forms in Africa into forest and non-forest forms. "Differentiation of the lowland forest biome from the savanna biome is probably the oldest on the continent (at least as old as the Miocene)" (Moreau, 1952). The diagram below shows the cleavage expressed as percentages of species within each order represented in East Africa.

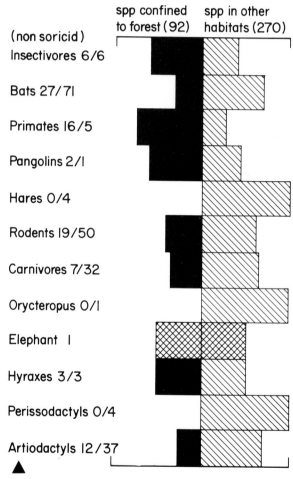

East African mammals restricted to forest or found in other habitats. Expressed as proportions of species in each order.

Moreau has also brought attention to the mountain forest relics of East Africa and their faunas. He has suggested that for most of the northern glacials, generally cooler conditions would have occurred and forests might have been extensive in wetter areas, particularly during the severest period of the last glaciation and he has proposed that a "montane biome", with a lower limit of about 450 m above sea level in contrast to the present 1,500 m, might have allowed some sort of ecological connection between Ethiopia and southern Africa, and between the Cameroon highlands and eastern Africa. For the reasons mentioned earlier, at Olduvai (altitude 1,000 m) there is no evidence of a forest fauna throughout the Pleistocene, and the location and relative age of African forests and their faunas during this period have to be deduced at present from zoogeographic and phytogeographic comparisons.

A comparison of mammal species confined to forest with the species from all other habitats reveal that 92 species (i.e. a quarter of the total) are restricted to the very small and scattered East African forests (see diagram). This figure strongly suggests that forests played a very much more important part in East Africa's past than they do today.

It has been customary to regard forest animals in East Africa as West African intrusives and, although these undoubtedly occur, this regional name, attached because the animals were first collected in West Africa, blurs vital distinctions if it is used indiscriminately for all African forest fauna, and it is as well to remember the historical caprice that made for this title.

There are today numerous important differences in the forest mammals of the Upper Guinea area, the Cameroon-Gaboon area and Central Africa. The number of endemics confined to each of these three areas is surprising, considering the minor nature of the present geographic barriers and the ecological homogeneity of the African forest belt. Furthermore, there are very few endemic mammals in the central Congo basin south of the River

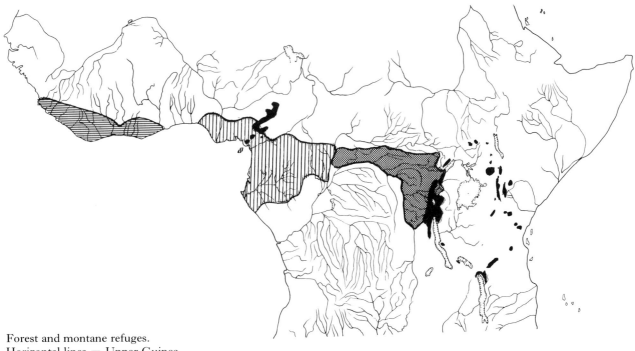

Forest and montane refuges.
Horizontal lines = Upper Guinea
               Refuge.
Vertical lines   = Cameroon Gaboon
               Refuge.
Stipple   = Central Refuge.
Black   = Montane areas.

Congo which, although clothed in forest today, is the site of Kalahari sands which are considered to be the evidence of former desert conditions between 75,000 and 52,000 years ago. Subsequent shorter arid periods are believed to have affected parts of the western Congo basin 40,000 and again 10,000 years ago (Moreau, 1966).

Subject to these vicissitudes, forest would have retreated to the west and to the east, creating widely separated refuges for forest fauna, and a corridor of drier conditions would have connected the northern and southern savannas in between. In this connection there is an interesting anomaly in the distribution of the race of an elephant shrew, *Petrodomus tetradactylus tordayi*, which occurs in the central Congo basin, although the range of this typically Southeast African species is otherwise outside true forest. Instead of retreating as other animals might have done as the climate changed, it looks as though this population may have adapted itself to the forest. It should be noted that this elephant shrew is only found south of the River Congo and not in the forest refuge areas to the west and east. The distribution pattern for the vervet, *Cercopithecus aethiops*, also suggests that this species might have invaded the northern savannas through a Congo corridor (see p. 209).

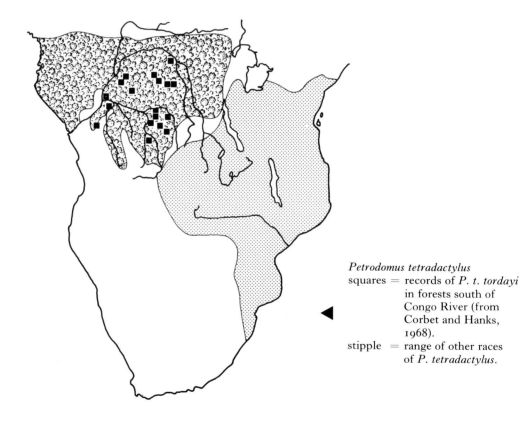

Petrodomus tetradactylus
squares = records of *P. t. tordayi* in forests south of Congo River (from Corbet and Hanks, 1968).
stipple = range of other races of *P. tetradactylus*.

The evidence for older periods of separation may be deduced from distinct mammalian genera confined to West and East Africa. During arid periods a western refuge would have centred on the Cameroon-Gaboon, with a second forest refuge beyond in Upper Guinea. We may regard the otter shrew, *Micropotamogale lamottei*, the mandrill, *Mandrillus*, the golden potto, *Arctocebus*, the anomalurid, *Zenkerella*, and the pigmy hippo, *Choeropsis*, as well as many species and races of squirrels, duikers and monkeys as endemic to the West African refuges.

The Central Refuge (which has been called "Ituri-Maniema" by Misonne, 1963) would have been forest in the general vicinity of the Ruwenzoris and Kivu. It would have provided a larger and ecologically more varied environment than the Western Refuges, with the additional advantage of being physically accessible from every direction (see map, p. 65). This area contains the largest number of forest species, 15% of which are restricted to this area (Bigalke, 1968). Forest species endemic to the Central Refuge are the Ruwenzori otter shrew, *Micropotamogale ruwenzorii*, the hero shrew, *Scutisorex*, the owl monkey, *Cercopithecus hamlyni*, the giant genet, *Genetta victoriae* and many others (see Rahm, 1966).

It can be asserted with confidence that past climatic changes, repeatedly isolating and then reuniting the forest areas of Africa have been a major mechanism in the speciation of forest mammals. Monkeys, duikers, squirrels and other rodents show morphological differences that frequently coincide with the forest refuges. It is sometimes possible to perceive differences in activity, behaviour or ecological preferences which follow serial steps indicative of directional evolution or orthoselection. Too little is known at present of the detailed natural history of these species to allow much more than guesses as to the behavioural and ecological correlates, but the field holds great opportunities for future research. A single example can be drawn from the *Malacomys-Colomys* (rat) group, where differences cutting across "subspecies", "species" and "genus" show a more or less neat coincidence with the refuges. Furthermore, the differences, although varying in degree, show a steady direction of evolution, in which the most superficial and obvious trends are a more velvety fur and greater contrast of colour, correlated with an ecological and adaptive development towards aquatic habits (see Vol. II). As the murid mice are thought to have appeared on the African scene during the late Pliocene, the invasion of the forest by this group is probably a Pleistocene development.

This series is a reminder of the commonplace that categories are the creation of classifiers and not nature.

UPPER GUINEA REFUGE	CAMEROON—GABOON REFUGE	CENTRAL REFUGE
Malacomys longipes cansdalei ———	*Malacomys longipes longipes* —— *Colomys bicolor* ———	*Malacomys longipes centralis* *Colomys goslingi*

Mountains and Refuges

The smaller Upper Guinea Refuge lacks a well-marked topography but both the Cameroon-Gaboon and Central Refuges include high mountain massifs. These refuges therefore include several distinct biomes, which have persisted in spite of climatic fluctuations that must have altered greatly the relative proportions of the afro-alpine (Paramos) zone, the montane forest and the lowland forest. Altitudinal distinctions in many birds and a few mammal species suggest that some highland forest areas might have been physically separate from lowland forests during dry periods. A widespread adaptability on the part of forest mammals with regard to altitude and temperature, and the rarity of very clear distinctions between montane and lowland forest mammals may in itself be indicative of the demands put upon them by "Montane Biomes" as postulated by Moreau.

Mammals that seem to be strictly confined to low-lying forests (*Manidae* (pangolins), *Haemoschus* (chevrotain), *Neotragus batesi* (pigmy antelope), *Cercocebus albigena* (mangabey)) reflect in their limited distribution the very restricted extent of low-altitude forest in eastern Africa. Low-lying forests of any great extent might have been maintained only in the vicinity of the Central Refuge and along the northeastern margins of the Congo basin. Moreau has pointed out that the Semliki and Ituri Valleys might have provided a stable lowland refuge for birds throughout the periods when montane conditions were very extended. In spite of the small area involved, a very large number of mammal and bird species exists hereabouts (an area of which Bwamba is a part), bringing its peculiar fauna within the compass of this work.

The sharp distinctions that exist between montane and lowland forest birds in Africa are not as common in the more adaptable mammals. However, an altitudinal distinction is apparent between two related sympatric squirrels occupying Uganda forests. The range of one species, *Paraxerus boehmi*, is more extensive both in area and altitude, being found at altitudes up to 2,000 m and as far south as Malawi. The other, *Paraxerus alexandri*, is found only in lowland forest below 1,300 m in Uganda and northeastern Congo. This boundary agrees with that of other mammals and birds, which tend to divide at around 1,500 m.

A white-bellied sun squirrel, *Heliosciurus ruwenzorii*, is exclusive to the montane forests of the eastern Congo, Kigezi and Ruwenzori. Another squirrel, *Funisciurus carruthersi*, is restricted to montane forests between Ruwenzori and Lake Tanganyika.

The mountain monkey, *Cercopithecus l'hoesti*, in common with many other "typically montane mammals" is also found in lowland forest. There are other montane species that extend their range above the forest; the harsh-furred mouse, *Lophuromys woosnami*, is found from 1,800 m above sea level to 2,700 m on the Kigezi Mountains and Ruwenzori. The tree-hyrax, *Dendrohyrax*, and the golden mole, *Chrysochloris*, are dominant species in the high altitude paramos that are also found in montane and lowland forest. There are also species that have become confined to the alpine and subalpine zones above the forests on several widely separated mountains. The duiker, *Cephalophus nigrifrons*, is a rare duiker on the Cameroon mountains and in lowland forest in the Cameroons and Gaboon, but it is a successful and common colonist of the higher altitudes on the Bufumbira volcanoes, Ruwenzori, Elgon and Mt Kenya. The species shows subspecific differences in every case and the nature of their variation suggests that the group may have originated in the Cameroons (an area with many species of red duiker). Perhaps during a period of lower temperature the intervening country became suited to the duikers' expansion, yet, whatever the past climatic situation, it must have been very different from today; for over its eastern range the species is strictly limited to an altitudinal stratum above 3,400 m. On Elgon, it shares the paramos habitat with a thick-coated race of the bushduiker, *Sylvicapra*, while the forest below 3,400 m is occupied by another duiker, *Cephalophus callipygus*.

The situation for the relic species *Otomys denti* and *Rhabdomys* on mountain refuges was mentioned earlier. A group of archaic shrews, *Myosorex*, shows similarities in pattern. The genus is well-known from fossils at Olduvai, and the most primitive living member of the genus, *Myosorex schalleri*, has been found in lowland forest in the eastern Congo. In the isolated Uluguru mountain block, *Myosorex geata*, is known to range from 600 m to 2,300 m. Two other species, *Myosorex caffer* and *Myosorex varius*, occur at low altitudes in South Africa, where the generally lower temperatures of the southern latitudes and the geographical position as a continental cul-de-sac, may be important determinants. Otherwise the genus is confined to high altitude grassland; *M. preussi* on Mt Cameroon, *M. blarina blarina* on Ruwenzori, *M. b. babaulti* on the Bufumbira, *M. zincki* on Kilimanjaro, and *Myosorex* (subgenus *Surdisorex*) *pollulus* on Mt Kenya, *Myosorex* (*Surdisorex*) *norae* on the Aberdares. This last subgenus has become specialized and it fills a similar niche to that of the golden mole, *Chrysochloris*. Perhaps the absence of the latter from Mt Kenya has hastened the development of this specialization, for *M. pollulus* is the only endemic species of mammal on Mt Kenya. Mountains provide specialized habitats, particularly at higher altitudes, and they also become refuges for older, formerly widespread species in retreat; these two factors are inseparable in the case of *Myosorex* (Heim de Balsac, 1967).

Northern and Southern Forests

The Central Refuge shows differences that are not determined by altitude. There are a number of mammals with a northerly distribution which do not occur to the south, where there are different species or races.

There is the possibility that earlier forest blocks in East Africa had a generally more southerly distribution than they have today. As theories of past climates rest on assumptions as to whether the equator was stable or not, the literature on equatorial instability is considerable. It is thought that the equator might have oscillated as much as seven degrees to the north and south during the Pleistocene alone. A shift in the distribution of East African forests may therefore be a reflection of climatic alterations caused by these terrestrial movements. The problems raised need not be gone into here, they have been summarized by Moreau (1952).

The species and races in the northern forests are sometimes ecologically equivalent to those in the south and the "watershed" between forms is often perceptible in the vicinity of Lake Kivu and in the East Congo.

The blue monkey, *Cercopithecus mitis stuhlmanni*, is widely distributed both east and west of Ruwenzori, but to the south of Lake Edward it is largely replaced by *Cercopithecus mitis doggetti*. The black and white colobus, *Colobus abyssinicus*, occurs north of the Congo River and extends south as far as Lake Kivu. South of Kivu, races of *Colobus polycomos* replace this species (one race, *Colobus polycomos cottoni*, ranges still further north, where it is sympatric with *Colobus abyssinicus*). The range of numerous rodent species also extends no further south than the vicinity of this lake. An interesting variation of this pattern can be seen in the distribution of the giant duikers: *Cephalophus spadix* ranges from Ruanda southwards, where it is found east of Lake Tanganyika in scattered montane forests as far as the Usambaras and Kilimanjaro. The larger *Cephalophus sylvicultor* occupies the remaining forest areas: West Africa, the Congo basin, Kenya and Uganda forests (see map, p. 71). As a larger and more elaborate species, this duiker probably evolved more recently and may have replaced *Cephalophus spadix*. If so, the pattern

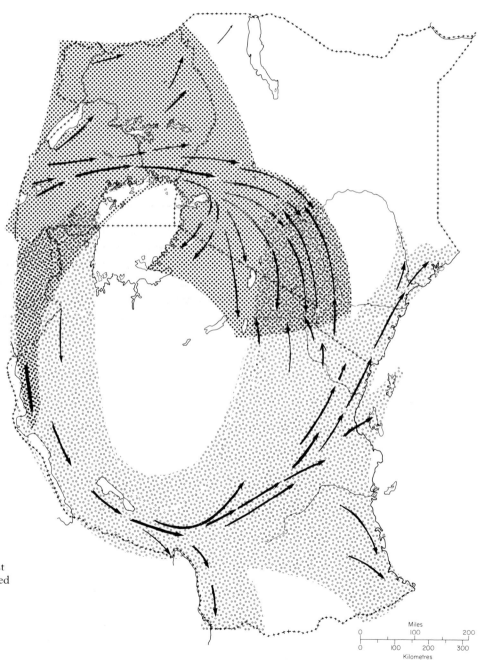

"Northern" and "Southern" forest faunas in East Africa with suggested dispersal routes (arrows).

suggests that the southeastern forests were colonized by forest species at an early date, but a climatic shift closed this dispersal route to more recent radiations of forest species.

A rather similar pattern may be seen with the pouched rat *Beamys*, a rare forest genus with a relic distribution restricted to eastern Africa. One species occurs in parts of Malawi and the Southern Highlands of Tanzania, the other along the Kenya littoral and in North Tanzania. This primitive genus is probably closely related to the ancestral stock of the very widely distributed and successful *Cricetomys* which occupies both forest and dry country habitats over much of sub-Saharan Africa (Hanney and Morris, 1962). A rare and little known species of mongoose, *Rhynchogale melleri*, also seems to have a very scattered distribution restricted to southeastern Africa.

The north-south division of fauna that is perceptible in the eastern Congo becomes very much more marked further east.

The East African forests are predominantly montane and are all now in isolated blocks, some of them very small. Forest mammals have colonized these forests by two different routes, one connects Ruwenzori and the forests of the Congo with Elgon and the Kenya Highlands, the other passes down the eastern shore of Lake Tanganyika to the montane areas of the Southern Highlands and on to the coast. Mammal populations along the northern route are generally indistinguishable racially from the animals in the eastern Congo. Although earlier expansions and contractions of forest may have established a few mammals on the mountains, the greater part of the present mammalian forest fauna of western Kenya has been recently in continuous connection with the fauna of the Central Refuge. The lack of older forms suggests that these forests were formed and colonized by mammals very recently, perhaps during a prolonged wet period about 30,000 years ago. This period followed a phase of tectonic activity that contributed to the present topography of the rift valley in Kenya. The connecting forests would have followed the northern shores of Lake Victoria, which is also thought to be of mid-Pleistocene formation.

To the southeast of the Central Refuge, forest mammals are found in scattered montane forest relics down the eastern shore of Lake Tanganyika, through the Southern Highlands and the "African Ghats" to the geologically ancient mountain blocks of the East Coast, the Uluguru and Usambara Mountains and on to the coast, where the last traces of forest fauna peter out at about the Somalia border. In the extreme south of Tanzania there are forest relics near Liwale and on the Rondo Plateau, while fauna with forest origins is found still further south. The forests along the southern route are now very small and fragmented but there are some endemic species, for example the suni, *Neotragus moschatus*, the red duiker, *Cephalophus adersi*, the squirrels, *Paraxerus vexillarius* and *Paraxerus lucifer*, and the elephant shrew, *Rhynchocyon chrysopygus*.

These phylogenetically ancient groups probably represent relics of formerly widespread types and also species that have evolved within this area, so that the forests which harbour them, even if of very small extent, must be of great age. There is a likelihood here of a succession of disconnections and reconnections with the Central Forest Refuge which might have augmented or changed the fauna from time to time; the red colobus, *Colobus badius*, the

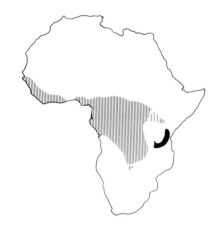

Distribution of giant duikers:
vertical lines = *Cephalophus sylvicultor*.
black = *Cephalophus spadix*.

pied colobus, *Colobus polycomos*, the mangabey, *Cercocebus galeritus*, the monkey, *Cercopithecus mitis* and the tree-hyraxes, *Dendrohyrax arboreus* and *Dendrohyrax validus*, can hardly all have come east at the same time. Indeed these species suggest that representatives of successive faunal eras are superimposed on one another in this area.

A radiation of soft-furred, grey-bellied mountain squirrels, endemic to the ancient mountain blocks of Tanzania, Malawi and Mozambique shows that within this region there was probably an expansion and contraction of forests which did not involve connection with the main African forest block but which did join up the isolated montane islands of the present, causing the speciation of four distinct mountain squirrels, i.e. *Paraxerus byatti*, *Paraxerus vexillarius*, *Paraxerus vincenti* and *Paraxerus lucifer*. This radiation is discussed and illustrated in detail in Vol. II.

The differences between northern and southern forest faunas become more apparent when a comparison is made of the known mammals of two forest areas well to the east of the Central Refuge, the Kaimosi-Mau and the Southern Highlands.

SOUTHERN HIGHLANDS	KAIMOSI-MAU
Chrysochloris stuhlmanni	
Rhynchocyon cirnei	
Petrodomus tetradactylus	*Manis tricuspis*
	Perodicticus potto
Cercopithecus mitis moloneyi	*Cercopithecus mitis stuhlmanni*
Colobus polycomos	*Colobus abyssinicus*
Colobus badius gordonorum	*Cercopithecus ascanius schmidti*
Heliosciurus gambianus	*Heliosciurus rufobrachium*
Paraxerus lucifer	*Protoxerus stangeri*
Paraxerus byatti	
	Atherurus africanus
	Hylomyscus stella
Otomys anchietae	*Otomys typus*
	Hylochoerus meinertzhageni
Cephalophus harveyi	*Cephalophus callipygus*
Cephalophus spadix	*Cephalophus sylvicultor*
Neotragus moschatus	*Boocerus eucerus*
	NOTE : All species on this list are indistinguishable or very closely related to western Uganda populations.

At least twenty mammals are common to both areas but almost all of these are regarded as being subspecifically distinct.

Both areas are montane forest except for Kaimosi (1,500 m) and this lower altitude may determine the presence of *Manis tricuspis*. Both lists are no doubt incomplete, but the species in the Southern Highlands are surprisingly numerous considering the small size of the relic forests. The faunas will be seen to be distinct at almost every level; there are species in both areas that are not found in the other, while those that are shared are nearly all different subspecies.

The difference between the northern and southern forests are not absolute and in the few scattered relic forests that occur in the intervening 1,000 km there is some overlap of fauna. Mammals typical of the northern forests, *Cercopithecus ascanius* and *Protoxerus stangeri*, reach as far south as the Kungwe massif, half way down the eastern shore of Lake Tanganyika, while those characteristic of the southern forests are found as far north as the Sango Bay forests where, instead of the northern *Colobus abyssinicus* and *Dendrohyrax dorsalis*, one finds *Colobus polycomos* and *Dendrohyrax arboreus*. This forest has many species of trees (*Podocarpus*, *Ilex*, *Trichocladus*) which are usually restricted to montane forests; they may have been established on the seasonal flood plains of Lake Victoria by means of the Kagera River. It is, on the other hand, much more likely that Sango Bay is an ancient forest surviving under peculiarly favourable edaphic and climatic conditions. It furnishes some indirect evidence for Moreau's "Montane Biome", representing as it does many elements of a montane community at 1,100 m above sea level (although typically montane birds have not been found there). Under a climate not much cooler or more humid than the present one, forests of this sort might have linked Ufipa and the Southern Highlands with the Central Refuge through Kungwe and the Lake Tanganyika drainage system.

The animals dispersing along the northern route have faced fewer and less formidable barriers. Nonetheless, a number of the mammals found in western Uganda failed to get as far east as Kenya (or conceivably might have died out or retreated). According to present knowledge, the Nile is the eastern boundary along this route for the elephant shrew, *Rhynchocyon*, the squirrels, *Funisciurus pyrrhopus*, *Paraxerus boehmi* and *Paraxerus alexandri*, the rats, *Deomys ferrugineus*, *Hybomys univittatus* and *Malacomys longipes*, the tree-hyrax, *Dendrohyrax dorsalis*, the hero shrew, *Scutisorex*, and the black mangabey, *Cercocebus albigena*. The river is of course a formidable and ancient barrier, perhaps more so for small terrestrial animals like *Rhynchocyon* and *Scutisorex* than for others, but some typical mammals of the Central Forest Refuge are limited to its immediate vicinity without very obvious barriers to their expansion; for example, the pigmy antelope, *Neotragus batesi*, the forest mongoose, *Crossarchus alexandri*, and the rat, *Aethomys longicaudatus*. Moreover, there must have been many subtle interactions involving climatic tolerance and competition with other species, which have been operating over long periods of time with considerable fluctuations of temperature. These factors are by their very nature obscure.

Some of these species probably represent populations that have spread east from West Africa. There could have been a lengthy time-lag between the initial forest expansion into Kenya (taking Central Forest Refuge mammals with it) and the arrival of successful West African colonists. *Dendrohyrax dorsalis* looks rather like a "late arrival"; it may perhaps have failed to cross the Nile because the formerly broad forest front had contracted by the time it reached its present limits. This species appears to be in the process of supplanting and hybridizing with the East African tree hyrax, *Dendrohyrax arboreus crawshayi*, which preceded it in Uganda and is the typical Kenya forest hyrax. Another species following the northern edges of the forest that may also be a "late arrival" is *Cephalophus rufilatus*, a West African species with a distribution that goes no further than the banks of the river in West Nile district.

For *Funisciurus pyrrhopus*, *Deomys ferrugineus*, *Hybomys univittatus*, *Malacomys longipes* and other species the factors delaying or limiting their eastward advance remain to be sought. Meanwhile the physical barrier of the Nile remains a convenient if rather inadequate explanation.

The distribution of *Cercopithecus neglectus* has a curious pattern. This monkey is absent from the south-eastern forests but has an extensive and very patchy distribution in the Congo basin, Uganda, western Kenya and southern Ethiopia. The pattern implies a very successful expansion at some time after the southern forests had become isolated, but also a more recent "withering away" over much of its range. It is suggested in the profile of this species that its declining status may be due to displacement by a lately-arrived monkey that had originally evolved in West Africa, *Cercopithecus ascanius*.

Some species from the outlying parts of the northern forests have sub-speciated, for example the black and white colobus east of the Gregory Rift (*Colobus abyssinicus*), *Colobus a. kikuyuensis* and *C. a. caudatus*. While this may seem to imply that these monkeys have been established there for a very long time, it is more likely to be the result of relatively recent isolation, for the same species has minimal variation over the rest of its range, where there has presumably been little interruption of gene-flow.

One ancient species that occurs east of the Nile is a golden mole, *Chrysochloris stuhlmanni*, which is found on two older mountain blocks, Elgon and the Cherenganis. It is interesting that the only definitely ancient relic species east of the Nile is a subterranean form. Perhaps its cryptic habits allowed it to survive the period of great tectonic activities of the Pliocene and Pleistocene, when volcanic ash and lava might have destroyed the earlier fauna of the Kenya Highlands over a very wide area; at any rate, this species is somewhat of an anomaly in the fauna of Kenya.

The faunas of the northern and southern forests overlap to some extent at their outermost limits, but the mammals of the southern forests meet a minor "Wallace Line" between the Usambara Mountains and Kilimanjaro, since the volcanic mountains east of the rift valley are largely inhabited by mammals coming from the west by the later northern path. However, some of the mammals colonizing these volcanoes have been drawn from those populations already established in forests to the south-east of these mountains. *Cephalophus harveyi*, *Neotragus moschatus* and races of *Cercopithecus mitis* are notable examples. Other southern mammals, *Dendrohyrax validus*, *Paraxerus byatti* and *Cephalophus spadix*, colonized no further than Kilimanjaro. The crested rat, *Lophiomys*, probably colonized Mt Kenya from the north and a local, dry country species, the hyrax, *Procavia johnstoni*, colonized the alpine rocks.

The coastal forests and part of the Usambara forests are low-lying and in many cases they have escaped the full rigours of dry periods because of their

74

proximity to the sea. They are however very impoverished faunally and only contain about 10% of the species found in western Uganda. The isolation of Zanzibar was discussed in Chapter II.

The coastal forests north of Mombasa are of special interest as there are indications that they may have suffered a longer isolation than even Zanzibar. Populations of forest species in this area are generally distinct from those to the south. Examples are listed below:

COASTAL FORESTS NORTH OF MOMBASA (Southern boundary Kilifi creek)	RELATED FORMS
Rhynchocyon chrysopygus	*Rhynchocyon petersi* on Coast, Zanzibar and Usambaras.
Colobus badius rufomitratus	*Colobus badius kirkii* on Zanzibar, *Colobus polycomos* south of Mombasa.
Cercocebus galeritus galeritus	*Cercocebus galeritus* populations in Congo basin.
Paraxerus palliatus tanae	*Paraxerus palliatus suahelicus* on the Coast. *Paraxerus palliatus frerei* on Zanzibar Island.
Cephalophus adersi	*Cephalophus adersi* also on Zanzibar Island.

The differences in the two red colobus populations could be of considerable significance, for the Tana populations have a close resemblance to the Uganda *Colobus badius tephrosceles*, while *Colobus badius kirkii* is very different and has an intermediate related form in the Southern Highlands.

It is possible that there were two ancient connections with the western forests. If this was so, Pleistocene volcanics may have been responsible for the absence of any trace of the older forest fauna west of the eastern rift in Kenya.

In summary, the following broad pattern appears in the distribution of forest fauna in East Africa.

1. The principal reservoir of forest fauna, the Central Forest Refuge, just includes East African territory in Bwamba, where the low-altitude rain forest is part of the Ituri-Maniema Forest. (This area is discussed in detail in Appendix II.)

2. A relic forest fauna occurs in some of the coastal (lowland) forests. Differentiation in some species suggests that these forests have been isolated for a great period of time and it is possible that the tectonics of the late Pliocene and early Pleistocene played an important part in breaking the link with the western forests; the general level of the interior of eastern Africa was raised by 600—1,200 m (Cooke, 1968).

3. There appear to be two principal routes for forest populations moving eastwards. The southern route follows the "African Ghats"; the northern route crosses Uganda and the Highlands of Kenya.

4. The differences in these faunas suggest that the southern route retains many older forms, while the presence of some endemic species on mountain massifs is indicative of isolation over a considerable period. This route is

unlikely to have been continuously linked with the central African forest block for a considerable period of time.

5. The northern forests have a broadly homogeneous fauna and were clearly in recent and continuous contact with the Central Forest Refuge. Many successful and widely distributed forest species in this zone have been unable to reach the southern forests or the coastal forests because of isolation by intervening dry country.

6. The absence of older faunal elements in the forests in the vicinity of the eastern rift could be due to the destruction of fauna by volcanics, while the late formation of Lake Victoria may have been an important factor by encouraging the growth of a broad, bridging belt of forest. The presence of two distinct red colobus on the coast may indicate that both routes are of ancient standing, but nonetheless the "newer" character of the northern forest fauna is striking.

Paths out of the Forest

Earlier it was asserted that the isolation of fauna in the three principal refuges was of major importance for the speciation of forest mammals. This process led to a proliferation of types and the filling of numerous niches within this fundamentally stable biotope, which has been on the whole a closed system.

On the other hand, the forests of eastern Africa, small as they are now, are numerous and have been subjected to fragmentation and isolation probably since before the Pliocene. From extensive galleries of riverine forest in woodland, wooded grassland or uplands they may graduate by stages under drier conditions to thicket or scrub. The general pattern seen today may be closer to an arid extreme but it comprehends the same range of natural vegetation types as have existed for millions of years, and it is the proportions and not the composition of the mosaic that have greatly altered. With drastic climatic changes, populations of forest mammals isolated within this mosaic must usually have perished or hung on in sheltered corners. They were unlikely to have been fed back into the forest system in the way western isolates generally were, but sometimes mammals may have started upon an adaptive path that led out into drier and more open habitats. In the species adapted to dry country it is difficult to parallel the radiations observable in the forest species, as the habitats have not exerted the conservative and localizing forces found in forests. Nonetheless, some savanna species have certainly derived from a forest stock and, in a few mammals, an invasion of savanna can be traced through a cline. A montane forest squirrel mentioned earlier, *Paraxerus byatti*, has a southern Tanzania form, *Paraxerus byatti laetus*, which shows characters anticipating a forest squirrel, *Paraxerus vincenti*, which is isolated on Namuli Mountain in Mozambique. This squirrel in turn has a very close resemblance with *Paraxerus sponsus*, which inhabits the coastal forests of Mozambique and Zululand. This species shows a clinal gradation into *Paraxerus palliatus*, which is distributed along the entire East African littoral, often in dry scrub and savanna. (This radiation is discussed and illustrated in colour in Vol. II.)

Another instance of adaptation can be seen in the squirrels. *Paraxerus boehmi* is a small striped species found in West Uganda, on the eastern shore of Lake Tanganyika and in the Southern Highlands. This animal is exclusively arboreal and confined to well-developed forest. In the vicinity of the Uluguru Mountains there is a slightly larger and more grizzled squirrel, *Paraxerus ochraceus*, with distinct but less contrasting stripes; this species lives in riverine forest and *Brachystegia* woodland. To the north, varieties of this squirrel are found in forest, woodland and thicket from Karamoja to Mt Urguess and seven subspecies are described from Kenya alone. In Somalia and in the northern frontier district of Kenya, a very pale form, *Paraxerus ochraceus ganana*, has lost the stripes and is not infrequently seen on the ground in dry acacia scrub. This may be a superficial and possibly relatively recent development, but the process has been operative for many millions of years, and there is less chance of finding living relics in older groups, particularly unstable intermediate forms. Savanna species that represent an early stage in the radiation of their group have generally survived the competition of later types only by specializing or becoming modified to fill some peculiar niche. This may be the case with the neotragines, where a somewhat broken series of forms show considerable specialization (i.e. the klipspringer, *Oreotragus*). Nonetheless, this series of small antelopes is suggestive of an adaptation to drier and more open conditions by ancestral forest neotragines not unlike the suni, *Neotragus moschatus*.

An interesting case concerns *Galago demidovii*. This species ranges throughout the forested areas of the continent, with a distinct increase in size in the most easterly population, *Galago demidovii thomasi*. In the montane forest relics of southern and eastern Tanzania, along the coast and on Zanzibar, there is an interesting galago exactly intermediate between *Galago demidovii* and *Galago senegalensis*. This galago is described in the profiles under its original name *Galago zanzibaricus*.

Fossils of *Galago senegalensis*, inseparable from the contemporary species, are found throughout the Pleistocene deposits at Olduvai, so that if this species emerged from an isolated eastern population of *Galago demidovii*, as is suggested in the present work, this development could only have occurred in the Pliocene, with the implication that the forests had been drying out and breaking up during this period and had already been isolated for a considerable period. Miocene deposits in Uganda and West Kenya have yielded teeth of a very small galago, *Progalago minor*, which are indistinguishable from those of *Galago demidovii*, so that this species might represent an ancient stock of prosimians.

Paths into the Forest

Some savanna mammals have also invaded or retreated into the forest. The tree hyrax, *Dendrohyrax*, the bongo, *Boocerus*, and the okapi, *Okapia*, are examples. Moisture conserved by valleys and rocky outcrops encourages thicket growth in the rock hyrax habitat; the exploitation of this source of food would seem to have led to more arboreal habits and ultimately to emancipation from the shelter that rocks provided. The behaviour of *Heterohyrax* exhibits the awkward situation of an intermediate form, for when these animals are disturbed by a predator they tumble out of the trees and race for the rocks, while the truly arboreal *Dendrohyrax dorsalis* freezes or, at the most, scuttles for a hollow branch nearby.

For many species of savanna ungulates, the thicker, richer vegetation around rivers or along forest edges provides food and shelter during the dry season. Riverine forests or thickets are extensive in all but the driest areas and were even more widespread for long periods in the past. They are also natural foci for the growth of forest as climatic conditions change, and those animals able to take advantage of this habitat would not, therefore, have been driven out when forest expanded. There are several mammal species that favour an ambiguous forest-savanna habitat; the potto, *Perodicticus potto*, and the greater galago, *Galago crassicaudatus*, are animals typical of the forest edge and of gallery forest; the latter species is also common in woodland. The bongo, *Boocerus*, also favours the forest edge in areas uninhabited by man.

Among the murid mice there are forest species that have clearly derived relatively recently from savanna species. As a whole, the murids in Africa have a degree of resemblance that suggests that they have not diverged a great deal, a condition that is confirmed by fossils. Before the Pleistocene, murids are unknown, but cricetines and still earlier phiomyds and anomalurids are known to have occupied most of the rodent niches, and it is presumed that murids only entered the African scene at the end of the Pliocene.

The murid genus, *Aethomys*, has nine savanna species which are principally restricted to the southern savannas, with a marked radiation in the general vicinity of Angola and central Africa. The main African forest block is occupied by one species, *Aethomys longicaudatus*, which presumably invaded the forest somewhere in the Congo basin. A further forest species, *Aethomys defua*, occurs further west in the Guinea forest. The invasion of forest may be very recent, for *Aethomys longi caudatus* has not colonized the East African forests and this species is only known from our area in Bwamba. The situation of a forest race of *Petrodomus tetradactylus* was mentioned earlier (p. 66). These species may resemble the weaver birds' (*Malimbus* species) that have adapted to forest (see Moreau, 1958).

Many minor climatic fluctuations have certainly occurred that were too slight to have linked up all the forests, so that the opportunity for specially adapted forest animals to move in was limited. This perhaps helps to explain the wide range of animals that flourish equally in forest and savanna, such as the buffalo, *Syncerus caffer*, the bushbuck, *Tragelaphus scriptus*, the elephant, *Loxodonta africana*, the bushpig, *Potamochoerus*, the leopard, *Felis pardus*, the mongoose, *Herpestes sanguineus*, the ratel, *Mellivora capensis*, the dormouse, *Graphiurus murinus*, and the elephant shrew, *Petrodomus*.

Forest Barriers

Savanna fauna is often neglected in discussions of climatic change because so many species range from the Cape to the Sudan and then west to Guinea and Senegal. Although savanna is now continuous along this arc of nearly 10,000 km, the presence of a solid belt of forest across East Africa in the late Pleistocene would have virtually isolated northern populations from southern ones. If the forests lasted for a long time, differences could be ex-

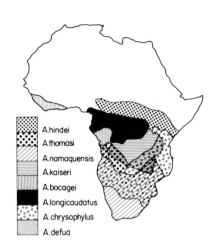

A.hindei
A.thomasi
A.namaquensis
A.kaiseri
A.bocagei
A.longicaudatus
A.chrysophylus
A.defua

Distribution of *Aethomys* species in Africa.

pected to develop and indeed there are differences between the fauna of the northern and southern savannas. Davis (1962) has described the line dividing these faunal subdivisions (running from the Congo basin to the Tana watershed), as the "Sclater Line".

The distribution of the savanna monkey, *Cercopithecus aethiops*, roughly follows this division. The monkey shows a variety of races within East Africa. One is a small, pale grey form, *Cercopithecus aethiops arenarius*, found in the more arid areas of northern Kenya and Uganda, while a larger, darker, greenish form, *Cercopithecus aethiops tantalus*, is found in wetter western Uganda, and the commonest form, *Cercopithecus aethiops pygerythrus*, occupies the rest of East Africa. Superficially the extremes in size and colour seem to be simply a response to climatic and ecological conditions, but the characteristics of the northern savanna form, colloquially known as the tantalus monkey, and the southern savanna form known as the vervet, are distinctive enough to have led some taxonomists (Dandelot, 1959) to assign them separate specific status. In southern and western Uganda there is a broad zone of "hybridization" between *Cercopithecus aethiops tantalus* and *Cercopithecus aethiops pygerythrus*, and among the predominant "hybrids" there are individuals that appear to be pure breeds of both parent stocks. The area in which "hybridization" has taken place is broadly coincidental with plant communities that have been variously described as "pseudo savanna", "forest savanna mosaic" and "communities derived from forest", it is also the area through which, in the recent past, the forests of western Kenya and the Congo must have been connected.

Hybridization is generally the result of an expansion of previously isolated groups and is common after periods of climatic change. In this instance, since a barrier of forest isolated the populations, climatic change augmented by cultivation and settlement has created conditions suited to both populations which have moved in. By contrast, the northern Kenya, *Cercopithecus aethiops arenarius*, has so adapted to arid conditions that difference in habitat would seem to be sufficient to cause a sharper geographical boundary and to discourage interbreeding with the tantalus. In other words, an ecological barrier tends to separate these races. Thus within a single species there appears to be a different outcome on the two fronts where formerly isolated populations meet.

It is interesting to examine other species for which forests are barriers. The distribution of two well-defined races of hartebeeste, *Alcelaphus buselaphus cokii*, in southern Kenya and northern Tanzania, and *Alcelaphus buselaphus jacksoni*, in Uganda and northwest Kenya, suggests that they were separated by forest. At some time in the past a belt of suitable habitat to the east of Mt Kenya and in the drier eastern rift valley provided opportunities for the two populations to meet, and these areas became hybrid zones. The northern race, *A. b. jacksoni*, has southern extensions of range which are well within or beyond the former forest belt. The colonization of these areas may have been assisted by climatic change, but it is worth noting that the western extension is along the migration routes of the pastoral Bahima coming from the north in the thirteenth century (see Dale, 1954), and that grazing conditions have been encouraged by the pastoralists burning the range for their stock, which incidentally might have suited the hartebeeste.

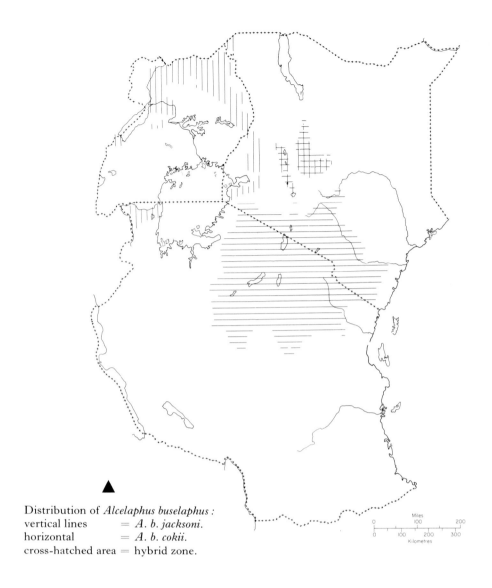

Distribution of *Alcelaphus buselaphus*:
vertical lines = *A. b. jacksoni*.
horizontal = *A. b. cokii*.
cross-hatched area = hybrid zone.

The giraffe also has distinct races separated by the "Sclater Line" and, although there may have been some southward extensions of range in Uganda in recent times, this animal is not found in the "pseudo savanna" of central and southern Uganda.

The vegetation of the northern savannas, sandwiched between forest and the Sahara Desert, forms serial belts which have migrated north and south, but have not been complicated by great changes of altitude, nor chopped up and isolated in the complex mosaic pattern found in southern and eastern Africa. Perhaps it is for these reasons that the northern savannas have not provided the same opportunities for speciation as the southern ones and are much poorer in number of species. The simple homogeneous nature of the northern savanna belts may therefore discourage their colonization by some southern savanna species.

80

A list of some mammals endemic to the northern and southern savannas respectively is indicative of the imbalance of species in what are broadly comparable habitats.

A LIST OF SOME TERRESTRIAL MAMMALS
BROADLY ENDEMIC TO THE NORTHERN AND
SOUTHERN SAVANNAS

Northern Savannas	Southern Savannas
Cercopithecus patas	
Xerus erythropus	*Paraxerus cepapi*
Tatera robusta	*Tatera boehmi*
	Tatera leucogaster
Cryptomys ochraceocinereus	*Heliophobius argenteocinereus*
	Pedetes capensis
Hystrix cristata	*Hystrix africae-australis*
Aethomys hindei	*Aethomys chrysophilus*
	Saccostomus campestris
Mylomys dybowskyi	*Pelomys fallax*
	Thallomys paedulcus
	Zelotomys hildegardeae
	Otomys angoniensis
	Rhabdomys pumilio
Mungos gambianus (limited range)	*Poecilogale albinucha*
	Rhynchogale melleri
	Equus burchelli
	Hippotragus niger
Redunca redunca	*Redunca arundinum*
Cephalophus rufilatus (marginal species)	*Connochaetes taurinus*
	Alcelaphus lichtensteini
	Aepyceros melampus
	Raphicerus campestris
	Raphicerus sharpei

Most of the northern savannas species are isolates of widely distributed genera and have related species in the south. Two notable exceptions are *Cercopithecus patas* and *Cephalophus rufilatus*; both species have emerged from the forest and become adapted to the northern margins of the forest block, the latter is still a forest edge species, while the peculiar condition of the former species is discussed on p. 210.

In spite of the dangers inherent in deducing past events from present-day distribution patterns, ecological or geographic faunal assemblages taken as a whole and compared with other assemblages, suggest some meaningful evolutionary patterns, some of which have been discussed in this chapter.

The postulation of a general geographical locality for the emergence of distinct populations is often possible, but with the present state of knowledge, attempts to relate the origin of any particular species or race to a climatic event in the past are speculative.

In conclusion eastern and central Africa has been a region peculiarly suited to speciation because:

1. Tertiary rifting, mountain building and general uplift in eastern Africa have led to the formation of geographical features that have constituted considerable, but not absolute physical barriers.

2. The climatic vicissitudes of the Pliocene and Pleistocene have effectively and repeatedly isolated populations of animals adapted to dry or to wet conditions.

3. Geographic features such as deep rift valleys or lakes and isolated massifs have had their efficacy as isolating mechanisms, reinforced by the great ecological diversity which is associated with differences in altitude and climate; in the case of some mountain habitats ecological and physical isolation are inseparable factors.

Isolated and ecologically peculiar habitats can induce speciation, which sometimes leads to special adaptations. Fluctuating climate may also have encouraged the acquisition of greater ecological plasticity by a wide range of mammals. Fossils have shown that mammals apparently occupying limited ranges or specialized niches today may formerly have been widespread, but have presumably become restricted to remote corners or less favourable habitats by later competitors.

The broad suggestions contained in this chapter have been illustrated by reference to specific distribution patterns. There is generally a more detailed discussion in the profiles of the species concerned.

Appendix II
Anatomy of Mammals

It is impossible to describe mammals, let alone discuss their structure, without being aware of form and function. J. Z. Young has called the terminology of modern biology "tool language", it is a language that is inescapable when we try to formulate ideas, whether we contemplate the animal alive in its natural surroundings or its disarticulated skull in a museum cabinet. It is clearly a mechanism, albeit an infinitely finer one than any that men can make. Indeed, mammals in common with other organisms have been defined in Young's tool language as self-regulating machines. This definition is a useful one—in particular as it helps to abolish some man-made compartments that the study of animals has been separated into—embracing as it does all aspects of the animal's life and structure. D'Arcy Thompson (1917) used tool language when he compared the most fundamental part of the mammalian body, the vertebral column, to the Forth Bridge and classified some mammalian structural types into "families of quadrupedal bridges", according to features which corresponded to certain mechanical conditions and functions. However, the analogy may be taken too far; a static bridge is a far cry from a genet racing through the tree tops, but the terminology of bridge-building remains useful when we examine this basic unit of the skeleto-muscular system of mammals.

In a typical quadrupedal mammal the vertebral column is arched between the supporting legs (which need hardly be called piers), the function of the column is analogous to the compression member of a bridge, the tension

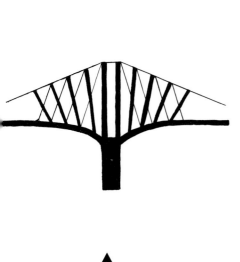

Diagram of a cantilevered girder
as used in bridge building.

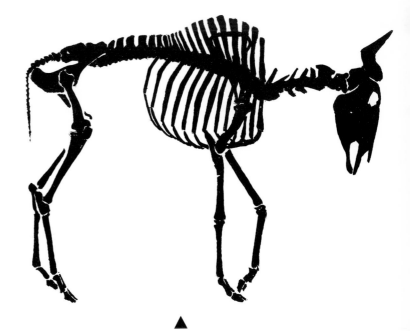

Skeleton of gnu,
Connochaetes taurinus.

being taken on the muscles and ligaments of the back which serve as tension members. The vertebrae are generally heaviest and the muscles are greatest in the small of the back, where the load is not spread over ribs or limbs. The vertebral spines and the interspinous ligaments represent a web of struts and ties to give both strength and lightness to the back. The spines vary in height, the tallest being where the greatest bending moments are. In many ungulates the head is cantilevered out on a long neck, and as much as three-fifths of the weight may be taken on the forelegs. The spines are generally longest over the shoulders, where they serve as struts, taking the weight of the head through the tension members, the *Ligamentum nuchae* and *Sacrospinalis*.

Variations on this pattern are illustrated by the elephant, camel, giraffe, cheetah and rhino.

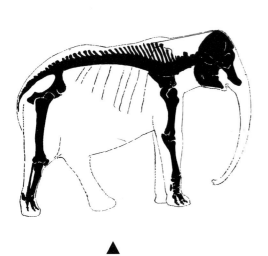

▲

A tuskless elephant or "Buddu".

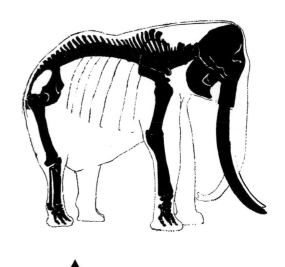

▲

An old tusker.

The elephant has tall thoracic spines sloping back to take the strain of the heavy head, their length and form are correlated with the weight of ivory carried (hence the characteristic stance of an old tusker) and the back of the skull has an enlarged area for the attachment of the big neck muscles. In the old elephant the arrangement may be likened to balancing scales with the forelegs acting as a fulcrum, whereas in the younger or tuskless animal a bridge is more apparent, as the head is relatively light and most weight is spread behind the front legs.

In the camel the hanging neck and head represent a heavy forward extension, weight is so balanced at the forepart of the body that the hindlegs almost become mere propellers and carry very little body weight.

The giraffe relieves some of the heavy vertical load of the neck through its sloping back, but most weight is taken on the forelegs and the balancing of the head on top of the cervical vertebrae makes for a columnar neck which

▲

Giraffe

▲

Camel

is under compression when standing. When the animal runs or browses on low bushes the head and neck are cantilevered out on a highly developed and very powerful tension member the *Ligamentum nuchae*.

The cheetah represents a lightly built extreme of the quadrupedal bridge with flexibility of the back at a premium, the back tends to hang slightly when standing.

Heavier quadrupedal bridges are carried in a distinct arch and a greater proportion of weight may be taken on the hindlegs as in the rhino. The double demand for strength and flexibility in the rhino's shoulder is shown in the broad blade-like thoracic spines, which have slots to accommodate the spine behind when the vertebrae are compressed. Where most weight is taken on the hindlegs, as in the aardvark, the lumbar region carries high vertebral spines.

▲ Cheetah   ▲ White rhino

The pangolins have a vertebral column with the greatest power and weight behind the sacrum. The tail acts as a protective sheath, a strut when digging and a principal limb in the climbing form.

In the spring hare, *Pedetes*, which is bipedal the lumbar and sacral spines are the longest and strongest due to the overhanging load of the fore part of the body; this animal can be mechanistically described as a single balanced cantilever.

▲ Spring hare   ▲ Giant pangolin

Another biped is man but his erect posture puts a heavy vertical compression down the vertebrae making a columnar form with head balanced on top. The pelvis has a basin-like shape to which muscles bracing both the legs and the spine are attached. Gravity is one of the forces influencing animal form, an example can be seen in the ribs of many mammals which hang from an approximately horizontal vertebral column in the form of a single vertical hoop usually narrowing towards the sternum. D'Arcy Thompson suggests that gravity has played a role in moulding the form of the human rib-cage, "each pair of ribs forms a hoop which droops of its own weight in front, so flattening the chest, and at the same time twisting the rib on either hand near its point of suspension".

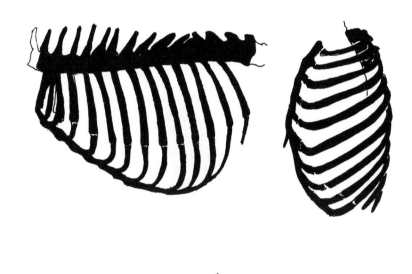

The rib cages of lion and man compared.

The potto, *Perodicticus*, which moves with rather reptilian side to side flexion of its back and may even travel hanging from its hands and feet, has a long backbone made up of rather simple undifferentiated vertebrae which have increased in number, some individuals having as many as nine sacral vertebrae. The cervical spines have been relieved of their original function and are employed in socio-sexual rituals (see p. 284).

The otter shrew, *Potamogale*, which swims with lateral movements of the body, has unusual mobility in this plane. In wholly aquatic mammals the body must work against the resistance of water to move, stiffness in the vertebral column is needed and the uniform column is like a flexible tapered rod. In both dugongs and whales movement is in a vertical plane.

Lastly I must mention the extraordinary backbone of the hero shrew,

Scutisorex (see drawing). This animal is a compulsive scent marker and the back is employed to rub scent on objects and possibly other shrews, during which ritual the body is flexed into unusually contorted postures (see Vol. II).

The muscles throughout the vertebral column act as ties or tension members. The dominant muscle of the back is the *Sacrospinalis* (commonly eaten as fillet). The thoracic part of this muscle is called the *Longissimus dorsi*. The tail is operated by the *Abductor caudi* muscles, their development and detailed functions vary from species to species as the tail may do service as balancer, fly-whisk, flag, club or limb.

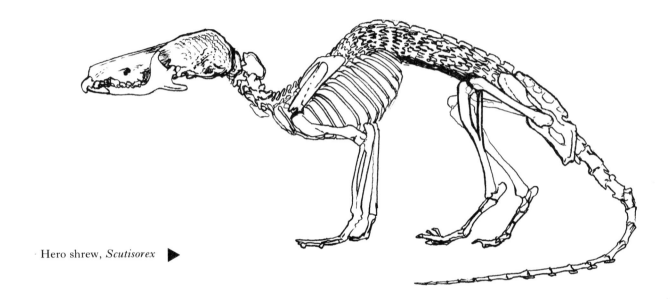

Hero shrew, *Scutisorex* ▶

The skull is held by an elaborate variety of neck muscles attached to thoracic and cervical vertebrae: the tension is taken on the *Ligamentum nuchae* in large mammals with heavy heads. The *Semispinalis*, *Longissimus capitis*, *Complexus* and other muscles act on the cervical vertebrae which make the neck very mobile most particularly near the head, due to the peculiar structure of the large atlas and axis. The atlas articulates with the skull to allow a nodding movement, the skull and atlas rotate on the axis to allow movement in all directions.

The mammalian limbs, while acting as levers and supports for the body, have assumed a great variety of secondary functions such as seizing and holding, feeling, digging, hanging, flying and in some mammals swimming. The body is supported and levered along by a variety of means, on toes or toe, on the soles or even on the elbows in moles and "running" bats.

The limbs are balanced by forces exerted by the body weighing down through the limb against the ground and the disposition of weight within the limb itself. In movement the strains are constantly altered and are met by the muscles which act as adjustable braces and ties. Each muscle has an antagonist, which may be another muscle or a ligament the action of which may vary; those for speed of action being long and parallel, those exerting much

88

force being bundles of short-grouped fibres. The bones act as struts being under compression. The attachments between muscles and bones may be direct or through tendons; the leverage may be exerted according to any of the three orders of levers. The pull and the load may act on either side of the fulcrum or hinge. Alternatively, an asymmetric loading on one side of the fulcrum may be counterbalanced by a pull between the load and the fulcrum, or the pull may be exerted at the end of the lever beyond the load.

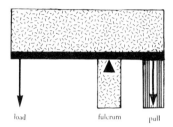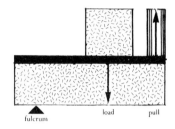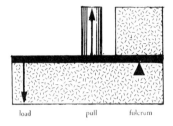

Diagram of three Orders of levers—1st Order, 2nd Order, 3rd Order.

Bones and muscles in the limbs alter from species to species, but in the less specialized types the arrangement indicates some of the mechanical adjustments that were necessary to convert lateral reptilian limbs into the supporting levers of mammals. When in phylogeny the elbow swung back into line with the body the "hand" had to rotate forwards; this is shown in the twisting of the radius round the ulna and the accommodation of muscles to this modification. In the hindlimb the knee swings forward, phylogenetically this was achieved simply by a bending of the orientation of the femoral head (see Romer, 1945; Young, 1957). The forelimb "floats" on the thorax attached by a wrapping sling of muscles. A clavicle joining the scapula to the sternum is found in primates, bats and digging mammals and in some primitive types, its vestigial remains are found in other mammals. This bone serves as a tie, stopping compression of the two limbs across the thorax. Both the clavicle and the scapula are modified remnants of the reptilian shoulder girdle. The scapula transfers the stress of the body's weight to the limb or, in running, digging, flying and swimming, may transfer and spread the stress of that activity onto the barrel of the body. The slinging of the trunk onto the scapula and hence the limb is largely achieved by the *Serratus anterior* (see drawing overleaf). The scapula is held to the body on its upper edge by the *Rhomboid* bound over by the more superficial *Trapezius*. The *Levator scapulae* is a strap running from the metacromium of the scapulae to the upper cervical vertebrae and the skull.

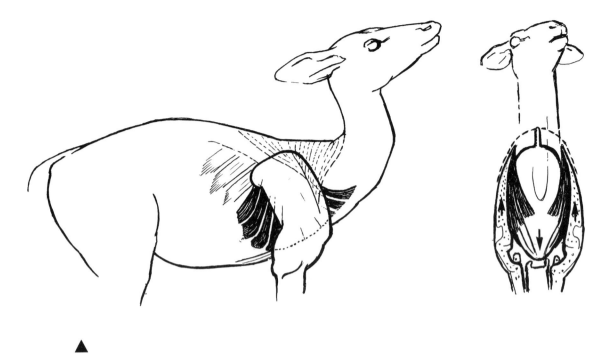

Diagram of the *Serratus anterior* slinging the trunk
onto the forelimbs (bushbuck).

Control and bracing of the shoulder is achieved by a number of muscles
attached round the joint. The *Pectoralis* or chest muscles (very often hidden
in the dissection drawings which are drawn from a sideview) attach the
shoulder to the median line of the chest. The *Sterno-cleido-mastoid* is a sheet
muscle running from the back of the skull to the shoulder joint, and the
Basihumeral a strap from the skull to the humerus and clavicle. A fan-shaped
sheet muscle from the humerus to the fascia of the back is the *Latissimus
dorsi*. The scapula and humerus are tied by a strong ligamentous connection
and also by the *Supraspinatus* and *Infraspinatus*. The *Deltoid* is also a tie and
an agent in moving the foreleg, it is attached to the *Spinatus*' fascia, scapular
spine and the humerus. The scapular-humeral joint is flexible and these
muscles with the *Biceps* and *Triceps brachi* and *Teres major* give it stability
and strength (particularly in digging animals).

The *Biceps* and *Triceps* run from the shoulder to the elbow and are the
principal flexors and extensors for the whole leg, the *Triceps* joins the olecra-
non process which is part of the ulna (in many ungulates this has become fused
with the radius to make one bone). In ungulates the muscles below the elbow
become semi-tendinous and then extend as tendons to the toes or toe, thus
reducing the weight that has to be moved in running. In primates and other
mammals the more primitive condition is retained, the radius carries the
hand and by rotating round the ulna achieves a great mobility; stresses coming
from the hand or from the body are evenly distributed, as the bones are very
firmly bound by the interosseus membrane. Where rotation is possible the

Brachioradialis, *Pronator quadratus* and *Pronator teres* hold the hand prone, with the finger nails to the front. When the hand is twisted so that the palm or wrist face forward, this supination is achieved by the *Supinator* and *Biceps*. The bones of the hand or forefoot have an original arrangement of three proximal carpals, a central carpal, five metacarpals followed by five digits of two or three phalanges, but there are numerous adaptive patterns—the single toe of the zebra, the wing of a bat, the dugong's flipper, and clawed, hoofed or nailed feet. The hindlegs are sometimes less specially adapted than the forelegs and in the main their function is propulsive.

The pelvis is a light, strong, balanced structure, sometimes fused onto the vertebral column but sometimes capable of flexion. This bone forms a ring through which sexual and excretory ducts pass, and the ball and socket joints on the outer side allow very considerable movement; at the hip joints this movement is operated by *Gluteus* muscles. The most powerful action involved in movement, that of "kicking off", initiating or maintaining momentum, is caused by the thrusting forward of the whole body by the muscles in the upper part of the hindleg. These muscles are appropriately large. They are the *Biceps*, *Gracilis*, *Semimembranosus* and *Semitendinosus* attached to the hindermost part of the ischium. Reduction of bulk and weight in order to decrease inertia in the recovery phase is achieved in lower legs by the muscles becoming tendinous. This begins on the tibial semitendinous *Gastrocnemius*. The extension of the knee is achieved mainly by the *Quadriceps femoris*, the muscle in front of the femur. This is largely covered by the superficial *Tensor fasciae latae* and the *Gluteus maximus* which helps stabilize the hindleg. The bent position of the hindlegs calls for a certain degree of permanent tension in the upper muscles to prevent collapse and a balancing function can be added to the propulsive one. The lower leg and foot in their original arrangement resemble the foreleg and hand, they are subject to a similar range of special adaptations.

In some more primitive vertebrates than mammals the head is merely a concentration of sensory, neural and feeding apparatus at the anterior end of the body. In mammals the head has become a clearly demarcated structure in which various activities are contained in one organized whole. The head accommodates the brain, eating apparatus, sensory organs for seeing, hearing, smelling and balancing, the nostrils and nasal passages. The skull may also grow horns, tusks or special teeth for defence or catching prey. Superficial features may become elaborate for sexual or social communication and the highly specialized sonar systems of bats involve an astonishing variety of modifications to the nostrils, ears, facial muscles and skull.

The basic form of the skull is usually determined by the relationship of the cranium to the jaws and teeth, but some species have developed one sense enormously, often at the expense of other activities. The relative importance of a sense can be assessed by the proportion of the skull it takes up, thus the development of one activity may lead to a single allround enlargement of the area concerned with that activity, one of the best examples being the swollen cranial vault of man. Others are the nasal apparatus of the porcupine and the eyes and their orbits in galagos. In these cases the areas concerned have been free to enlarge rather as a fruit would, becoming more or less globular in form. Other animals enlarging a special sense have to accommodate the area between other structures or the form of their skull may be subject to other special limitations. The nasal enlargement of the porcupine can be compared with that of the aardvark, *Orycteropus*, and the dik-dik, *Rhynchotragus*. Unlike the porcupine, the enlargement is restricted in these animals by other special considerations both physical and genetic; in *Orycteropus* the form must be contained within a tubular snout and in *Rhynchotragus* the nasal bones have retreated to allow the main structure of the nasal apparatus to become fleshy or cartilaginous.

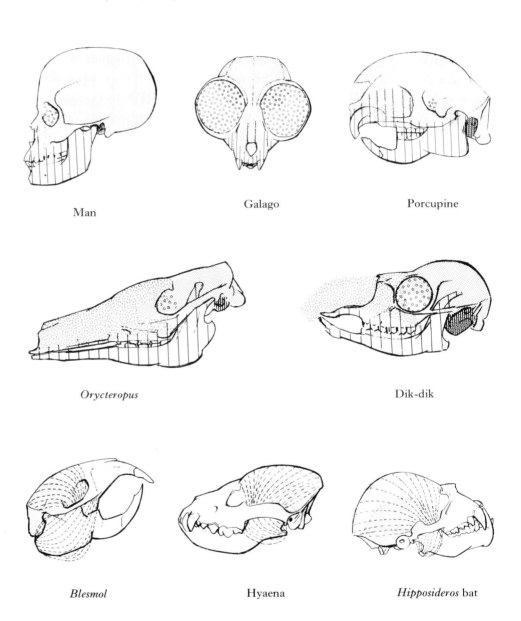

Man

Galago

Porcupine

Orycteropus

Dik-dik

Blesmol

Hyaena

Hipposideros bat

Whenever the jaws and teeth are specialized for crushing hard materials, the whole skull is modified by the need to buttress the teeth and strong bony bases to the teeth are apparent. Appropriately large muscles must operate the jaws and these must also have extensive bony anchorage; extreme examples of this are the mole rat, *Heliophobius*, the hyaena, *Crocuta*, and the giant leaf-nosed bat, *Hipposideros commersoni gigas*. The form of the skull in these species is dominated by their peculiar diets and habits. The structure associated with the other activities, the brain case, eye sockets, hearing and smelling apparatus have been displaced or relegated to a relatively inconspicuous position. Similarly the shape of the cranium is often partly determined by the areas of the brain that are most developed, visual centres swelling the hindbrain in one form, and olfactory ones the front in another.

There are examples where the form of the head is moulded by a special development or by the elaboration of some special feature; the horns of the lelwel hartebeeste, *Alcelaphus*, are an example of the latter. The bat, *Platymops*, lives in cracks between rocks and the head and body in this species have become very flat and broad and the form of the skull betrays this special adaptation. The white rhino, *Ceratotherium*, has a skull that is dominated by the stresses imposed by its own weight and the extraordinary horns on the nasal bones.

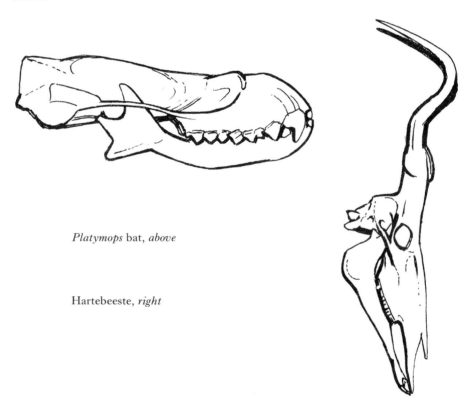

Platymops bat, *above*

Hartebeeste, *right*

The adaptive changes wrought at a phylogenetic level can be paralleled within a species. Ontogenetic development or the absence of some normal pressure reveal interesting structural differences between normal and aberrant animals or between adults and juveniles. Earlier in this chapter the mechanical differences between tusked and tuskless elephants were mentioned. A comparison of the skulls of elephants with and without tusks reveals that the heavy-tusked skull needs strong buttressing to hold the base of the tusks and also it reveals that the weight imposes stresses and muscles are developed so that the whole form is altered in almost every particular. The elephant's tusks are considered to function as counterweights (J. Z. Young, 1950) necessary for the proper balance of the animal. The architecture of the elephant skull has clearly evolved to bear the weight of the tusks and the tuskless condition is aberrant. The development of a very short and high back to the skull is necessary for the attachment of the neck muscles carrying the weight of the head, while the front area allows effective attachments for the jaw muscles. This special arrangement is achieved by air sinuses in the bone surrounding the brain; these start to appear some months after birth so that juvenile skulls are quite different to those of adults.

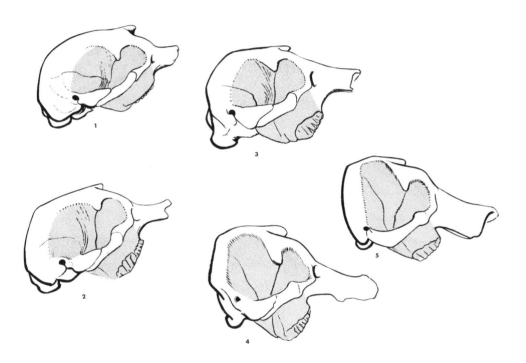

Elephant skulls, 1. infant, 2. small juvenile,
3. half grown, 4. adult tuskless, 5. adult tusker.

In most species several separate but equally important activities interact within the form of the head and the structures associated with the different activities frequently overlap. For instance, an important region of the skull may at one and the same time provide attachments for muscles connected with eating, bridging material between the eyes and the nose and also roof over the nasal apparatus or buttress the teeth.

The activities contained within the skull seldom occupy neatly separated areas, but the spring hare, *Pedetes*, the cheetah, *Acinonyx*, and the klip-springer, *Oreotragus*, are species in which the separate activities are more clearly defined than usual.

94

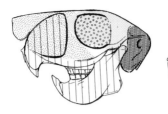  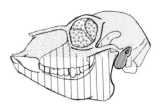

Spring hare   Cheetah   Klipspringer

The description of skulls and their attachments involves a terminology that identifies particular bones in relation to those of other vertebrates, defines homologies and subdivides the whole. The bony components of all mammalian skulls have a common origin and the identification of homology is a worthwhile task, but in appreciating the form and function of the head a sectional approach is not always helpful. For example, the nasal bone may be a long, thin component in a tube-like structure, or it may provide part of a heavy base for a horn or a small prominent knob to help carry the muscles of a trunk or it may disappear altogether. In all cases it is inseparable structurally from the specific rhythms of the head and the particular adaptive tendencies that the rhythms are directed towards. The illustrations of bones and skulls in this work do not usually include the representation of sutures, and the tracing of homology will not normally play an important part, instead the form of heads and bodies and the functional meaning of the altered form will be a major consideration.

A brief summary of the larger superficial bones of the skull may be given here. The teeth are carried in the premaxillary and maxillary bones of the skull and in the mandible below. The brain is roofed over by pairs of bones, the frontals, parietals and occipitals and a single bone the interparietal (which is of diagnostic significance in Hyrax). The sphenoid bones, the squamosal and periotic (containing the auditory capsule) are arranged around the sides of the brain, while the floor of the cranium is formed from the basisphenoid, presphenoid and ethmoid bones. A perforated bone around the olfactory lobes of the brain is called the cribriform plate. Above the maxilla are the

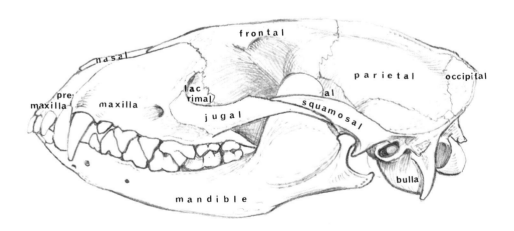

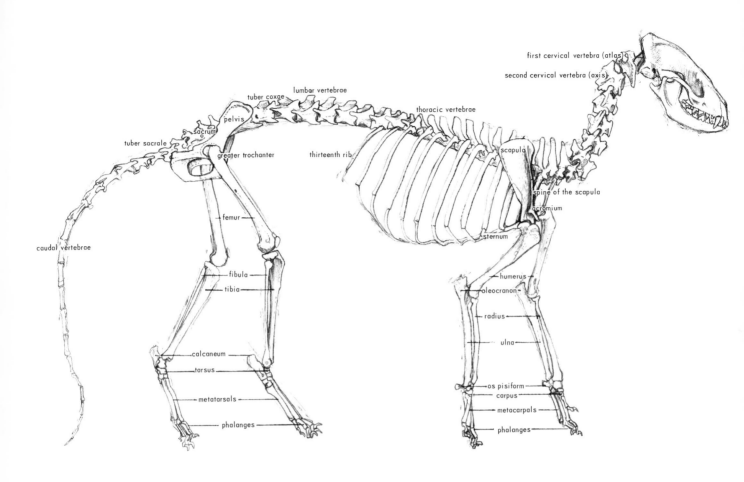

first cervical vertebra (atlas)
second cervical vertebra (axis)
lumbar vertebrae
tuber coxae
thoracic vertebrae
pelvis
sacrum
tuber sacrale
greater trochanter
thirteenth rib
scapula
spine of the scapula
acromium
caudal vertebrae
femur
sternum
humerus
oleocranon
fibula
radius
tibia
ulna
calcaneum
tarsus
os pisiform
carpus
metatarsals
metacarpals
phalanges
phalanges

Superficial dissection of civet cat.   ▶

Deeper dissection of civet cat.   ▶

paired nasal bones. The rest of the nasal tube is made up of the premaxilla and maxilla with the palatine bones flooring the back, behind these are the pterygoid processes for the attachment of muscles to move the jaw from side to side. The cranium and the maxilla are generally connected by a bridge of bone that is formed by processes of the squamosal and the maxilla and a connecting bone, the zygoma. In some mammalian groups a zygomatic process on the frontal may join the zygoma to form a second bridge and thus create a more or less circular orbital cavity. In the advanced primates the orbital cavity has become a selfcontained compartment.

The principal muscle attachments are on the back of the skull where the neck muscles, *Splenius, Trapezius, Semispinalis, Obliquus* and *Rectus capitis* hold up the head and join it to the body. On the skull itself the only large muscles are those connected with chewing, the *Temporalis* and the *Masseter*.

96

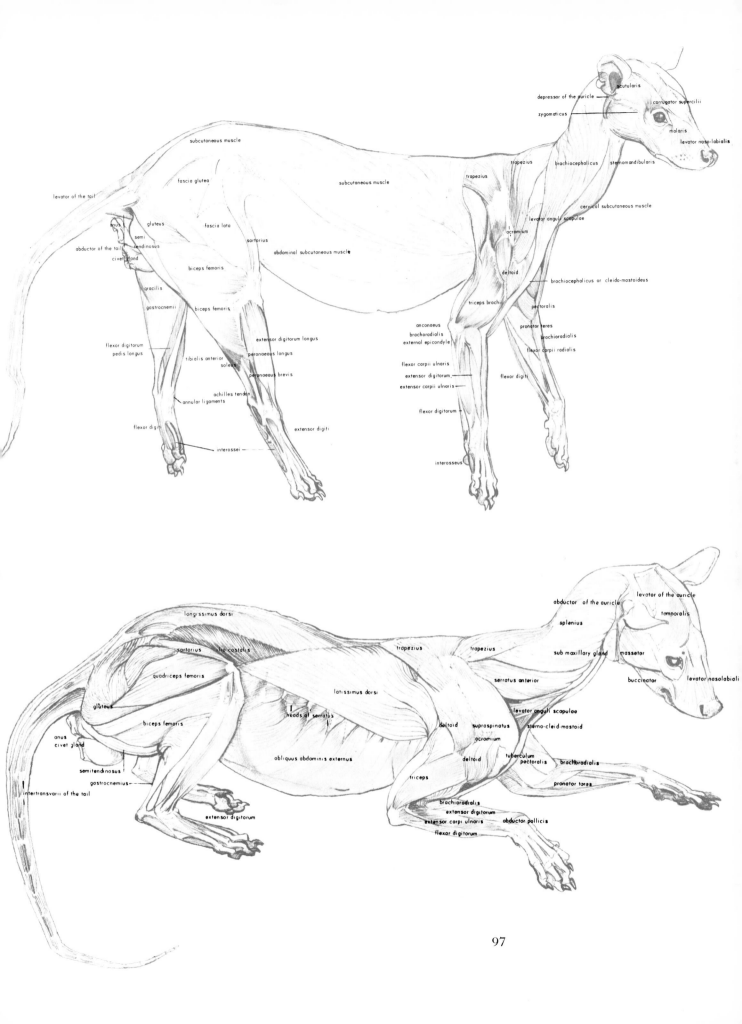

97

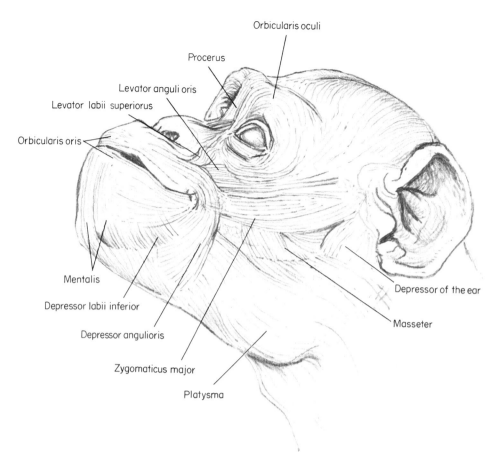

Facial muscles of chimpanzee.

Numerous smaller muscles are found on the face but some of these may become major muscles in specialized mammals like the elephant and the dugong. Muscles moving the lips and nostrils are the *Levator labii*, *Levator nasolabialis*, *Pyrimidalis nasi*, *Zygomaticus* (or *Malaris*) above and the *Quadratus labii inferioris* below. Lining the sides of the mouth is a sheet muscle the *Buccinator* and surrounding the mouth itself the circular *Orbicularis oris*. The eye is surrounded by circular *Orbicularis palpebrarum* and may be attached to thin subcutaneous muscles, the *Corrugator supercilii* (or *Temporalis*). A thin, subcutaneous sheet muscle is the *Cervical panniculus* or *Platysma*. The facial muscles are differentiated in higher mammals, particularly in primates, into muscles of expression. For the many mammals with large ears, the *Abductor*, *Levator* and *Adductor* of the ear are important and conspicuous muscles. Tactile vibrissae are found in the majority of species and tend to be concentrated round the mouth, although smaller clumps or even single hairs may be found on various other parts of the body.

This review is of some of the more obvious and superficial features of mammalian anatomy which are represented in the dissections and drawings of East African mammals. The inadequacies of this description will be obvious and the reader is referred to J. Z. Young *The Life of Mammals* (1957), for a comprehensive introduction to the anatomy and physiology of mammals.

98

The Primates

PROSIMII
ANTHROPOIDEA　　**Cercopithecoidea**　　**Lorisidae**
　　　　　　　　　　　Hominoidea　　　　　　**Galagidae**
　　　　　　　　　　　　　　　　　　　　　　　Cercopithecidae
　　　　　　　　　　　　　　　　　　　　　　　Pongidae
　　　　　　　　　　　　　　　　　　　　　　　Hominidae

Living members of the order primates are remarkable not only for their variety but for representing a series which, in their graduated scale of complexity and physical organization, approximates to the evolutionary progression of their order as a whole.

Some of the more "primitive" of these primate species may be taken as representatives of groups that have had their day. These species may survive as relatively unchanged relics in island refuges where they have avoided the challenges of competing species, or the rigors of ecological fluctuations. Among the primates this is best illustrated by the tarsier of the Indonesian Islands. Alternatively, they may adapt in some special way and avoid thereby the competition of "advanced" primates. Prosimians, which were a numerous and varied group in the Eocene, have only survived in Africa by acquiring special habits and becoming nocturnal; the modern representatives of this ancient suborder are reduced to a very few species.

Several varieties and numerous fossil remains of apes are found in Miocene deposits of East Africa, suggesting that they were a varied and successful group 20 million years ago. Today only the declining gorilla and the chimpanzee survive, while the aberrant *Homo sapiens* is highly successful having become an ingenious and specialized tool user. The generalized cercopithecus monkeys presumably now fill most of the niches occupied by the smaller Miocene apes.

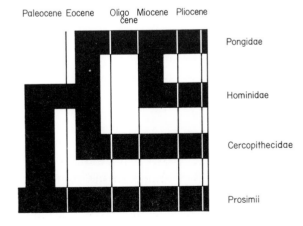

99

A remarkably representative cross-section of primates occurs in East Africa, comprising 4 families and 22 species. Primates span a wide range of types so that there are rather few obvious features common to all. Typically primates have a generalized anatomical plan and their principally arboreal life has necessitated a singular mobility of the limbs and hands, and the retention of some primitive features such as five digits and a clavicle. The eyes and the associated areas of the brain are well-developed. Stereoscopic vision, which is one of the hallmarks of modern primates has been achieved in more than one way; the eyes and orbits of the potto have tilted upwards without reduction of the nose, a device that is also seen in the Eocene prosimian *Adapis*. In more advanced primates the sense of smell has less importance and stereoscopic vision has been accompanied by a reduction of the nose; there is also a corresponding reduction of the olfactory lobes of the brain.

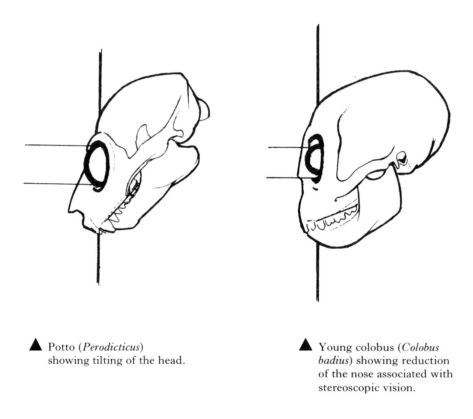

▲ Potto (*Perodicticus*) showing tilting of the head.

▲ Young colobus (*Colobus badius*) showing reduction of the nose associated with stereoscopic vision.

The general expansion of the brain started earlier and proceeded further in primates than in any other group of mammals. The earliest primates known are Palaeocene and Eocene prosimians from Eurasian and North American deposits. A variety of divergent forms have been found and both lemuroids and tarsoids had already radiated into many specialized forms on these continents. In Africa, the earliest primates known are anthropoids (*Aegyptopithecus, Parapithecus, Apidium, Propliopithecus* and *Aeolopithecus*) found in the Oligocene beds at Fayum.

100

These fossils suggest that the anthropoid stock originated in Africa and that the African primates had already started to diverge into distinct branches. The principal division of the living *Anthropoidea* into *Cercopithecoidea* and *Hominoidea* has been made on the basis of cusp patterns and other peculiarities in contemporary species, but the origin and meaning of the divergence are still a matter for speculation.

At the end of the Oligocene or during the early Miocene, African primates entered Eurasia and it is probable that the ancestral stock of contemporary Asiatic primates migrated at this early date. By the Middle Miocene, dryopithecine apes were widely distributed in Europe, Asia and Africa and are particularly numerous in East African deposits. This group may be ancestral to man as well as modern African apes. Known types ranged between animals the size of a gorilla, *Proconsul major*, to the monkey-sized *Proconsul africanus*.

Contemporaneous and presumably sympatric with the early Miocene apes were small cercopithecines which are known so far only from fragments. Later in the East African Miocene recognizably colobine monkeys are known.

Both anthropoids and prosimians are deeply involved in the epidemiology of various virus diseases that are transmittable to man, also numerous parasitic diseases are shared by man and monkeys. Primates have therefore become very important in medical research, principally as experimental animals.

The characteristics of primates and general reviews of their biology, taxonomy and anatomy are available in several excellent reference works (Le Gros Clark, 1959; O. Hill, 1953—1966; Napier and Napier, 1967; Morris, 1966).

Hominoids

Hominoidea

Ever since Darwin the relationship between man and the apes has excited extraordinary interest. Yet, even now, after a multitude of fossil hominoids have been excavated, the point at which the ancestral stock of man parted company with that of the apes remains a matter for debate.

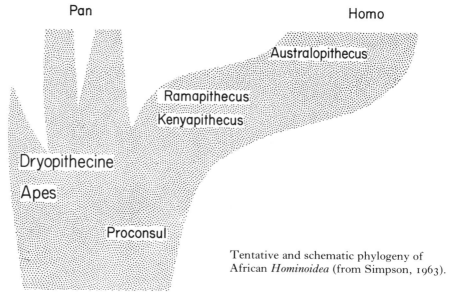

Tentative and schematic phylogeny of African *Hominoidea* (from Simpson, 1963).

Recent research has shown that man's blood serum proteins and chromosomes are very similar to those of African apes, a discovery which shows that, in spite of his development of brain and erect posture and his use of tools,

Hominoid chromosomes: representative karyotypes (from Klinger, 1963).

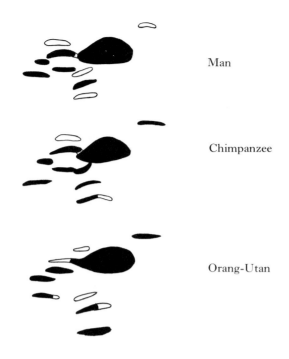

Man

Chimpanzee

Orang-Utan

Comparison of serum proteins. Diagram of 10
faster migrating components in starch-gel
electrophoresis patterns (from Goodman, 1963).

man retains a biochemical system that has scarcely altered since his divergence
from the same stock as that of the African apes. Curiously enough, the Asiatic
orang-utan, *Pongo*, differs more from the African apes in the patterns of its
blood serum proteins and chromosomes than man does, in spite of the orang-
utan's greater anatomical resemblances with the African apes.

These paradoxical resemblances and differences accentuate the conservative nature of the apes' adaptive anatomy, and also the degree to which man
has diverged in appearance without marked changes in his biochemistry.

The faunal evidence suggests that the ancestral stock of the orang-utan
entered Asia during the earlier Miocene but opinion is divided on whether
the hominid and dryopithecine stocks parted at about this period or earlier.

Until more fossils of Miocene and Oligocene hominoidea are turned up
no conclusive reconstruction can be made of man's affinities with the modern
apes.

NOTE ADDED AT FIRST REPRINTING 1978

Hominid fossils 1968–1977

This volume was written in 1967–68, before the Turkana (former Lake Rudolph) deposits were discovered by Richard Leakey and explored by teams directed from the Kenya National Museum and before the discovery of abundant new hominid remains in the Turkana area, in Laetolil and in the Omo and Hadar valleys in Ethiopia.

It is impossible to overrate the importance of the abundant new fossils that have come from these richly productive sites. These finds and others from Eurasia have produced a complex and often puzzling picture of human evolution.

The geographic and temporal range of *Ramapithecus* (see pp. 105–107) has been greatly enlarged with ages from 20 to 7 million years attributed to specimens which have been discovered in Pakistan, Greece, Turkey and Hungary as well as Africa and India. However, there are still no satisfactory post-cranial remains or complete skulls.

An abundance of robust *Australopithecus* specimens have been discovered in eastern Africa including several fairly complete skulls. These exhibit great individual variation and make it less likely that *Australopithecus robustus* and *A. boisei* are separate species. An upper molar that might be australopithecine has been found at Ngorora (9 to 12 million years) but the oldest confirmed *Australopithecus* is a gracile form from Lukeino (Lothagam) and has been dated 6·5 million years. No *Australopithecus* fossils have been found outside Africa yet. It is possible that cut stone tools from 3 million year old beds in the Omo Valley were used by *Australopithecus*. Artefacts aged 2·6 million years in east Turkana are from a horizon containing two hominid species.

Reconstruction of a flayed *Australopithecus*.

A most important new development has been the discovery in east Turkana, Laetolil and Hadar of large numbers of remains belonging to Early Pleistocene hominids that are likely to be close to the main line of human evolution.

The earliest specimens are 3·7 to 4 million years old in Laetolil and some resemble the gracile *Australopithecus* (in some there is even an affinity with *Ramapithecus* in the structure of the premolars) but all show an evolutionary advance and anticipate characteristics of *Homo erectus*. They are emphatically different from the robust *Australopithecus* with its enormous cheek teeth, heavily molarized premolars and relatively small front teeth.

These fossils have often been assigned to "*Homo habilis*" but current usage has tended to refer to catalogue numbers because generic attribution is still uncertain and earlier palaeontologists all too frequently had to retract after hasty attributions and too promiscuous a nomenclature.

These hominids had short faces, an enlarged cranial capacity, relatively large incisors and canines and reduced molars and premolars. Fragments of a 3 million year old skeleton that combines primitive australopithicine and human attributes, fondly named "Lucy" by its discoverers have been found in Hadar. Seven individuals, whose ages at death ranged between twenty-five and seven years, have been found in a group together in the same area. Although very short in stature these hominids are more advanced than Lucy in spite of being somewhat older. The remains of many vertebrates occur on living floors of this period and stone implements of Oldoway or "pebble culture" type are apparently associated with these hominids.

From east Turkana deposits aged 2·9 million years comes an almost toothless skull, "ER 1470", which Richard Leakey has suggested might have an affinity with the type specimen of *Homo habilis* (which was described from an Olduvai mandible). This large-brained hominid has a flat broad face unlike any other known form. Another east Turkana hominid skull, "ER 1805", has the unique combination of sagittal and nuchal crests, a relatively large cranium and small teeth. The gracile *Australopithecus africanus* seems to be most nearly represented in east Turkana by a cranium and palate (ER 1813).

A nearly complete skull of *Homo erectus* (ER 3733) has been found in east Turkana deposits aged 1·3 to 1·6 million years. An earlier specimen (1·9 million years) is known from Indonesia and others have turned up in Algeria, Morocco, Germany and Spain as well as Java and China. It is possible that populations of *H. erectus* became isolated in Australia, parts of Indonesia and southern Africa and that this type of human survived up to 30,000 or even 10,000 years ago.

In all late Miocene, Pliocene and Pleistocene sites rich associated faunas have been discovered.

The wide range of new fossils from eastern Africa have ensured that this is incomparably the most important region in the world for the study of our early ancestors and their cultures and environments. The map on p. 56 shows the age and location of all the major East African fossil sites.

Man and Pre-man

Hominidae

The earliest evidence of an ape that might belong near the ancestry of man is a fossil from Fort Ternan, Kenya, from deposits that are 14 million years old. Here and in much later Pliocene deposits in Eurasia a number of jaw fragments have been found, all of which have been described as *Ramapithecus* (syn. *Kenyapithecus*) (see Simons and Pilbeam, 1965). These fragments exhibit a form of palate very much more like that of a man than any previously known fossil apes. The teeth of *Ramapithecus* differ from the dryopithecines in the reduction of both canines and incisors, and the less complicated, shorter and more widely spaced cusps on the molars. Correlated with this reduction of the front teeth the face is less prognathous and the mandible slighter. These dental characteristics mark a departure from the typical dryopithecine pattern, but until more fossils are available it is impossible to guess how long the lineage of *Ramapithecus* had been distinct from the dryopithecine apes which are so numerous in the East African Miocene deposits. Although *Ramapithecus* fragments only consist of teeth and jaws the reduction of the canines suggests that the functional role of these teeth had become obsolete, at the same time chewing movements were probably no longer restricted to up and down champing as the canines no longer locked the upper and lower jaws. From this, changes in diet and changes in behaviour and in habitat have been inferred.

It has been assumed that the principal ecological difference between *Ramapithecus* and the dryopithecine apes was a greater degree of adaptation to the open savanna country which became widespread in the Later Miocene. This is probably a safe assumption, as the remains of *Ramapithecus* are scattered over perhaps as much as 4 million years and vast distances, having been found in Europe, India, China and Kenya, suggesting that for a primate it was an exceptionally mobile and successful animal. Indeed, there are rather few animals today that are found over so vast a range, and it may be useful to list some contemporary mammal species that have had or still have a wide African and Eurasian distribution: hare (*Lepus*), porcupine (*Hystrix*), jackal (*Canis aureus*), ratel (*Mellivora*), hyaena (*Hyaena*), caracal (*Felis caracal*), leopard (*Felis pardus*), lion (*Felis leo*), and cheetah (*Acinonyx*).

Each species has special characteristics and limitations peculiar to itself, to its family and to its order. *Ramapithecus* too must have been unique and specially adapted in some respects while being typically ape-like in others. However, it must have shared a common preference (or at least a common adaptability) to allow it to live in dry, relatively open country. In this way it was clearly not very ape-like.

There are other characteristics which it might have shared with these widely distributed species. The majority are carnivorous and some are scavengers. Most are capable of ranging over a large area and the majority have resorted to burrows or caves particularly for breeding. All are capable of protecting themselves or can run. Most of these species may sometimes be seen in family parties or small aggregations, but excepting lion and cheetah none can be termed more than slightly social.

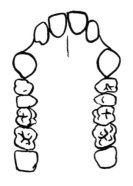

Proconsul major palate.

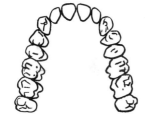

Reconstruction of *Ramapithecus* profile.

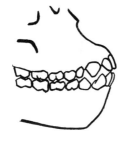

Ramapithecus palate.

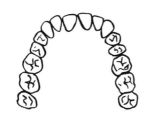

Homo sapiens palate.

Outlines of palates (in part after Eimerl, 1966).

The only modern ape known to venture into relatively open country is the chimpanzee and there is valuable comparative evidence from observations of East African chimpanzees that an open habitat induces smaller units of more consistently mixed or family-like composition and with a far greater range (Nishida, 1967; Suzuki, 1969). It is possible that this social trend was even more marked in *Ramapithecus*, which was probably omnivorous. *Ramapithecus* may have fed on similar foods to those of the animals listed above, that is: small vertebrates, fruit, roots and perhaps carrion, and it is possible that the use of caves as refuges may also have accompanied its adaptation to open country, and that it may have fallen prey occasionally to large felines.

How could *Ramapithecus* flourish in an exacting and dangerous environment? The small defenceless primate must have been vulnerable to a wide range of carnivores, so that the reduction of the canines would appear anomalous and a reversal of the situation found in the two present-day open country primates, the baboons and the patas monkeys. These species effectively display their enormous canines in threat and ultimately are capable of using them in defence. These monkeys (most particularly the patas) also rely on running, for which their generalized quadrupedal gait is rather well-adapted. Whatever locomotive pattern *Ramapithecus* adopted it was unlikely to have been capable of exceptional speed. The brachiating of modern apes may well have become exaggerated in phylogeny but a tendency towards this type of locomotion has probably been characteristic of apes for a very long time, and Napier (1963) has suggested that *Proconsul africanus* was a semibrachiator.

A defensive device that these small animals might have resorted to could have resembled that evolved by the savanna-dwelling social mongoose, *Mungos mungo*, which, when threatened, close in on one another and "bunch" thereby giving the appearance of a single, large and aggressive animal instead of numerous, relatively helpless small animals (see Vol. III). Elephant poachers are reputed to resort to a similar ruse sometimes when caught in the midst of a large excited herd of elephants.

It is not known whether *Ramapithecus* was bipedal or not but one inference that can be drawn from the reduced canine teeth is that an alternative means of defence or threat was available and this was likely to have involved some degree of emancipation of the hands for wielding weapons.

Wild chimpanzees are known to have used stones to break open nuts, they also throw objects or carry branches when making a threatening display. Gorillas likewise uproot and throw vegetation, sometimes accompanied by bluff charges on the part of the males. Indeed a most marked characteristic of the African apes is their intimidating display of hooting, roaring and thumping together with slapping, dragging and throwing of vegetation. The two activities of loud noise and violent slamming or manipulating of objects are often linked and seem to be a common expression of heightened excitement; at night chimpanzees hoot and scream at the sound of a leopard coughing, gunfire, thunder and earthquakes. (In cinema-going humans behaviour reminiscent of this is elicited by a failure of the lights.)

Noisy behaviour would appear to have some functional advantage for modern apes, its further development in *Ramapithecus* might have had even greater benefits and it may be worth mentioning that throughout East Africa, villagers drive off dangerous carnivores and crop-raiding ungulates or elephants with nothing more than a hullabaloo and the waving of sticks and branches. Furthermore if the ritualistic ape-like pattern of behaviour was to develop a directed and co-operative character at all similar to that seen among contemporary villagers, there might have been a shift in the significance of sounds from being a communal expression of an emotive state, as appears to be the case in apes, to a functional condition where communication was more controlled. It is possible to postulate an initial selective advantage for simple social threat behaviour by apes as a response to predators in open habitats. The advantages in controlling sounds to direct behaviour might have had wider uses and might conceivably have provided a basis for the development of speech.

Discussion of the problems associated with predation may well obscure the primary source of any species' success: its capacity to utilize an abundant and sustaining natural resource effectively. In this respect climatic changes in the Miocene and Pliocene are thought to have converted vast areas of land to drier more open vegetation types. There is no evidence that such habitats were less rich in food resources for primates than forests even if these resources were more dispersed. Apes were the dominant primates of the period so that there is nothing essentially problematic about the groups' invasion of savanna.

For the present there is little alternative to speculation about *Ramapithecus*; by contrast the australopithecines provide adequate evidence of their anatomy, as extensive remains have been found in South and East Africa. The structure of the feet and the pelvis was very like that of man and it is clear that these animals were adapted to erect walking. Furthermore, three very different types are known. Two, *Australopithecus robustus* and *Australopithecus boisei*, were large, heavy-jawed animals that may have been primarily vegetarian, and the third was a smaller, probably carnivorous type, *Australopithecus africanus*. Several races or varieties of *Australopithecus africanus* have been found and it was probably a more advanced population of this type that gave rise to *Homo erectus*. The larger *Australopithecus boisei* survived until only three-quarter of a million years ago and appears to have been sequentially sympatric with both *Australopithecus africanus* and *Homo erectus*, whereas it appears that *Homo erectus* came to replace the former species. Replacement is a widely spread phenomenon and is found in all groups of mammals where there is direct competition.

Remains of *Homo erectus* have been found in many African localities, in Europe and in the Far East, it was clearly a highly successful polytypic species. It is probable that among the many regional varieties an advanced population arose and became *Homo sapiens*. This species eventually eliminated its own ancestral stock, and it is further probable that subsequent phases of "competitive intolerance" eliminated other types of early man. Having eliminated his own ancestors, *Homo sapiens* is now eliminating the environment from which he emerged and with it the mammalian communities which his ancestors must have known so intimately. It is in this context that a reconstruction derived from the numerous fossil remains of *Australopithecus africanus* (from various sites) is offered here for its special interest as an important Pleistocene mammal and one that would probably still roam East Africa were it not for competitive intolerance from its descendants.

Modern Man

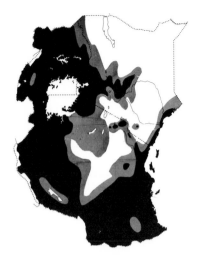

Reliability of Annual Rainfall
(from Royal Commission
Report on E.A.,1954)
■ A good prospect of 20in.
■ Fair prospect of 20in.
□ Chances of failure to receive 20in.
   30-100 years in 100 years.

The unique properties of our species have been extensively explored and are
so well-known as to need no discussion here. Indeed it may seem rather foolish
in a work such as this to include man as another species of mammal, but in
the first place the fate of most of the other mammal species is now dependent
on man and in the second place his specialization and peculiarities have been
evolved by fundamentally the same processes that have determined the
peculiarity of other mammals, both these facts bear discussion.

Until relatively recent times man interacted with other mammal species
in the simple role of one more particularly effective predator, but his population and his range were limited in much the same way as any other predator;

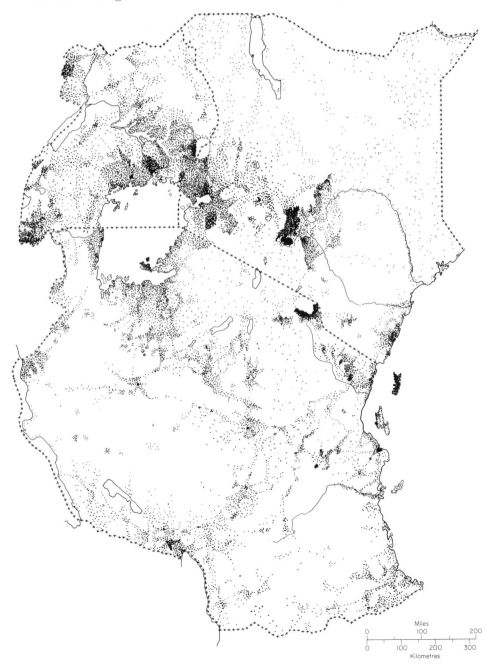

■ Areas in which Tsetse Fly
   vectors of Humans & Animals
   Trypanosomiasis are absent
   or have been eliminated
□ Distribution of Tsetse

Miles
0   100   200
0 100 200 300
Kilometres

in short the distribution of man was controlled primarily by several biological factors differing not at all from those controlling any other animal, namely water supply, food, intra-specific competition and disease.

The role of the hunting and food-gathering peoples and their influence on the natural environment stretches back in time to the first hunting bands of hominids. Modern peoples living in this way are now rare. The Hadza around Lake Eyassi, small bands of professional hunters and honey gatherers in Tanzania and Kenya and a few Bambuti and Batwa pigmies in West Uganda are the last people pursuing this most ancient way of life. Otherwise agriculture and pastoralism have provided man with the principal means of feeding himself and his distribution in East Africa has reflected the limitations these occupations place on him. Agriculture capable of sustaining a large population needs a reliable rainfall, and the largest populations in East Africa generally occur within areas that can rely on a 75 cm annual rainfall that is reasonably well-distributed through many months of the year. Malaria has been a severely limiting factor on populations living in low-lying areas. Pastoralism does not generally support such high human numbers, but it spreads man's influence and the primacy of his interests over large areas. The principal limitation on this way of life in East Africa has been the tsetse fly, as bovine and human trypanosomiasis carried by the fly precludes stock and usually man from the areas where this insect flourishes. The fly has also discouraged agriculture as people are seldom prepared to live without any livestock.

It is interesting to correlate a map of the areas of greatest human density with fly distribution in East Africa and the areas of good reliability of a 75 cm rainfall (see maps on facing page). These maps indicate that unreliable rainfall and disease (most particularly trypanosomiasis) have been the most important limitations on the spread of human populations or their stock. Improved agricultural methods, boreholes and irrigation are now changing this pattern and making it possible for man to utilize many marginal areas. During the past 60 years East African governments have evacuated populations from areas where human trypanosomiasis is common, so that the disease is now very rare, but the eradication of tsetse has until now been far too expensive an undertaking for East African governments to carry out on a large scale out of their own resources. This is also changing, as foreign aid and development are providing the economic and technological means of "reclaiming the bush" ("bush" being an emotive word to describe a wide range of natural vegetation types, which are in some way or other inimical to man, his stock and his fields). These natural limitations on man in tropical Africa and the presence of numerous and diverse mammal species are linked phenomena; the present wild life of East Africa owes its existence largely to man's past inability to destroy it or replace it. This inability is now being replaced by an ample capacity for the extermination of the majority of larger species if man wishes it, or the extermination may well happen by default. In the profiles of species it will be seen that human activities are increasingly affecting the great majority of mammals. It is most important to realize that the continued existence of many of the mammals discussed in these volumes has been up to the present time a matter of accident, but that their future is entirely a matter of choice for mankind. Mammal species will continue to

◄ Distribution of human population in East Africa (in 1962).
Derived from map N6 *Ann. Ass. Am. Geogr.* Vol. 56, 1966
by P. W. Porter. *NOTE:* In this map dots are only intended
to give an impression of relative density.

survive only because man deliberately allows them to, which means that legislation and planning with that specific aim in mind is necessary.

A major biotope that might have been threatened at an early date in East Africa is forest: the small and vulnerable forests in this region generally grow within areas of reliable rainfall and on fertile soils, but the economic value of timber led to the early creation of forest reserves, which have allowed the existence of unique and rare species to continue as an incidental by-product. The contemporary need to speed up the turnover of timber resources may lead to the replacement of many of the remaining indigenous forests by exotic soft woods and monoculture plantations, this policy would lead to the ultimate disappearance of many species of forest mammals.

The future of mammals in East Africa will therefore depend on the decisions of people. Many species have very limited ranges and, for some of these, the decisions cannot be delayed much longer. This work is the first comprehensive inventory of East African species and the first attempt to map their overall distribution. It is my earnest hope that it may help to provide a basis for deciding whether certain mammal species will be allowed to remain or not.

> Accuse not nature, she hath done her part;
> Do thou but thine, and be not diffident
> Of wisdom, she deserts thee not, if thou
> Dismiss not her. (*Milton, Paradise Lost*)

In spite of being a commonplace, the statement made earlier that man's former limitations in tropical Africa are linked with the presence of so great a variety of mammal species can be profitably amplified in terms of evolution. As a dominant animal, man replaces the other wild animals wherever he is able, turning over to his own use the natural resources which the animals formerly used. In doing this he scarcely differs from any other ascendant species. Huxley (1943) stated that "Evolution may be regarded as the process by which the utilization of the earth's resources by living matter is rendered progressively more efficient". This sounds very like a definition of human progress, the demands of which we have already seen are taking over the habitats of many species of wild mammals.

Organic evolution has a genetic base needing immense periods of time for its operation, in this it differs absolutely from human progress which relies upon more speedy means of diffusion, but there are numerous analogies between the two processes, indeed one school of anthropology suggests that the workings of the two processes are identical (Sahlins and Service, 1960). "Progress" derives from man's employment of tools, of true speech and from his tendency to migrate and it depends upon a more or less continuous human tradition, yet with every step of progress, man's history, his origins and his earlier environments are annihilated or obscured. Almost everywhere reminders of the past embarrass the myths of the present. The capacity to make things, to develop them and then make them obsolete is a central characteristic of human traditions and human behaviour. Development is often a form of elaboration and it is human elaboration that often obscures simpler sub-human or animal traits.

Gorillas and chimpanzees in the wild drape themselves with bits of vegetation in what appear to be attempts to elicit positive social reactions from their fellows, in other words they "show off". This playful hominoid trait is also common in children and has become elaborated in the personal adornment of adults. In spite of displaying specifically human skills in the making these adornments often also act as identification marks, and they seem to have converged in their function with some of the genetically determined territorial "marking" or "signalling" devices found in lower mammals. In the human, a playful origin might allow the degree of plasticity needed for the complex and variable needs of the human species. For instance, with cercopithecus monkeys both the patterns on their faces and backsides and the monkeys specific responses are genetically determined. Instead for the human a mark or signal may be changed from generation to generation. It is only in responding positively to the display of distinctive marks or patterns that man and monkey resemble one another.

Human societies all over the world have devised distinctive marks that identify the members of a group. These marks may take the form of tribal scarifications, skin painting, tattooing, etc., or may be developed further into cult ornaments, clothing or material symbols of social status. The marks are only discarded with great difficulty and are always replaced by new ones. The creation of distinctive forms to which the human group is deeply attached becomes extended from body ornaments to fetishes, sculpture, etc. and perhaps ultimately to the many symbolic artefacts that characterize a human culture and come to be called the signs of a cultural tradition. From these artefacts the member of the group derives so deep a sense of satisfaction and security, that threats to the objects or symbols of his group generally stimulate a violent defensive reaction. A mark, whether it is a simple scarification, a pendant, an elaborate fetish or the more sophisticated forms of developed cultures is only fully meaningful to those that created it, or those that grow up with it, and the range of the cultural marks' effectiveness or "meaning" will therefore extend no further than the boundaries of the group's territory.

The role of a distinctive mark is not initially unlike that of a territorial scent-marker in many mammal species. A group member "knows where he is" whenever he sees or smells another fellow or object properly marked, and is insecure in the absence of the familiar marks of his group. Marks therefore enhance solidarity within a group's range, but for humans they become meaningless outside it.

Human marks have probably derived from simple beginnings and may have become complex symbols by a process of elaboration. The form and the choice of a site for a distinctive mark may be found in a wide variety of human artefacts, showing that the choice of a vestment for it may be unimportant. The symbolic and phenomenal origin of the mark is also quickly lost and is of less significance than the meaning of the elaborative process itself, which has become a necessity for man, paralleling the territorial marking behaviour of other mammals. This behaviour is most exaggerated in a newly formed group, an example of this was the government of pre-war Italy; the fascist emblem was marked on every building, on every school book, and on every person (in the form of a badge). One is forcibly reminded of social mongoose, *Mungos*, this species uses its anal glands to mark vigorously every object and every fellow mongoose in the vicinity of the home burrow.

However, by their very nature, human marks tend to obscure animal origins rather than reveal them, for the elaborative process enhances the peculiarity of each and every group, and generally involves the display of distinctively human techniques.

It has long been recognized that the earliest "marks" scratched or rubbed by men onto the walls of their caves, on bones, rocks or clay fall into two quite distinct stylistic categories which have been called the *organic* and the *geometric* (Read, 1936), an origin for the latter type of "art" which is most characteristic of the neolithic has been suggested above. The other form of "art", typified by palaeolithic cave paintings generally represents animals. In spite of temporal and regional stylizations, the animal species can usually be recognized immediately, and often life-like postures or silhouettes highly characteristic of the animal species are faithfully represented. Rock paintings and engravings of animals, whether by Aurignacian or Magdalenian men or by African Bushmen, probably derive from a process that is fundamentally different to that involved in the making of a symbolic, geometric mark. Palaeolithic man's dependence on the animal was absolute. Hunting

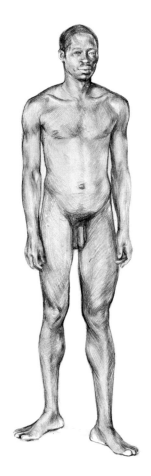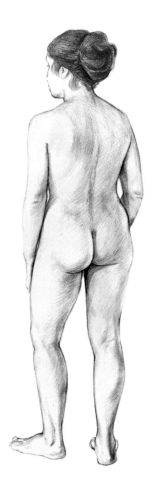

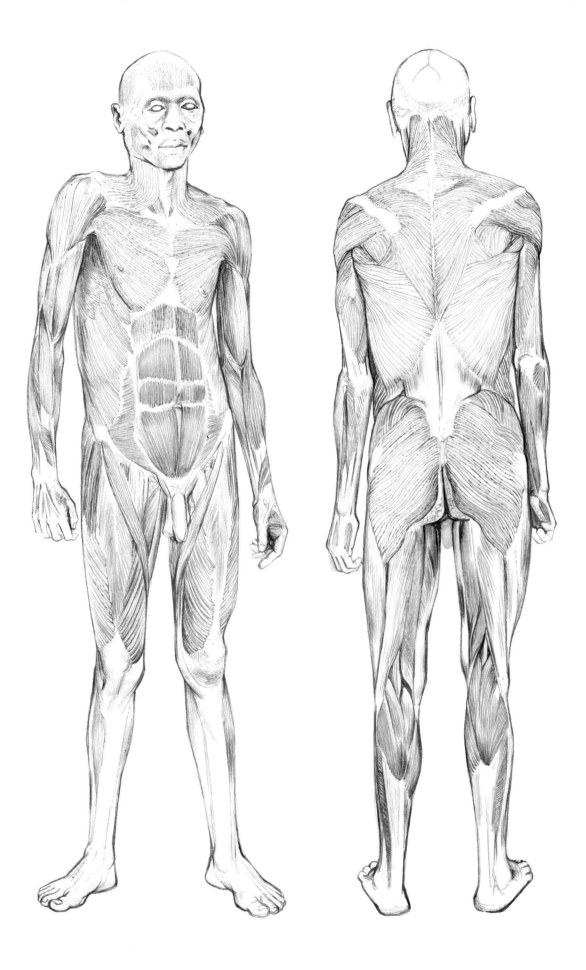

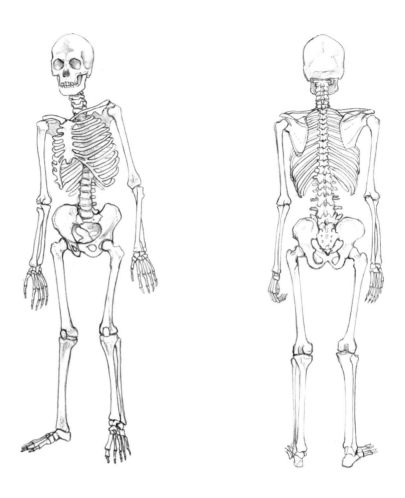

probably engaged all the men's energies and, during difficult times, great
distances must have had to be covered. In the intensive search for animals,
particularly in wooded country, a man's visual acuity must have played a
major role in finding the animal before the animal saw the man. During
periods when game was scarce, the capacity to recognize the form of an animal
from a distance, in spite of dappling, countershading and all the cryptic
devices of animal colouring, was crucial and the search for the significant
form of the animal amidst the shifting kaleidoscope of vegetation, sky, rocks
and ground must have preoccupied the eyes and the brain of the hunter for
long sustained periods. It is possible that each time an actual animal was
seen, an intense visual pattern became printed on the brain. This dominant
mental model or pattern might subsequently have reasserted itself in an
analogous manner to a retinal after-image, while the man was looking at a
suggestive shape on a rock surface in the home cave. The emphasis of those
shapes by additional marks (see Giedion, 1962) was a small matter for men
that had developed a high degree of manual dexterity and skill. Once this
initial step had been taken a specifically human delight in technical elaboration probably took over.

One interesting display of primitive marking behaviour, where plainly mammalian traits are scarcely obscured by human ones, is related by Armstrong (1964): "When I was real little, us kids would pound up a bucket of red brick dust and sell it to the prostitutes on Saturday mornings, make maybe 50 or 70 cents. Every Saturday they'd scrub their steps down with pee and then they'd throw the brick dust on the sidewalk in front, and that brought them luck Saturday night. That was their superstition." The mark in this instance has involved a direct appeal to both the senses of sight and smell and "elaboration" is associated with an explicitly sexual connotation. Galagos have an imperative need to scent their home range with urine and the smell of its own scent seems to reassure an individual animal. Working girls in a factory have been noticed to gain visible satisfaction from their own body odour when they were tired.

In the wild, the sexual behaviour of the gorilla is not particularly conspicuous, the greater part of the animal's time being taken up by gathering food. In zoos, where food is supplied in bulk and the physical environment of the cage is barren, erotic behaviour, including masturbation and variations on the copulatory positions of the animal have been observed which have not been seen and are not likely to occur in the wild. Recourse to excessive sexual activity is found in humans as a symptom of boredom, stress or deprivation, but ingenious elaboration of sexual activity is often taken much further than in gorillas (i.e. "Bunny Clubs").

Sometimes however the great apes elaborate behaviour under natural conditions. Loud vocal and drumming displays are occasionally sustained by chimpanzees for periods greatly in excess of the usual minute or so; in these performances heightened excitement and the release of tension seem to oscillate and so prolong the display. The beginnings of deliberate rhythmic "dance" were observed by Kohler (1925) in captive chimpanzees: "Marching in a circle, one behind the other, the big animal stamping its foot violently at every step or every other step and the others exaggeratedly accentuating the marching movements."

Human drumming, Spanish dancing, war dances and other activities involving rhythmic noise and vigorous movement exhibit many of the characteristics of the chimpanzee behaviour. The components of drumming and singing are noises depending upon motor patterns that scarcely differ for man and ape. The stimuli or conditions of excitement in which these performances occur also show some resemblances, obscured perhaps by the elaborated character of most human activities. Nonetheless, in this instance the resemblances could derive from behaviour patterns with phylogenetically ancient roots, which are common to apes and man.

It has been difficult enough for individuals belonging to specific cultures to recognize the common roots that their customs and rituals have with other cultures, it is still more difficult to recognize the animal origin of some of our responses to visual, oral and olfactory stimuli, particularly when we know less and are even less interested in the behaviour and customs of mammals than we are in those of people from far away places.

An appreciation of the essential unity of mammalian life, of which our own species is a part, enhances our enjoyment in studying mammals. For this reason I hope that extensions of this discourse, which may have taken liberties with both human and animal behaviour, will be indulged.

Apes

Pongidae

Apes are found in the forested areas of tropical Africa and Asia (see world distribution map, p. 53).

The two African species are anatomically very similar, but the chimpanzee is smaller, more arboreal, more active and of a livelier temperament than the gorilla. Anatomical differences such as the gorilla's cranial and neuchal crests are principally due to the greater size and weight of the animal and the development of massive jaw muscles.

Although their divergence could be ancient (possibly as early as the Miocene), these African species are nonetheless much closer to one another than they are to the orang-utan, *Pongo*, a fact which is not reflected when the three apes are given equal generic status. Simpson (1963) states: "Placing all the African apes in *Pan* permits classification to express the clear fact that they are much more closely related to each other than to any species of other genera."

Features by which apes are commonly distinguished from monkeys are the patterns on the molar teeth, the apes having five cusps on the lower molars to the monkeys' four, the tail is absent and there are differences in the skeleton, the apes having a larger head, longer neck and limbs (particularly the arms), a broader thorax and a long flat pelvis below a short lumbar region. (Apes generally have 4 lumbar vertebrae to the monkeys' 6 or 7.)

The apes' long limbs allow a compact tubby body to be efficiently levered about through the trees or over the ground. Lengthening of the limbs is an adaptation to give a longer stride whether the "stride" is made by the legs or the arms, on the ground or swinging through the branches. It is a locomotory system capable of modification to special conditions, illustrated by such extremes as the heavy terrestrial gorilla and the very long-armed arboreal gibbon. However, it is probably a form of locomotion with mechanical limitations and its adoption by the apes may have a bearing on their status.

The scattered distribution of the very few species of modern apes is both a measure of their specific success relative to the many extinct ape species and also the measure of a general failure of the group in relation to monkeys.

The long-backed monkeys can spring, run, climb and leap through the trees or over the ground, grasping with both hands and feet and using their tail as an acrobat's balance, whilst in the most perilous situations. In contrast to the apes, the monkeys' legs are slightly longer than their arms. Such a body architecture is decidedly superior for the exploitation of the slender branches of tree tops, where fruit, flowers and leaves are most abundant. It is interesting that the only surviving ape that is of comparable size to the monkeys, *Hylobates*, has had to specialize the brachiating habit to a quite exaggerated degree, while monkeys have retained a more generalized pattern.

The locomotory limitations of the apes are offset by their brain being much larger; their manipulative skills have developed further, and their capacity for facial expression and social communication is greatly superior to that of monkeys.

116

There is an interesting and very extensive literature on apes, notably by the following authors: Yerkes, 1929; Zuckerman, 1932; Kellog, 1933; Kohler, 1925; Carpenter, 1942; Le Gros Clark, 1959; Goodall, 1965; Napier and Napier, 1967; Morris, 1966, 1967; Schaller, 1963.

Gorilla
(Pan (Gorilla) gorilla)

Family   Pongidae
Order   Primates
Local names

Gorilla, Makaku (Swahili), Ngagi (Lukiga).

Measurements
height
1,750 (1,400—1,850) mm
weight
140—275 kg (males)
70—120 kg (females)

Gorilla

The gorilla and the chimpanzee show many resemblances, indeed the gorilla differs most in those features related to its great size and weight. Yet the two species are sympatric over much of the gorilla's range and have never been known to hybridize, so that any recent common ancestry is not at all likely. As the main radiation of apes occurred during the Miocene, since which time they have declined, and as the fossil Miocene apes found in East African sites show similar differences in size, it is possible that the ancestral stock of the gorilla and the chimpanzee may have been distinct at this early date, which speaks a lot for the conservative frame of the great apes.

Like many other African forest mammals, the gorilla is split into two populations on either side of the Congo basin. The situation can be initially ascribed to climatic changes that desiccated the central Congo basin in the past, but the status of the gorilla is very different to that of the chimpanzee that has a very much more considerable range and seems to be a generally more adaptable and successful species. Schaller (1963) shows that rivers are barriers to the heavy gorilla, which is apparently unable to swim and that the succulent green growth on which they feed limits their distribution to humid forests with plenty of herbaceous undergrowth.

In the light of East Africa's past vicissitudes and the gorilla's limitations it is perhaps not surprising that its distribution in East Africa is limited to two very restricted localities in Kigezi, where the total population probably does not exceed 200 individuals. The prospects for their survival in the Kigezi "Gorilla Sanctuary" are discussed in the last report issued by the Uganda Game Department (1962): "Encroachment by cultivators, bamboo cutting and other human activity continues to cause considerable disturbance and unless it can be brought under control it is certain that this rare species will eventually disappear from the area altogether." The area concerned is only 23·3 sq km and much of it is now unsuitable habitat. The area of forest reserve in which the gorillas are found was reduced to this size from 34 sq km in 1950. Considering that a troop's range is 26—39 sq km it is not surprising that the gorilla are consequently no longer permanent residents in this area. Population pressure round the Kayonza forest will probably in the near future make similar inroads on this last permanent refuge of the gorilla in East Africa.

The habitat in these two areas is very different. At Kisoro the forest is mostly bamboo with some small patches of mountain woodland. The greater part of Kayonza is montane forest rising to 2,500 m, the northern section is at a lower altitude—1,350 m and this area is one of the few places in East Africa where montane and lowland forests merge in an unbroken succession. The country is very hilly and there is a very rich undergrowth providing the gorilla with a rich supply of food; the principal plants eaten here are the leaves and pith of the vines *Momordica foetida*, *Mikania cordata*, *Urera hypselendron*, the shoots and pith of tree ferns *Cyathea* and other ferns and the pith and leaves of *Aframomum*. At Kisoro, the gorilla's favourite foods are the central stems of wild celery, *Peucedanum* spp., the root of the herb *Cynoglossum*, the pith and blossoms of the shrub *Vernonia*, the hook-leaved

Map of western Kigezi showing gorilla distribution in Kayonza and Impenetrable Central Forest Reserve and in Mt Muhavura "Gorilla Sanctuary". The map shows the pressure of cultivation and the ease of access to the area. Less than 100 gorillas live in East Africa.

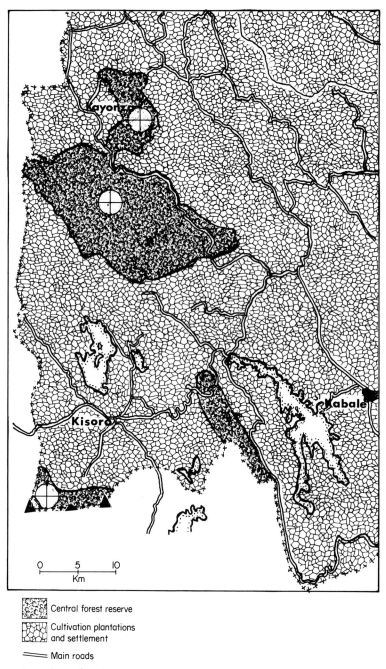

Central forest reserve

Cultivation plantations and settlement

══ Main roads

++++ International boundary

vine *Galium*, which is rolled into balls, the stems and leaves of nettles, *Laportea*, the leafy stems of the everlasting *Helichrysum*, the pith of *Senecio*, the berries and new growth of blackberries, *Rubus*. Between October and December and again in March the bamboo put up new shoots on which the gorilla feed.

There is no evidence that the gorilla ever eats any animal matter and the animal has the placid disposition of a herbivore. Although there is great individual variation in the males, gorillas as a whole and particularly the females do not have much curiosity and are not as excitable as the chimpanzees, in spite of their noisy displays and mock charges, which punctuate what is otherwise a sober and leisurely existence.

The animals have sharp eyesight and hearing, particularly once they are alert to a disturbance in the neighbourhood and Schaller has evidence of gorillas smelling a man downwind at 20 m. In Kayonza, the smaller animals climb and make their nests in trees, but whenever they do climb they are very cautious and the older animals spend almost all their time on the ground. They walk on all fours, frequently stopping to eat, when they sit or stand on three limbs while picking stems with the free hand. The weight is taken on the knuckles of the hand and on the entire sole of the foot.

As much of their foraging takes them through thick, prickly growth they often walk with the head turned sidewards. Normal communication is by grunts, by means of which the troop keeps together and there is an alarm note that sounds like a very loud, deep, bushbuck bark. They leave a broad trail that is easily followed and excrete large quantities of pungent horse-like dung at frequent intervals. While on the move they usually walk in file but while feeding spread out within easy hearing of one another. A typical day consists of leisurely feeding during the morning followed by a rest period, during which a day nest may or may not be made. If the troop moves any distance in the course of a day this is generally accomplished in a short active period in the late afternoon after the rest. The nests are made at dusk and may be

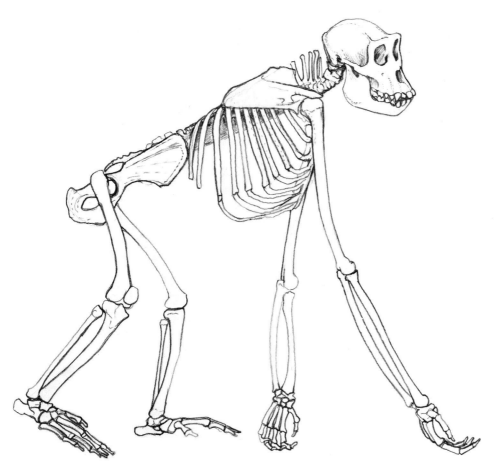

anything between a huge bed of bent and broken branches or bamboos to a mere rim around the curled up animal.

The troop contains between 5 and 27 individuals, but temporary associations of troops may lead to larger numbers. A fairly typical troop might consist of two or three silverbacked males (silverback males are probably over twelve years old), two blackbacked males, eight or nine females, three juveniles and six infants and it is not unusual for nearly half the troop to be immature. Troops range over 26 to 39 sq km, unlike chimpanzees they do not rely on well-known tracks but they know their home range well and frequently revisit favourite areas. The troop is of relatively stable composition, although many males may come and go and solitary males have been observed. Schaller saw a meeting between two troops in which the females exhibited no interest or excitement, while the leading males made displays and then settled down amicably only to part shortly afterwards.

Display is elicited by the proximity of other troops or man and in this there are resemblances with the ritualized hooting and tree-beating of chimpanzees, which also seem to be a release from tension. Schaller has described the display in detail; briefly the gorilla pursues the following sequence, a preliminary hooting, in which the air sacs in the throat inflate, followed by leaves being snatched and put in the mouth, then rising on the hindlegs, tearing and throwing of vegetation, a rapid drumming on the chest and kicking followed by a violent sideways run with slapping, tearing at the vegetation and ending with a final thump of his palm on the ground. The display is relatively stationary but other gorillas get out of the way if the male comes in their direction. Other males in the troop watch the performance intently and may add their own tattoo to the chest beating, the sound of which may carry a mile. Schaller heard chest beating one night when two troops were sleeping near one another.

The display is distinct from fighting behaviour and in the rare quarrels between males, tension may start with a hard fixed stare, barking and screaming. Males also roar at alien animals, Schaller saw a mobbing raven roared at in this way and this frightening noise is not infrequently directed at the human intruder; other members of the troop tend to cluster behind the male when this call is made. Yawning is seen in essentially the same circumstances as in other monkeys and is commonest in the leading silverback males which are more alert and nervous than the females.

The gorillas' bluff charges and occasional attacks on humans have led to innumerable stories and the killing of many gorillas. Attacks are seldom thrust home and then usually when the human intruder fails to stand his ground. Merfield and Miller (1956) relate that among the Mandjim Mey in Cameroon it is a disgrace to be bitten by a gorilla for it is generally understood that only a coward would be attacked.

Apart from man, leopards are the only predators of the gorilla (the drawing opposite is of an adult silverbacked male killed by a leopard in February 1961). These occurrences must be extremely rare and this individual gorilla was crippled by his lack of fingers in one hand. Virus, blood, intestinal and other parasites are known in the gorilla, which suffers from many human diseases including arthritis, heart disease, pneumonia, cirrhosis of the liver, malaria and yaws.

No breeding seasons or peaks are apparent and young are thought to be born at intervals of three and a half to four and a half years. The female has a cycle of 25—30 days, but, unlike the chimpanzee, has no perceptible swelling to advertise her oestrus period which lasts for about three days. During this period the female "presents" to the males.

The animals appear to reassure one another by turning their heads sideways, and Schaller saw a captive male strut around the female watching her out of the corner of his eye. This is a "showing off" posture frequently seen in other contexts in the wild and in captivity.

Sexual behaviour and copulatory positions seem to be subject to great variety in captivity and these are discussed at length by Schaller. Gestation is 251—296 days. Females are capable of conceiving at about seven years of age, the male becomes fully mature at nine or ten years.

The new-born infant is unable to cling and is supported by the mother, it does not leave the mother's body at all until it is about two and a half months old. At four months the young gorilla is able to walk but the mother continues to be closely attentive to it up to the fifth month. By the time it is a year old the young gorilla is wandering about in the group, but it continues to be cared

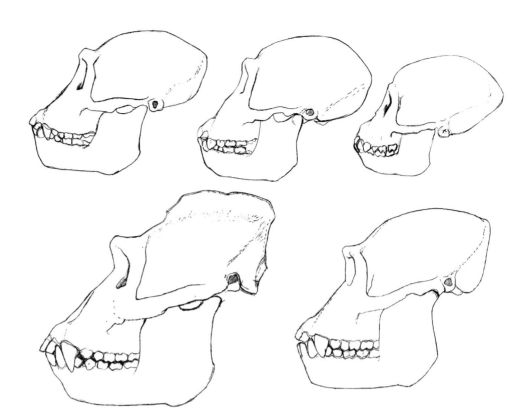

Skull series changes with growth in male gorilla. ▶

for by the mother for about three years. Play is largely confined to the young, in marked contrast to the stolid adults. Schaller describes their play most charmingly:

"An 8 months old bumbles around by the reclining dominant silverbacked male. With a wide overhand motion it swats the male on the nose, but he merely turns his head. The infant then runs downhill and turns a somersault over one shoulder and ends up on its back, kicking its legs in bicycle fashion and waving its arms above the head with great abandon. A ten-month old infant watches these proceedings while propped against the rump of the male. Suddenly the ten-month infant rises, hurries to a sitting juvenile, and pulls the hair on its crown with one hand. When this brings no response the infant yanks at the hair with both hands, but the juvenile remains oblivious. The infant desists, sits briefly, suddenly rolls forward over one shoulder, and with arms and legs flailing like a windmill rolls over and over downhill and disappears in the vegetation."

"An infant, $1\frac{1}{4}$ years old, sits and handles a leafy lobelia head. It bangs the leaves on the ground several times and swings them around in an arc. Finally it places them upside down on its head and remains motionless for several seconds under the green hat."

"Four infants . . . play in a vine-covered *Hypericum* as high as 20 ft above the ground. By themselves they swing back and forth on lianas. Then, when one races along a branch, all others follow. In a row they climb up some vines and slide down others, grabbing a mouthful to eat in passing or pulling and pushing each other. One stops on a branch, and, holding on to a vine, leans far out. The vine snaps and the infant tumbles head first through the tangle. . . . The others immediately jump and join the other, tumbling and sliding carelessly in a shower of dead leaves."

Chimpanzee
(Pan troglodytes)

Family Pongidae
Order Primates
Local names
Soko, Soko mtu (Swahili), Kitera
(Lunyoro), Ekikuya, Ekibandu,
Kinyamusito (Lukonjo), Esike
(Rutoro), Empundu (Runyankole),
Mkiba (Kuamba).

Measurements
head and body
635—940 mm
height
1,000—1,700 mm
weight
45—80 kg

Chimpanzee

The chimpanzee does not need description, yet its familiarity makes it the more difficult to look at objectively. The very name we give it, *Pan* (*Satyrus*) *troglodytes*, is a monument to our anthropocentric perceptions implying as it does some alienated, uncouth Caliban. Its disarmingly human features have made it a model for countless idiotic toys and cartoons. Its clumsiness when made to walk upright simulates the uncoordinated movements of a drunk, cast it as a clown in innumerable circuses and zoos, while psychologists rearing infants in nappies and cots and arming the animal with paint brushes have reinforced the public image of a caricature of humanity. Museum artists have used the chimpanzee as a model for their reconstructions of long extinct apes, which of course is to revert the image.

The ancestral stock of the chimpanzee is known from the Miocene of East Africa and the type was named *Proconsul* (after Consul, a London Zoo

Chimpanzee distribution
Legends suggest historic presence of chimpanzees here.

Miles
0   100   200

0   100   200   300
Kilometres

127

chimpanzee). This species, *Proconsul africanus*, was a smaller and more slender animal than the modern chimpanzee, which we may regard as a somewhat more specialized descendant of this dryopithecine ape. It has been suggested that the largest form of the three *Proconsul* species known might be ancestral to the gorilla to which it approximated in size.

The present distribution of the chimpanzee stretches across the forest zone of Africa from Guinea to the forests of western Uganda and Tanzania. These apes live over a wide range of altitudes up to 2,750 m and in many different types of forest. In the western region of Tanzania the chimpanzee lives in a mosaic of *Brachystegia* woodlands, thicket and riverine forest strips. Significantly most of this area is virtually uninhabited by man. The absence of large human populations along parts of the Lake Tanganyika littoral has undoubtedly allowed the chimpanzee to survive and possibly even to expand in this area, but the chief determinant for its presence anywhere is an abundance of food, particularly fruit. For instance, one adult animal has been seen to eat 620 bean-sized pods in six and a half hours, as well as many odds and ends of other food. It can be seen then that a pre-requisite to chimpanzee survival is quantities of food available throughout the year.

Various observers have catalogued the foods of chimpanzees and lists run

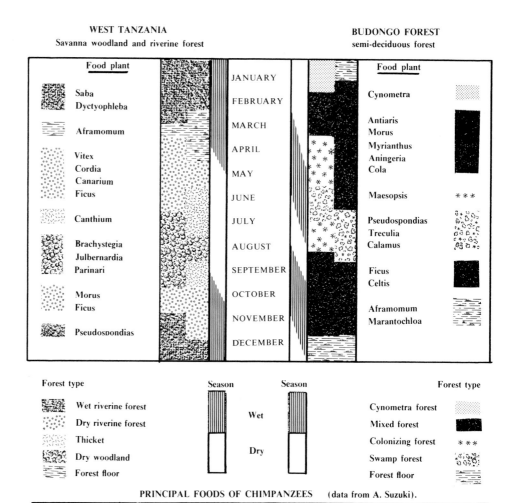

PRINCIPAL FOODS OF CHIMPANZEES   (data from A. Suzuki).

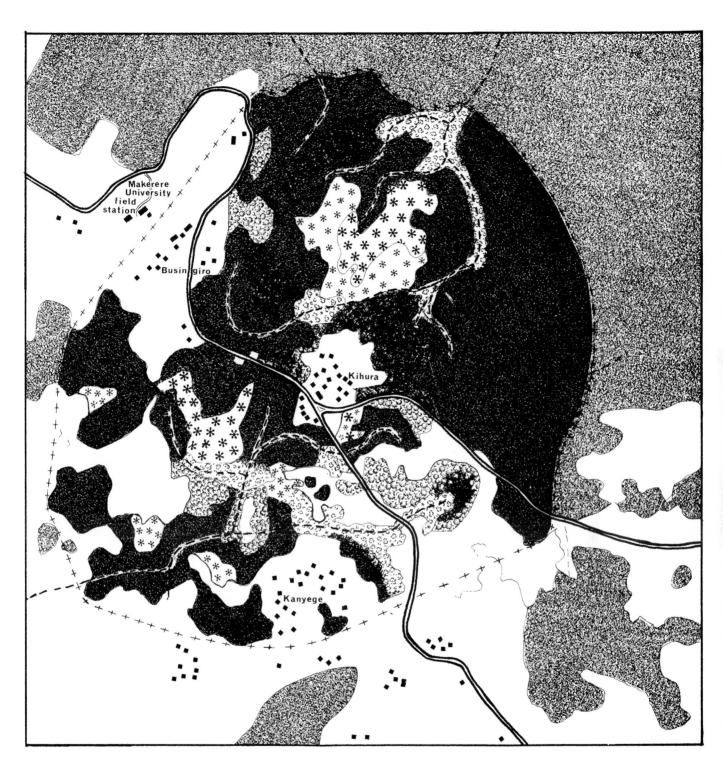

▲ The range of a regional population of about 80 chimpanzees
studied by A. Suzuki. (Map derived from Uganda Forest Dept.
Map B.N. 23.) Vegetation types as in food chart key:
white = grass, scrub or cultivation; *dashes* = rivers; *squares* =
houses; *grey* = forest beyond range of population. *NOTE*:
Four other regional populations overlap this range along N, E,
S and W peripheries. Population to the East (with no
ecological boundary) has the largest area of overlap.

from about 30 to over 80 different items. In all cases fruits constitute the principal diet; leaves, buds and blossoms, bark and resin are also eaten. Foods are gathered in a variety of ways and observers have noted that individuals may be recognized by mannerisms of feeding. Most fruits are picked with the hands and then eaten; early in the morning this is usually indiscriminate stuffing, but after a time feeding may become more selective and ripe fruit is picked in preference to the less ripe. There is a second feeding peak in the evening between four and six. Berries and seeds are usually eaten directly off the stem with the lips. Hard fruits are broken against a trunk. In West Africa, they have been seen breaking open nuts by pounding them between rocks (Beatty, 1951).

The type, locality and spatial distribution of food and its fluctuating abundance determine the chimpanzee's seasonal and day to day movements and also certain aspects of their behaviour and social structure. The area over which a troop ranges may be correlated with abundance of food. Suzuki has established that Tanzania troops of about 30—50 individuals ranged over 100—200 sq km of savanna woodland and small riverine forest strips near Uvinza, Tanzania. By contrast, troops numbering 70—80 individuals range over 7·5—20 sq km of Budongo Forest. I am indebted to Mr Suzuki for lists of the principal foods of chimpanzees in these two areas and the periods during which they were eaten. The range of the Budongo troop studied by Mr Suzuki is also mapped with the principal vegetation zones demarcated. In this way the temporal conditions for this troop's food supply can be correlated with the spatial distribution of food.

A variety of foods may be eaten on any one day, particularly when food is dispersed. This is a common situation for Tanzania chimpanzees, which in general walk greater distances and have a more varied diet. Forest chimpanzees by contrast often have a more concentrated and abundant source of food, and movement may be limited. Indeed groups of chimpanzees may sleep and feed in the vicinity of a large food tree or group of trees for several days on end. In addition to vegetable matter chimpanzees eat a variety of animal foods. They have been observed dipping twigs into a subterranean nest of honey (Merfield and Miller, 1951). In a similar fashion Goodall (1963) has observed termites being caught by the chimpanzee inserting twigs that have been stripped of leaves into the termitary. She also noted other insects being eaten: *Camponotus*, *Dorylus* ants and their nests, cocoons of *Megaponera foetens* and gall-fly larvae. Intensive study in recent years has revealed that chimpanzees in all habitats kill and eat meat not infrequently. Goats have been killed in Ankole and Bunyoro and young bushbuck have been seen to be dashed to the ground, the head and body being thumped on the ground until the animal was dead. The most frequent victims, however, are other primates. Young baboons, black and white colobus and blue monkeys have been killed. At Gombe Stream the commonest kill is red colobus. All the victimized species often feed close to the chimpanzees; solitary blue monkeys may accompany a troop and, in the case of the baboon, young individuals may even play together.

A cannibalistic episode has been seen and photographed by Suzuki when an adult male was found eating the legs off a living chimpanzee baby which screamed continuously. A female watched the performance intently but with-

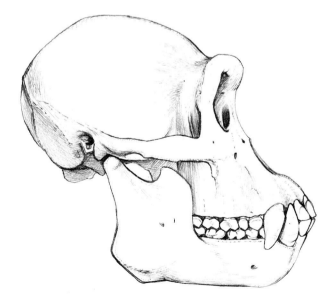

out moving or making a noise and other dominant males at one time came to touch and smell the baby. The occurrence followed a period of almost three months when *Celtis* had been the principal food (the acid juices of this fruit are sometimes wiped from the chimpanzee's mouth with leaves). This period of monotonous diet may have led to nutritional deficiencies or may have induced a craving. On the other hand, a characteristic feature of killings, both in the chimpanzee and the baboon, is that they occur quite suddenly and usually over a short period. Suzuki has correlated this trait with periods of excitement and tension which coincide with the decline or end of a specific fruiting season before a new, consistent supply of food has been found. Sugiyama (1969) also relates seasonal activity to strong agonistic interactions. This cannibalistic incident may not be as unique as it would appear, for a group which McKinnon had been watching, but with which he had lost contact, was heard screaming and crying. Later after coming up with the group again he found that a newly born chimpanzee seen earlier had disappeared.

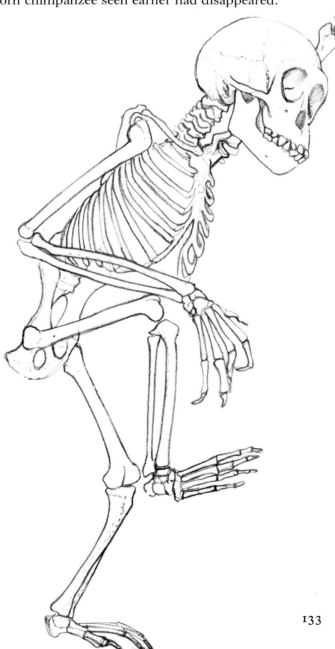

133

The killing of a half-grown *Colobus polycomos* in Budongo Forest was quite deliberate. The chimpanzees surrounded the colobus, which dropped to the ground, where it was killed by a large male chimpanzee which ate most of it, but allowed females to touch it and to take morsels. Even juveniles scavenged for bits of skin and fur which they sucked clean and later spat out (Suzuki personal communication).

An adult male in the Gombe Stream Reserve on sighting a bushbuck bristled all over; a neighbouring female seeing this male's behaviour became very nervous and crept off with her knees bent (Goodall). Killing does seem to be a prerogative of adult males so that their dangerous nature may be recognized by the females and young. The taste for meat, however, is found in both sexes and in the juveniles.

134

The adaptability of the chimpanzee to varied foods and habitats suggests that, were it not for direct competition from man, it could live over a wider range of country and of habitats than those in which it is found today. In Rungwe, there are legends of little black men of the mountain forests which may refer to chimpanzees and a single animal has been found still further south in a forest at Nkata Bay, Malawi, and there are other reports of chimpanzees in this area. The animals were reputed to have occurred in Mabira Forest in the 19th century. It is therefore not unlikely that the chimpanzee distribution in East Africa was more extensive in the immediate past than it is today. Recently a troop in Bwamba became isolated in a small forest area by expanding cultivation and human settlement and the chimpanzees were killed, having become a menace to the local people and their crops. This pattern of events might not have been infrequent in the past. Another factor controlling the chimpanzees may be disease. They suffer from many human diseases including malaria, and McKinnon has suggested that chimpanzees dying in the Gombe Stream Reserve may have picked up a disease in the fishermen's camp which they had taken to visiting.

The only reserve specifically created for chimpanzees is the Gombe Stream Reserve north of Kigoma, Tanzania. Otherwise the future of the chimpanzees in East Africa rests with the Forest Departments of Uganda and Tanzania and the Queen Elizabeth Park. Uganda chimpanzees will be secure for as long as West Uganda indigenous forests and their fauna are regarded as an economic and scientific asset, and so long as licences are withheld for commercial exploitation of the chimpanzee. Populations in East Africa are not large enough to stand any large scale interference. Heavy forest poaching in recent years, involving many hundreds of men and dogs and miles of netting have resulted, in one area at least, in the illegal capture of chimpanzees and serious disturbance of the habitat.

Serious consideration should be given to the creation of a National Park in west Tanzania in the area south of the Malagarasi River where chimpanzees are living in conditions probably comparable to those of the early hominids. It should be emphasized that this situation is quite unique and is of extraordinary interest to all students of mankind's origins.

The behaviour of the chimpanzee has been the subject of numerous studies, most of which are based on captives. Sometimes general conclusions have been based on single individuals, yet great differences in character are as apparent in these apes as in humans. No more than a brief outline of the social life of wild chimpanzees can be given here.

A troop or social unit of chimpanzees, which Sugiyama (1968) has termed "regional population",* may contain 25 to 80 or more animals and these will live within a limited area, from 18—21 sq km in forest and from 100—200 sq km in savanna mosaic. The home range in the forest has well-established paths which the chimpanzees follow. In savanna, game paths may be used. Within the main social unit there are smaller groups, these only amalgamate as an entire troop on rare occasions, usually when a very localized food supply brings them together. These sub-groups have a very loose and unstable composition in forest, where there is a great individual independence and the only stable relationship is that of a female with her young. There may be all male groups, adult groups, nursery or mother-young groups, mixed groups or solitary individuals. The size and individual composition of these groupings change constantly. By contrast in the drier, more open country of western Tanzania chimpanzees ranging over much larger areas of country have an overwhelming preponderance of mixed groups comprising one or more males with females and young. The stability of membership in these mixed groups is still unknown but the open habitat does seem to impose a particular type of social coherence on their nomadic life (see Suzuki, 1969).

Movement from place to place is along well-known paths on the forest floor. Here they can be very fast and quiet. They are generally cautious climbers and it is only by a human yardstick that they might be called adept or agile in the trees. They spend more time in the trees during the rains, when

* A controversial title as the long-term movements of individuals or groups outside their "region" has not been studied.

a day-nest may be made. Nesting is a typical feature of chimpanzee behaviour. Usually made in trees the nests may be at almost any height and are made by branches being pulled into a chosen spot whereon the chimpanzees sit. Other twigs and branches may be fetched from nearby.

A group of chimpanzees can be very noisy. Their bouts of hooting and particularly their drumming on tree buttresses can be heard over long distances. This intimidating noise probably serves as a directional indicator as well as being part of a threat pattern, the two functions are scarcely separable where territorial behaviour is concerned. The group bursts into a chorus of whoops which rise into a crescendo of screaming hoots and a drumming tattoo by a male or males, who beat with their hands on a tree buttress, or grip it while they hammer their heels on the resonating surface below. This performance is more prolonged and more frequent when two groups meet one another in an area where the range of troops overlap and on the peripheries of a troop's territory. Very occasionally the noise is kept up for an hour or more and there is about the whole performance an element of what could be called group hysteria in humans. The "drums" are mature buttressed species of trees, that have adequate resonating boards such as *Cynometra* or *Chrysophyllum*. The sight of favourite and well-known "drum trees" also seems to be a stimulus to drumming, which possibly helps to reinforce the boundaries of territories. The breaking of a thunderstorm, the rumblings of earthquakes and nocturnal gunfire also precipitate hooting and screaming even from the nests. If this is also a threat behaviour pattern, one might well be reminded of King Canute and the waves.

Chimpanzees are promiscuous and within a troop amicable, and all-male parties are not uncommon. Threatening behaviour is expressed by dragging and tearing at vegetation, banging and thumping with the arms and feet and screaming, meanwhile the hair stands on end. These expressions vary in context; they may serve to maintain a hierarchy within the social unit, to advertise territorial rights to other troops and also to intimidate other animals and potential predators.

A low-ranking or timid male of the group that visited Goodall's camp learned to rattle and bang an empty kerosene tin. This frightened his fellows, whereupon he rose to a leading position in the troop, a remarkable case of the self-made chimp!

Humans in the forest generally cause a rapid and silent decamp. The warning on such occasions is uttered by the first animal to spot the intruders. This is usually a very low, short but urgent moan which generally goes unnoticed by the human ear, but has an immediate impact on all chimpanzees in the vicinity, who stop feeding, slide to the ground and depart rapidly and silently, sometimes a louder alarm bark is heard.

On one occasion a single female chimpanzee was found feeding in the crown of a tree separated from her group by a path on which observers were standing; passing through the branches of the canopy this animal defecated and urinated on the party as she went and, without stopping, was seen to break off and toss down a short, stout branch with one deft movement of her right foot.

Kortland (1967) placed a stuffed leopard in the path of some wild chimpanzees and he observed threatening behaviour followed by an attack in which a stick was picked up and swung at the leopard. Hunters after killing a chimpanzee have described being threatened and screamed at by other members of the group, both from the trees and from ground level. I am unaware of any serious attacks on man by wild chimpanzees but on one occasion in the Queen Elizabeth Park a party of zoologists were threatened at close-quarters by a group of chimpanzees on the ground. There was no apparent provocation for this unusual behaviour (M. Delaney, personal communication).

The direct approach of a dominant male may elicit nervous panting on the part of inferior animals who take up a crouching posture or offer the palm in an appeasing gesture. The dominant animal may hug the submissive one or put his hand on its head, and this contact seems to have a reassuring effect on the animal. This reassurance by contact may be the ritualization of a juvenile trait carried over into adult life. It has obvious implications for humans who exhibit almost identical behaviour in social situations.

Sexual behaviour centres on the 36-day menstrual cycle of the female, who exhibits very large perineal swellings at oestrus. She is only receptive during this period and is mounted frequently, often by several different males who do not generally show any animosity towards one another. Gestation may last 202 to 261 days. When heavily pregnant, the female carries her back in a characteristic arc and is appreciably slower in her movements. The newborn baby does not suckle until the second day. The eyes open at about 20 minutes after birth.

For five months the baby clings tightly to the mother and takes its first steps at about six months. Until it is two years old it seldom strays far from the mother and on the least alarm seeks the reassurance of bodily contact with her. Other animals, particularly siblings and juveniles like to carry and play with babies. In his superb documentary films of chimpanzee life in the Gombe Stream Reserve, Van Lawick has sequences of many incidents centering around the growth and care of the baby. Among the many games he has recorded chimpanzees playing there are many which are quite indistinguishable from those played by children, "tag" and going round in circles until dizzy, are notable examples.

At ten months the young starts to groom. After the age of two it is independent of its mother for long periods and by four years of age it may walk and play quite independently; by this time the mother usually has another infant and the young animals may form adolescent parties. Sexual maturity sets in between eight and ten years of age, when females may become pregnant. At ten years the animal is fully adult and may live up to forty years, if zoo records are an indication.

Budongo Dec 5 '67

Cercopithecidae, Old World Monkeys

COLOBINAE   **Colobus**
CERCOPITHECINAE   **Papio**
                           Cercocebus
                           Cercopithecus

Published definitions and descriptions of the old world monkeys are numerous and need not be repeated here. The comprehensive monograph by Hill (1966) is the most detailed available.

The Cercopithecidae are very poorly represented as fossils and all accounts of their evolutionary past have been and must continue to be based almost entirely on speculation until more fossil evidence is at hand.

Simons (1967) has suggested an affinity between the Oligocene *Parapithecus* and the Cercopithecidae, but the earliest actual evidence comes from Lower Miocene beds in East Africa, where unmistakeably cercopithecid fragments have been found. As earlier fossils have been thought to bear most resemblance with the Colobinae it has been suggested (Jolly, 1966) that the initial divergence between Hominoidea and Cercopithecidae may have been due to a dietary specialization, the apes becoming fruit-eaters and the monkeys leaf-eaters. Of living monkeys the Colobinae certainly retain the most archaic features and are found over the widest range. It is also possible that some other anatomical features affecting locomotory habits led to the divergence.

Ancestral Cercopithecidae probably entered Eurasia with the earliest migrations of forest fauna out of Africa and the most numerous group in Asia today are the langurs, represented by 6 distinct genera, 19 species and 89 subspecies. Many of these forms occur on islands, where evolutionary change may be speeded up through isolation, equally well the islands act as refuges for relic species. In any case there is little doubt that the colobines have had a tenancy in Asia over a very long period.

The large number of Asiatic colobines contrasts with only two African genera, which exhibit a high degree of adaptation to leaf-eating. The black and white colobus are a particularly successful group with an extensive range in the tropical African forests. One can suppose that less specialized members of the group were eliminated by subsequent competition from other types of monkeys and indeed several extinct colobines are known from the African Miocene onwards.

If the divergence of the Cercopithecidae from the Hominoidea was indeed due to a leaf-eating specialization, the specialization might already have broken down at an early date, for a cercopithecine tooth has been found in the Lower Miocene site at Napak, Uganda, that lacks the high cusps typical of *Colobinae* (Pilbeam and Walker, 1968). At present the continental areas involved and the date of the divergence between the cercopithecids and the colobines are unknown. But the interchange of true forest species between Eurasia and Africa probably ceased sometime in the Miocene and the differences between the monkey fauna of Asia and Africa raise some interesting questions.

The macaques are typically Eurasian although they do occur in North Africa. The baboons, drills and mangabeys are African, although a fossil baboon, *Procynocephalus*, is known from the Pleistocene of Asia. The macaques, however, are clearly related to these African monkeys, and macaques, baboons, drills and mangabeys are sometimes regarded as a subfamily, Papioninae or Cynopithecinae, distinguished by relatively long, doglike muzzles in the males. The occurrence of *Macaca flandrina* in late Miocene deposits in North Africa hints that macaques might have come into Africa from Asia together with the hipparionids that also make their first appearance in Africa at this time, and fossils also show that the Cynopithecinae were already a distinct group by the end of the Miocene. The absence of fossil or living macaques in tropical Africa and the wide distribution of living forms in Asia suggests that they developed in Asia, possibly from an early colobine stock.

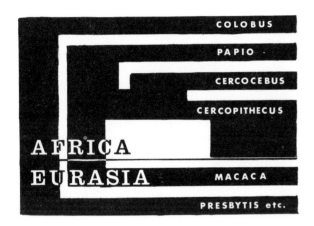

◄ Diagram of hypothetical relationship of some cercopithecoid monkeys.

In this connection it is to be noted that Asiatic colobines have radiated into distinct genera with widely differing habits and habitats. One interesting trend is most obvious in the proboscis monkey (*Nasalis*), although it is perceptible in other genera (i.e. *Simias*). In these colobines there is a lengthening of the nasal bones, a narrowing of the interorbital bones and a general forward growth of the jaws, which is particularly marked in the males and is accompanied by strong sexual dimorphism, adult males being about twice the weight of females. This trend is coincidental in *Nasalis* with the development of an extraordinary nasal appendage. This may act as a resonator for "warning" (perhaps territorial) calls peculiar to the male, or it might have evolved by sexual selection. Apart from the lengthening of the muzzle there is nothing about the skull of *Nasalis* to betray the presence of the aberrant nose; in fact the skull has a remarkable superficial resemblance with that of some macaques.

No direct relationship is implied by comparing this colobine with macaques, but the skull illustrates a most important fact; that the lengthening of the muzzle and the development of a marked sexual dimorphism are not directly connected with ground dwelling, for the proboscis monkey is a fairly typical tree-dwelling, leaf-eating colobine and the lengthening of the muzzle is not confined to the Cynopithecinae. It is possible that some early populations of similarly "aberrant" colobines, on taking to ground dwelling (as *Simias* is suspected to do), might have found an enlarged muzzle and sexual dimorphism an advantage.

A secondary change in diet and behaviour, together with a trend towards muzzle lengthening might have led to the divergence of colobines and the dog-faced macaques; a development that could only have occurred in the Miocene if it occurred at all. The example of *Nasalis* suggests or illustrates a possible trend and no ancestral position is assumed for forms which have continued to evolve and change. It is, however, worth mentioning that Verheyen (1962) found an almost complete overlapping of the allometric axes of African and Asian colobine skulls; as the two groups have probably been genetically separate since the early to Mid-Miocene, this implies an extraordinary degree of conservatism in skull structure for the Colobinae. As the macaques are recognized in the late Miocene, this genus too can be regarded as rather conservative although showing considerable variety of type.

Long muzzles are found in many macaques, but the development reaches its apogee with baboons and drills, where the male's heavy canine teeth and whole toothrow are cantilevered out from the cranium about as far as is mechanically feasible without structural weakness appearing. Considering the ontogenetic development of this feature, the infant male baboon begins with a relatively small muzzle distinguished by large and prominent milk incisors. Immediately in front of the orbits are swellings in the maxilla which are the buds of the upper canines. With their forward growth the canines describe prominent ridges, these channels of the growing teeth ossify and form the upper surface of the baboon's muzzle. The ridges are structurally important as they not only act as "flying buttresses" for the massive canines and incisors, but also provide support for the upper toothrows, which have lost direct cranial buttressing and rely on being supported between the nasal ridges and the base of the orbital plate. Direct buttressing above the toothrow is so inadequate that the roots are frequently exposed through their thin bony covering. The structure resembles the Late Gothic arrangement of flying buttresses that thrust the clerestory to ever greater heights in medieval cathedrals. In the lower jaw a modified buttress is also apparent but the teeth are well rooted in bone.

It is possible that special emphasis upon the front teeth might have had some influence on the development of cheek pouches. However, it is more likely that the advantage of pouches consists in increasing the amount of food that can be collected at one time and reducing the time spent at a food site. In this the Cercopithecinae resemble pouched rats (see Vol. IIB. pp. 546) but the spur to this development is the competition of other monkeys rather than exposure to predators. The Colobinae instead lack pouches, spend more time feeding and are less competitive over food.

The specialization of the incisors would seem to be a by-product of the extreme projection of the muzzle far beyond any functional requirement in feeding. This arrangement furnishes its owner with a visually emphatic signal which, in baboons, is enhanced by a bare face surrounded by a big

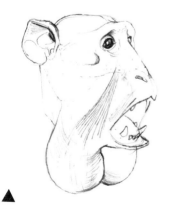

Dissection of *Cercopithecus mitis* head, showing distended cheek pouches.

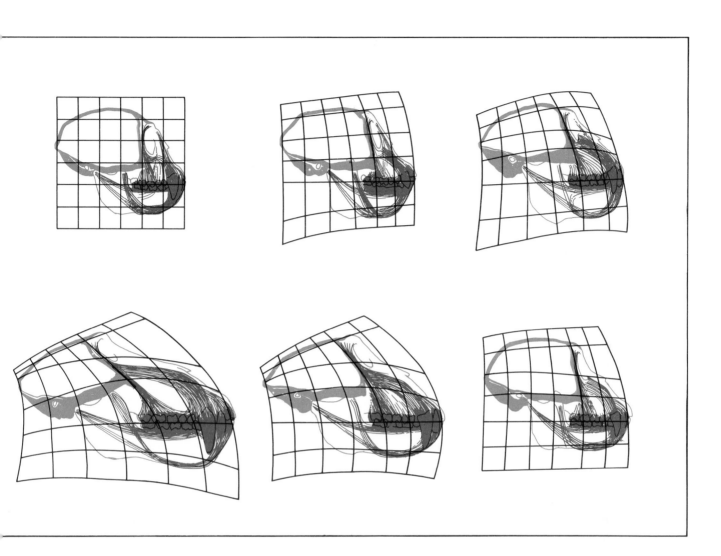

▲

Co-ordinates imposed upon skull profiles. *Upper left, Colobus badius kirkii ; middle, Simias ;
right, Nasalis ; lower right, Macaca silenus ; middle, Macaca nemestrina ; left, Papio
cynocephalus*. Profile with cranial region and zygomatic arch heavily outlined in blue.
Teeth and principal buttressing of the jaws in pink.

circular ruff or cape of hair. The canines also provide formidable weapons which may be used both in defence, in threat or for killing animals for food. In baboons, threat behaviour is frequently exhibited in tense situations and a social display of these features by a group of male baboons is rendered even more impressive by being punctuated with loud explosive barks; this performance is generally an adequate deterrent to potential predators.

Any primate attempting to live on the ground must have some means of escaping or averting danger. In the open, a tendency on the part of the males to threaten predators or scout the environment would either lessen the chances of survival, as predators would kill off monkey populations in which the males were actually defenceless or, if the male's aggressive bluffing or social threats proved effective, it would open up a new niche. In this the male's role as scout or protector would be enhanced by being unburdened by pregnancy or care of the young. It seems that a defence mechanism involving gross sexual dimorphism in structure, size and behaviour may have arisen out of a simpler and functionally different type of sexual dimorphism and muzzle elongation (such as that found in the proboscis monkey). This might have provided the basis on which natural selection could proceed.

The skull of the baboon can be compared with those of other cynopithecines and of various colobines by means of coordinates imposed on profile drawings of the skulls (Wentworth Thompson, 1917). A symmetrical grid of squares is imposed upon the outlined profile of a skull, in this case a typical short-faced colobine, *Colobus badius kirkii*. Using the line of the toothrow as a common axis, outlines of other monkey skulls have a similar network drawn so that the coordinates pass as nearly as possible through points that correspond with those of *C. b. kirkii*. The different orientation of features is reflected in the directions and distances between the lines. In the figure on the previous page the teeth and their principal buttressing system have been coloured in pink, while the zygomatic arch with the occipital and parietal crests are indicated in blue to stress the relationship between mandibles and cranium.

It is curious that the macaques, which are widespread in Asia and were in North Africa about 14 million years ago, are not represented even in relic pockets in tropical Africa. Instead, the most extreme dog-faced monkeys, the baboons, are represented throughout the African Pleistocene and are found in the Eurasian Pliocene in deposits often thought to represent dry country faunas. The physique and the social organization of baboons today seems particularly well-adapted to life in open country. The mobility and range of this primate which is distributed over the whole of the African savanna area may be some measure of the success of ancestral types, which were probably much more numerous than now both in species and numbers.

Within the group of dog-faced monkeys there are biochemical and anatomical differences between the baboons, *Papio*, and the drills, *Mandrillus*, which betray a considerable divergence (Jolly, 1967). There is similar evidence that the mangabeys, *Cercocebus*, and the geladas, *Theropithecus*, also represent divergent streams of a radiation within the cynopithecines. The principal feature common to all has been the presence of a long muzzle.

The mangabeys and mandrills have resemblances in the post cranial skeleton which may indicate some degree of affinity, but in skull structure there are considerable differences. The West African white-collared manga-bey, *Cercocebus torquatus*, lives in swamp forest but is frequently seen on the ground; it has a long muzzle but one that is nonetheless shorter than in any of the baboon types (see Rode, 1936). In common with the other mangabeys its skull is characterized by a suborbital fossa which is due to a buckling of the orbital plate, this seems to have been caused by the backward migration of the palate, whereby the maxilla has shifted back while the zygomatic bone has tended to retain an orientation appropriate to a more extended muzzle. This tendency for the palate to migrate back becomes more marked in other mangabey species. The most arboreal species, *Cercocebus albigena*, exhibits many baboon-like characters but there are significant differences in social behaviour which may be associated with a greatly reduced sexual dimor-phism. Dobroruka (1966) in a review of the species' structure of mangabeys thought that *C. torquatus* represented the most primitive species and suggested that the two other short-muzzled species groups derived from this older long muzzled type. If this is correct, the mangabeys represent a series illu-strating some ontogenetic developments. Considered together as a series, the ecology, behaviour and anatomy of mangabeys suggest that a return to the trees, together with a reduction in body size have rendered obsolete sexual dimorphism, the extended male muzzle and some associated behaviour.

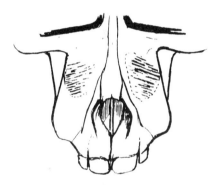

The shortest muzzle is that of the black mangabey, *C. albigena*. In the males of this species the vestiges of the "flying buttress" are discernible in diminutive "crepe" textured swellings above the bud of the upper canine, this is found in both juvenile and adult males but not in females. Black mangabeys have very thick incisors and a specialized diet which includes hard nuts. Correlated with this diet, the zygomatic bone has heavy muscle attachments (the anterior temporalis behind the orbit and the masseter on its lower margin). The zygomatic bridges are easily distorted and in this monkey not only is the zygoma bowed downwards, but the zygomatic process of the frontal bone has also been dragged down so that there is little trace of a bony bar over the eyes; the "peaky" narrow-faced expression of this mangabey derives from this downward drag of the facial bones.

Cercocebus albigena is a specialized monkey, not so *Cercocebus galeritus*; in this species the incisors are not exceptionally large, as the diet does not include very hard foods and, both the bones and the musculature of the jaws are altogether slighter; moreover, the zygomatic process of the frontal bone is higher, the zygomatic arch is slighter and less bowed and the suborbital fossa is less deep.

This trend is continued in an interesting swamp monkey, *Cercopithecus* (*Allenopithecus*) *nigroviridis*, from the Congo, which has conical mangabey-like teeth but which has lost the suborbital fossa which is the principal diag-nostic feature of the mangabey skull. In the skull of *C. nigroviridis* the zygo-matic arch and the suborbital plate regain the structural balance that has been disturbed in the mangabeys by the migration of the maxilla, it acquires thereby a typical short-faced cercopithecoid appearance. *C. nigroviridis* has other characters that are intermediate between the mangabeys and the cerco-pithecus monkeys (see Hill, 1966), and the species probably resembles the original stock from which all cercopithecus monkeys have derived. The difference between cercopithecus molars and those of mangabeys may be due to dietary differences, but it also might be correlated with the more direct and adequate buttressing of the roots, which followed the disappearance of the suborbital kink.

The co-ordinate diagrams (facing page) similar to those on p. 143 may be useful to compare the skulls of *C. nigroviridis* with those of mangabeys and baboons. The toothrow is again used as a common axis, but the cantilevered nature of the baboon skull suggests diagonal co-ordinates to coincide with the angle made by the orbital plate. The return to a more compact "square" form is well-illustrated in this way and the downward drag of the frontal bone in *Cercocebus albigena* is also particularly well-illustrated by the coordinates. The species figured are of course not ancestral to one another, but they probably have sufficient similarities with ancestral forms to illustrate in a graphic way an important aspect of the evolution of the cercopithecus monkeys from the mangabeys, an evolutionary development that represents the return of one branch of the dog-faced monkeys to a wholly arboreal existence. It would seem that the "terrestrial ordeal" gave this stock an edge over the older colobines. More active, more highly social and probably more aggressive, the guenons became the most successful monkey group in Africa.

The cercopithecus monkeys' status as a single genus and their strictly African distribution suggest that they evolved very recently. Their radiation, in which the climatic fluctuations of the Pleistocene have clearly played a major part, is discussed later.

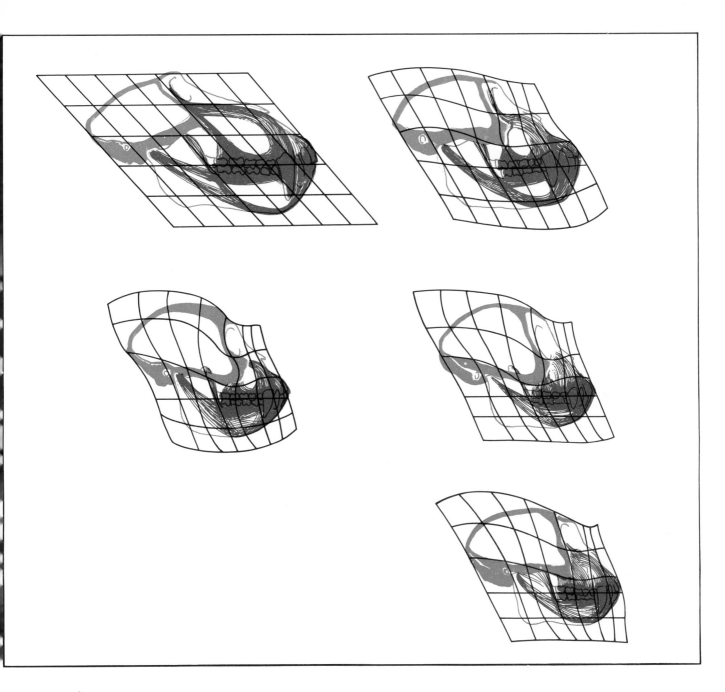

Co-ordinates imposed upon skull profiles. *Top left, Papio cynocephalus; right, Cercocebus torquatus; middle left, Cercocebus albigena; right, Cercocebus galeritus; lower, Cercopithecus nigroviridis.*

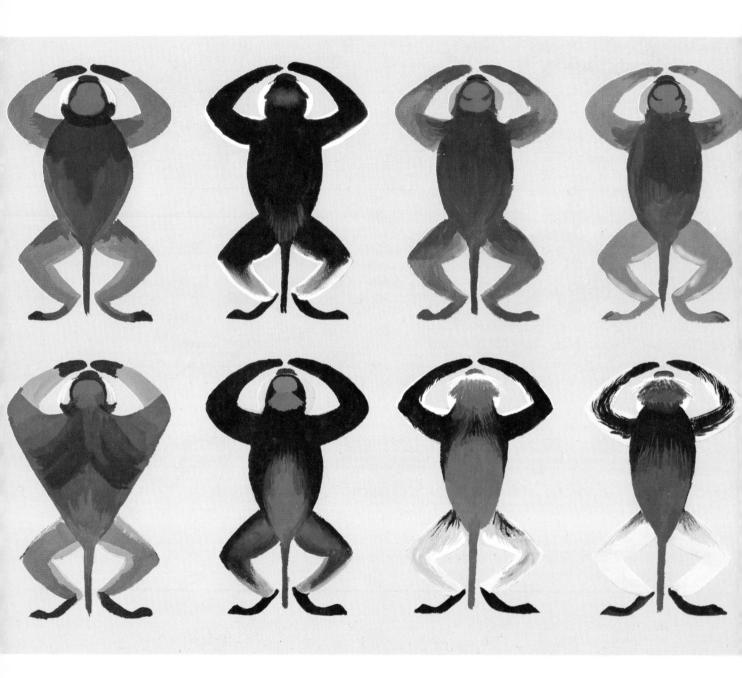

▲

Colobus badius. Some variations within local pops. *Extreme left* above and below,
C. b. tephrosceles (W. Uganda), *2nd left,* above and below, *C. b. gordonorum* (Iringa).
Right above, C. b. rufomitratus (Tana River), *right below, C. b. kirkii* (Zanzibar).

Colobus Monkeys

Colobinae colobus

The colobines are a very homogeneous group of monkeys with colobus in Africa, langurs and odd-nosed monkeys in Asia. There are numerous varieties (44 species and 112 races have been described). The subfamily occurs from Tze Chwan, South East Asia, Indonesia and India to Africa and on many isolated oceanic islands. Typically colobine fossil remains are known from the late Miocene of East Africa and from the Eurasian Pliocene. There are basic resemblances between the African and Asian colobines and Verheyen (1962) has concluded from a study of skull measurements that their common genetic patrimony has remained virtually unchanged for whatever period the two groups have been separated, which may amount to 20 million years. There are some more superficial resemblances across the Indian Ocean: the colouring of proboscis monkeys in Borneo is not very unlike that of some red colobus races, while contrasting fur and vivid white lips are features shared by the Zanzibar colobus and the Malaysian lutongs (*Presbytis*). It looks as if certain limited trends, albeit convergent ones, characterize the very conservative colobines. Considering these intercontinental resemblances it is not surprising that there are difficulties in assessing the status of colobus in Africa.

The Zanzibar red colobus poses a particularly interesting problem; Verheyen (1962) has given this form specific status, *Colobus kirkii*, and suggests that this is a primitive form near the stock from which both the red and pied colobus types developed. It is a smaller animal and has different textured hair to the western races of red colobus and, although the tinting and extent of colouring vary greatly, there are strong contrasts of colour and tone, there is also a distinctive white mark on the black almost hairless face (see drawings of *Colobus badius tephrosceles* and *Colobus badius kirkii*). However at Iringa, 850 km inland from Zanzibar, there is a race, *Colobus badius gordonorum*, that is intermediate between *C. b. tephrosceles* and *C. b. kirkii*. Some individuals of this population approach the range of variation of *C. b. tephrosceles* and all have the suggestion of a hair whorl characteristic of this race, but there is considerable variation and some individuals have the marked contrasts and rather lank, shiny hair characteristic of *C. b. kirkii*. This population might conceivably be the result of "hybridization" at a very ancient date, but it is much more likely that it represents the relic of a cline, nonetheless a cline that must antedate Zanzibar's separation from the mainland. Three principal factors might have combined to preserve the clinal nature of these populations over a very great period of time; first the genetic stability typical of colobines, secondly this species' conservative attachment to a locality and thirdly isolation and interruption of geneflow due to the climatic vicissitudes of eastern Africa during the Pleistocene.

It is extraordinary that *Colobus badius rufomitratus* from the coast north of

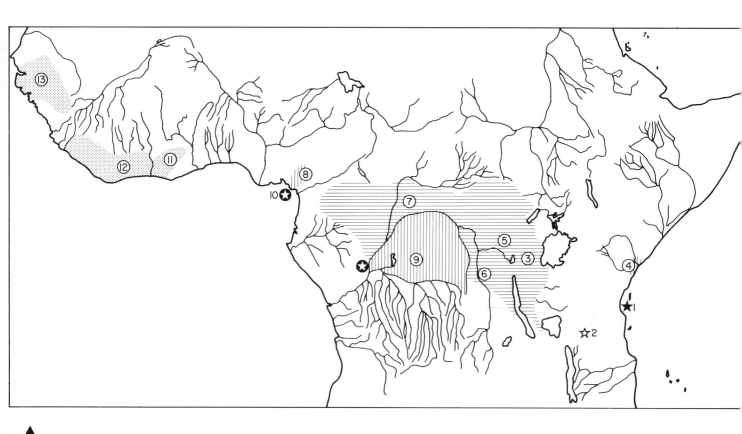

Total range of red colobus: 1. *Colobus badius kirkii*; 2. *C. b. gordonorum*; 3. *C. b. tephrosceles*; 4. *C. b. rufomitratus*; 5. *C. b. ellioti*; 6. *C. b. foai*; 7. *C. b. oustaleti*; 8. *C. b. preussi*; 9. *C. b. tholloni*; 10. *C. b. pennanti*; 11. *C. b. waldroni*; 12. *C. b. badius*; 13. *C. b. temmincki*.

Zanzibar is almost within the range of variation of *C. b. tephrosceles* but has no resemblance with *C. b. kirkii*, in spite of being its closest neighbour geographically; this anomalous situation is discussed in Chapter IV.

It may be relatively unimportant what taxonomic rank the Zanzibar colobus is given, but it does seem to occupy a special position intermediate in several respects between two very different trends in colobus evolution, the exterior signs of which are drab colouring in many races of the red colobus and extreme contrast in the pied colobus.

If Verheyen's assessment of the primitive status of *C. b. kirkii* is correct, the pied colobus might have had their origin in an extinct population bearing some resemblance to *C. b. kirkii*. This hypothetical population might have evolved black coats with white flashes, a tendency that is revealed in an individual *C. b. gordonorum* (illustrated in colour on p. 148). A polychromatic ancestor for the pied colobus is also hinted at in the skins of some West African individuals that have traces of red colouring which could be interpreted as reversion.

The pied colobus with the closest cranial resemblance to *C. b. kirkii* is *Colobus polycomos*, and this species is also less elaborate in pattern than the northern *Colobus abyssinicus*. Today, races of *C. polycomos* occur in isolated forests in the southern part of East Africa and across the forested areas of the continent as far as Senegal. In West Africa, these monkeys were subjected to isolation by climatic change which cut up the forests, a process which might have been irregular and repeated, so that today there are five races of pied colobus in West Africa. It is here that the black and white pattern must have become further elaborated and certain minor cranial differences appeared which find external expression in an altered profile and shape of the nose (see drawings). After a long period of isolation a change in climate allowed the western and eastern forests to join again. A western colobus population, which had become a distinct species, *C. abyssinicus*, crossed the Congo basin to the north of the river where it came to occupy the range of the older *C. polycomos*. Within this area, *C. abyssinicus* is today the more successful and numerous species and may be in the process of replacing *C. p. cottoni*; no hybridization is known to occur. The Congo River has apparently been a bulwark reinforcing the *C. polycomos* stronghold south of the river; further east, *C. abyssinicus* has also been inhibited from colonizing the southern forests through prior occupation by *C. polycomos*, but desiccation has probably been equally important and may have cut off any possibility of invasion. In northeastern Africa, the Late Pleistocene expansion of forest allowed *C. abyssinicus* to reach Harrar in the north and Kilimanjaro in the south. This successful extension of range contrasts with the apparent inability of *C. polycomos* (or *Colobus badius*) to do any more than hold their own, presumably they were unable to compete with *C. abyssinicus* wherever they were not already firmly established in large enough numbers.

The forests of northeastern Africa subsequently contracted and *C. abyssinicus* became isolated into numerous populations which have differentiated to some degree. The differences involve some adaptive features such as increased hair length in the cold highlands, but the trend towards more and more emphatic white contrasts is also apparent and this reaches its apogee in the most easterly population *Colobus abyssinicus caudatus*.

Black and white colouring occurs in many animals that form large troops, notably hornbills, storks, geese and zebra, none of which take pains to conceal themselves but instead frequently advertise their presence by loud noises such as honking and braying. Cryptic protection can therefore be discounted as a function for this colouring. Sexual dimorphism is very slight in the pied colobus and no displays that could be interpreted as sexual have been seen, so this too can be eliminated as an explanation. The suggestion that the colouring is of aposomatic or warning significance is also vitiated by the fact that crowned hawk-eagles and other predators kill colobus notwithstanding their black and white coats. The explanation for the development of this feature seems to centre on a most conspicuous aspect of colobus behaviour, which has been remarked upon even by the most casual observers. Pied colobus bounce about in the tallest trees of the canopy and make a great deal of noise several times a day, a practice that makes them both extremely conspicuous (particularly from above the tree tops) and very audible, and would seem an open invitation to the monkey-eating hawk-eagle. This display is associated with sleeping trees and the feeding area in particular, and is directed at and responded to by other colobus troops. There is an interesting parallel in the behaviour of the big pied hornbills, *Bycanistes*, which also make prolonged nasal fanfares in the early morning and evening, as they fly

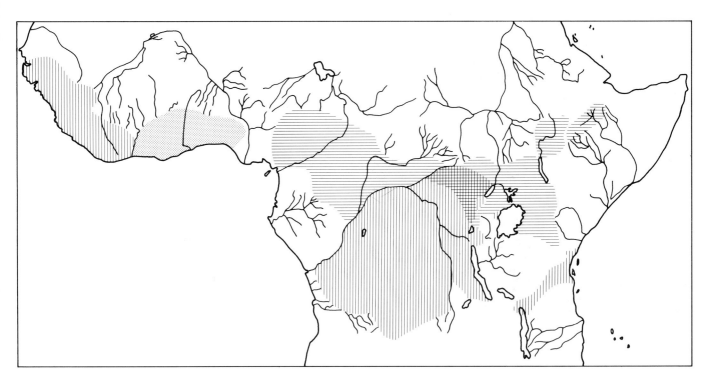

▲ Total range of pied colobus:
vertical shading   = *C. polycomos.*
stipple   = *C. p. vellerosus.*
horizontal shading = *C. abyssinicus.*

in and out from their social roosts and in their feeding trees. In both cases the behaviour must have advantages that more than offset the increased exposure to predators. The advantage for the species would seem to concern dispersal of populations throughout the habitat. It is not the immediate function of the display to "test" the environment, but rather to define territory by visual and auditory "marking" but the long-term effect may be to facilitate colonization. The flamboyant territorial display must quickly disclose which parts of the forest are occupied, so that as populations increase the entire available habitat can be effectively exploited.

The pied colobus is evenly and thoroughly distributed in forests where it has not been persecuted, and it is surely no accident that this colobus and the redtail monkey are dominant species in many Uganda forests, as both monkeys have the most highly evolved and elaborate patterns and signalling devices of their respective groups; each individual monkey becomes a visual and auditory marker for the troop's territory. By contrast, very local concentrations of red colobus are found within much larger areas of forest apparently equally suitable for colonization. This, I suspect, betrays a very limited capacity on the part of red colobus populations to disperse and move into unoccupied forests.

Colobus polycomos dollmani

Colobus polycomos vellerosus

Colobus polycomos adolfi-frederici (lowland)

Colobus abyssinicus uellensis (lowland)

Colobus polycomos sharpei (highland)

Colobus abyssinicus caudatus (highland)

Pied colobus: *top left, C. p. dollmani ; right, C. p. vellerosus ; left, C. p. adolfi-friederici* (low altitude); *right, C. a. uellensis* (low altitude); *bottom left, C. p. sharpei* (high altitude); *right : C. a. caudatus* (high altitude).

A difference that might involve a physiological superiority is the pied colobus tolerance for mature leaves whilst the red colobus mostly eats shoots.

A detailed comparative study of the social and territorial behaviour of colobus monkeys would be most interesting and would do much to illuminate the very different status of the various colobus species. Physically the red colobus is similar to the pied colobus, and the loud vocal rituals and colour pattern of the latter may only amount to exaggerations of features that are perceptible in some populations of the red colobus; in effect however, these "exaggerations" seem to be associated with a general superiority of the pied colobus over its relative.

C. b. tephrosceles (adult male).

Red Colobus
(Colobus badius)

Family   Cercopithecidae
Subfamily   Colobinae (Elliot)
Order   Primates
Local names
Ekajansi (Lutoro).

Measurements
(Colobus badius tephrosceles only)
total length
1,270—1,370 mm
head and body
610 (580—670 mm) males
585 (485—620 mm) females
tail
660 (600—730 mm)
weight
11 (9—12$\frac{1}{2}$ kg) males
7$\frac{1}{2}$ (7—9 kg) females

Red Colobus

Races

Colobus badius rufomitratus	Kenya Coast
Colobus badius gordonorum	Uzungwa Mountains
Colobus badius tephrosceles	Uganda and West Tanzania
Colobus badius ellioti	Bwamba
Colobus badius kirkii	Zanzibar

The red colobus is very variable in coat colour, both individually and among races (see colour plate p. 148). The Zanzibar race, *Colobus badius kirkii*, has lank glossy hair with bright strong contrasts. A trait which is detectable in the infants and juveniles of other races of *C. badius* as far away as the Ivory Coast. Adult *Colobus badius tephrosceles* and *Colobus badius rufomitratus* instead have a thick matt coat without very much contrast of colour or tone, while *Colobus badius gordonorum* is intermediate between *C. b. tephrosceles* and *C. b. kirkii*. Sexual dimorphism is marked in the western races and the males have a heavy head with a rather ape-like face.

Races of red colobus are found patchily from Gambia to Zanzibar across equatorial Africa, 15 races have been described. In East Africa, *Colobus badius ellioti*, is confined to a very small area of Bwamba, *C. b. tephrosceles*, has the widest distribution, occurring in scattered localities between Toro, West Uganda and Ufipa, Southwest Tanzania; *C. b. gordonorum* is known from only one locality near Iringa, but may turn up in the forests of the unexplored Luhombero massif. *C. b. rufomitratus* is found along the Tana River, at Dada, Ginda, Ozi and Wema and in the Arabuko Forest. *C. b. kirkii* is found in the Jozani Forest, and at Makunduchi, Muongoni and Popokaniki on Zanzibar Island.

The red colobus occupies forests at various altitudes but always in the immediate vicinity of permanent water. I have found *C. b. tephrosceles* feeding on mature leaves in the main canopy of dense, mixed lowland rain forest, in the undergrowth tangle of recently felled forest, picking up fallen fruit on the ground, eating the new buds on an isolated leafless *Albizzia* (in an area regenerating after logging), and they are also found in riverine forest, swamp forest and montane forest. They are primarily shoot-eaters but also take soft fruits.

Red colobus are adept jumpers and in broken forest climb up to the highest, outermost branches of trees and then leap the gaps with their arms stretched out to catch even the smallest twigs of the next tree, they also make vertical drops of as much as 20 m, landing with all four feet together on springy branches, which they allow to catapult them up and on again with the rebound. During rest periods, the younger animals form groups which chase one another and will drop one after the other into lower vegetation and then climb up in a leisurely way with their tails held vertically over their backs and repeat the process.

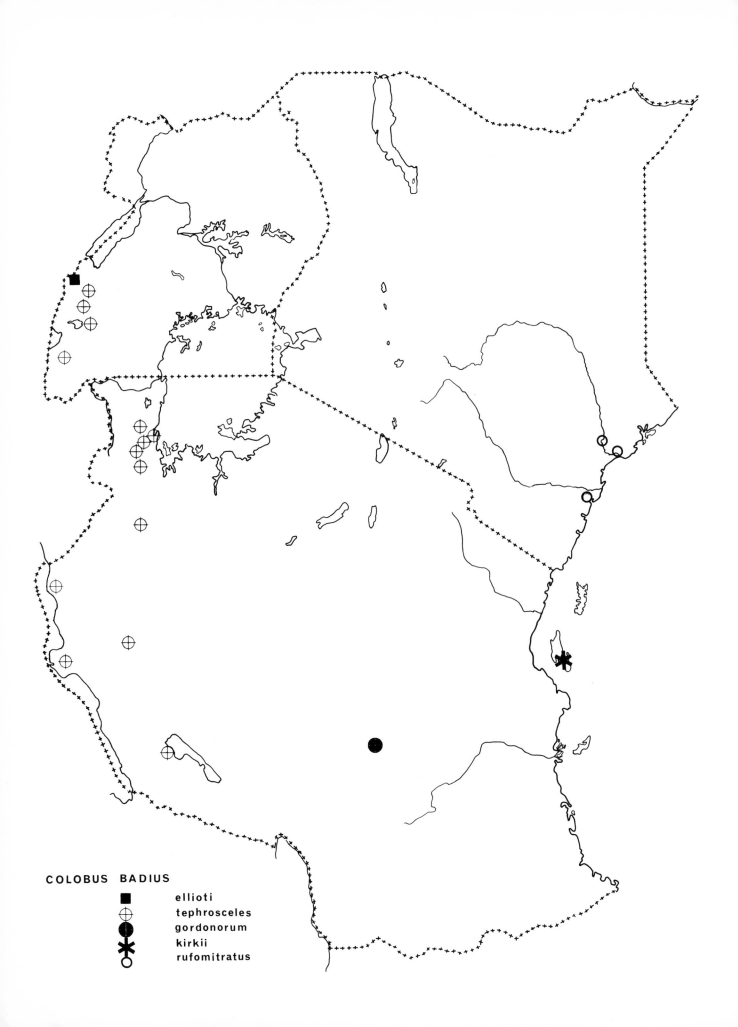

COLOBUS BADIUS

■ ellioti
⊕ tephrosceles
● gordonorum
✳ kirkii
○ rufomitratus

They are generally quiet and tame but the adult males make a sneezing high-pitched bark when alarmed and also make a short growling noise. Like black and white colobus they sometimes make a chorus in the early morning or evening and occasionally at other times, but their voice does not have the same resonance as other colobus species. They live in large, dense troops of fifty to a hundred individuals in a very localized home range; a troop can generally be found within a radius of a few hundred metres at any time of the year. The territories of several troops may abut, and a small area of forest can contain a very high density of red colobus. The relationship between

C. b. kirkii (sub adult).

troops and their interactions have not been studied.* The troop contains many large males and casual observation has not revealed signs of aggression within the troop. Adult males are often on the periphery of the troop, where they may feed together side by side. While feeding the troop may sometimes scatter over a small area of forest seeking leaves at all levels, but when sleeping and resting, or in response to heavy rain or alarm they tend to coalesce into what appear to be "family" groups containing a single male with several females and juveniles. The individuals in these groups stay very close together, even clustering on one small branch, while other groups may be on nearby branches and the entire troop may be present in a single tree.

They frequently walk by one another with their tail held vertically which allows an effective display of their genitalia. The female has marked perineal swellings which seem to be permanent rather than appearing only at oestrus, although the absolute size of the swellings may fluctuate. Furthermore, the male exhibits swellings round the anus which mimic the female's perineum very closely. This unusual development may enhance the close social cohesion that is the hallmark of the red colobus, by providing a permanent attraction to the adult males and by inhibiting their potential aggression. Like the black and white colobus, wrestling and hugging is common between adults of the same sex and also between mixed pairs, it probably serves to release tension. It may be worth mentioning that *C. b. tephrosceles* has a very pungent smell, this character is admittedly a subjective one, but it is one that obtrudes on the observer's senses and might be of some significance. The same race has an elaborate pattern of whorls in the red hair of the head, a feature which needs to be seen at close quarters to be appreciated. The absence of contrasting markings or sudden movements and the conservative and small home range can be correlated with close-range communication and the dense social organization of this colobus. The monochromatic drabness of some races must make the re-establishment of contact after dispersal more difficult, and it is noticeable that these red colobus do not scatter in alarm, but instead cluster close together, while males may threaten or take up the rear of the fleeing troop.

Red colobus mix with cercopithecus monkeys and they will associate with pied colobus amicably, feeding on leaves off the same tree and will even share the same branch. However, the pied colobus seems to avoid large red colobus males and will turn away from clustered groups.

Although remarkably tame in some areas (Pitman remarks: "when shot at they sit about and make a plaintive bark"), one troop I encountered fled immediately on seeing me, and it might be assumed that this troop had been persecuted. Their localized distribution, conservative habits and generally tame behaviour make the red colobus very vulnerable to man, and colobus are a delicacy for those who eat monkey. The species long term survival will probably depend on legislation and effective protection; *C. b. kirkii* is particularly vulnerable as its range is largely contained within a military training zone.

Booth (1957) reported that it was difficult to keep West African races of *C. badius* alive in captivity; lack of appetite set in and animals quickly succumbed to pulmonary or intestinal infections. Dekeyser (1957) also notes their failure to survive in zoos and their need for trace elements only found

Perineal area in female (top) and male anus (below).

* See postscript, p. 161.

in certain food plants. Similar experiences with captive *C. b. kirkii* suggest that the species is unlikely to do well in zoos unless kept in numbers and under semi-natural conditions with a nutritious diet. It may be of interest to note that a wild *C. b. tephrosceles* showed extensive grey spotting and partial collapse of the lungs.

In East Africa, the red colobus appears to have had difficulty in colonizing forest as it expanded, and present populations are almost certainly rather ancient relics. The dependence of this species on water and on minimal forest conditions make it a useful indicator of ecological and climatic stability, as these limitations suggest that the areas where red colobus are found have avoided any significant degree of dessication over a very long period.

The presence of a race resembling the Zanzibar colobus at Iringa, in the Southern Highlands is of particular interest, as it suggests that the forest relic in which it is found is part of a very ancient connection between the western forests and the coast, and that this forest has suffered isolation but never total dessication. Differences between the two populations found on the coast are also an indication that there might have been two very ancient forest "routes" across eastern Africa; this possibility was discussed in Chapter IV.

Postscript : This species is currently being studied by T. H. Clutton Brock and T. Struhsahker. They have determined an average sex ratio of 1 male to 3 females. They found a *C. b. tephrosceles* troop of 55 monkeys ranging over about 26 hectares of thick Uganda forest with numerous neighbouring troops nearby. In the less densely forested Gombe Stream N.P. a troop of 64 ranged over 50—80 hectares while a larger neighbouring troop of 81 ranged over 80—140 hectares. They did not find any consistent "family" groupings.

◀ *C. b. tephrosceles.*

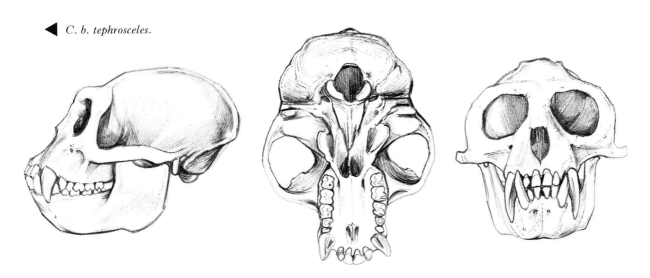

Colobus polycomos sharpei

Pied Colobus
Black and White
Colobus
(Colobus polycomos)
(Colobus abyssinicus)

Family   Cercopithecidae
Subfamily   Colobinae (Elliot)
Order   Primates
Local names
Mbega, Kuluzu (Swahili).
C. polycomos : Munyunga
(Luganda), Nkomo (Lukonjo).
C. abyssinicus : Mbega (Lunyoro),
Ngeye (Luganda, Kuamba), Dolo
(Lwo), Ekiremu (Lukiga),
Mongasiet (Sebei), Etepes (Ateso),
Echu (Karamojong).

Measurements
(vary with species, races and sex)
total length
1,220—1,660 mm
head and body
600—750 mm
tail
740—830 mm
weight
13—23 kg

Pied Colobus, Black and White Colobus

Species

Colobus polycomos, Colobus abyssinicus.
Colobus polycomos : White shoulder mantle, long white face whiskers, whorl on crown.

Colobus polycomos ruwenzori	Ruwenzori
Colobus polycomos adolfi-friederici	Sango Bay
Colobus polycomos sharpei	Southern Highlands
Colobus polycomos palliatus	Coast, Usambara and Uluguru Mts

The very minor differences (length and thickness of coat and extent of white markings) in these races can be attributed to altitude or the isolation of small populations.

Colobus abyssinicus : White cape over rump, woolly white beard, black bonnet, white brush.

Colobus abyssinicus uellensis	West Uganda
Colobus abyssinicus dodingae	North Uganda
Colobus abyssinicus matschie	Mt Elgon
Colobus abyssinicus percivali	North Kenya
Colobus abyssinicus kikuyuensis	Kenya
Colobus abyssinicus caudatus	Mt Kilimanjaro

Colobus abyssinicus uellensis, is a lowland form with a short, thin coat and the least extensive white marking.
Colobus abyssinicus caudatus, is a highland form with long thick hair and maximal white. The other forms reflect various intermediate conditions between these two.

C. a. uellensis.

The two pied colobus groups are geographically or ecologically distinct in East Africa although overlapping in the Northeastern Congo.

It is not yet possible to draw distinctions between the behaviour of the two groups, but a comparative study should be rewarding, as differences might correlate with the evolutionary gradient of which these groups represent two extreme wings. They almost meet in Bwamba, the Ruwenzori race of *Colobus polycomos* lives in the montane forests overlooking the Semliki Valley where *Colobus abyssinicus uellensis* range to the foothills.

Pied colobus are found in suitable habitat from sea level to over 3,000 m. As in other monkeys the length and thickness of the coat is correlated with temperature, so that the mountain races look very heavy and robust.

They are found in a very wide range of vegetation: bamboo, montane, lowland swamp and coastal forests, moist savanna, woodland and dry thicket forest on the northern mountains. They need green fodder and water throughout the year, and colobus may be found along wooded river courses in dry country if these requirements are met.

The food plants differ with the habitat; they generally prefer fresh growth to older leaves and troops can often be seen in deciduous trees of the forest canopy when there is a flush of new leaves. As the lower layers are usually evergreen with hard shiny leaves, the colobus are not often found at this level, except in more open areas, where sunlight has stimulated herbaceous plants and soft fresh growth. Where *Cynometra* forms solid stands, colobus probably have little variety to their diet. In montane areas, *Rauvolfia* leaves and fruit, *Juniperus procera*, *Podocarpus*, *Toddalia* and *Usnea* lichen constitute important items, in woodland and moist savanna, *Albizzia* and *Acacia* species are favourite food trees. In addition to leaves, which are the principal food, various fruit and grains, bark, twigs, palm nuts and insects are eaten.

I have on two occasions found colobus on the ground in newly burnt grassland adjoining forest, they seemed to be eating ash, other observers have found colobus eating *Haemanthus* bulbs and also earth. In common with other herbivorous animals they appear to relish a mineral supplement to their diet. They have been seen coming down to streams to drink, usually in the afternoon, but licking twigs and leaves in the early morning is often adequate. They feed by pulling branches towards them and sit picking the leaves off with their lips, the hands are rarely used to put food in the mouth.

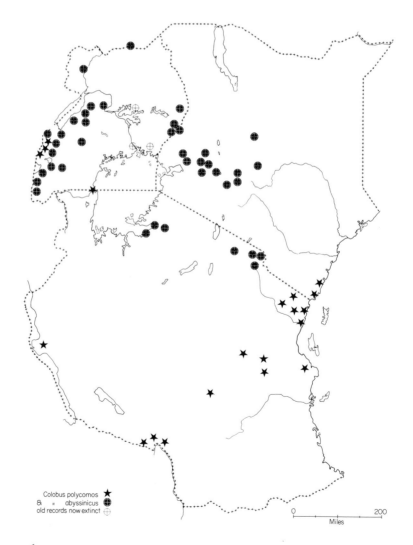

Colobus polycomos ★
& " abyssinicus ●
old records now extinct ⊕

0   200
Miles

Colobus abyssinicus uellensis.

The colobus is very tame and easy to watch in areas where it is not molested, but where it is hunted it becomes a very wild and wary animal. Their reaction to disturbance or danger vary widely and depend on the habitat and the enemy. I have seen a troop climbing into the highest branches of a tree when elephants were breaking trees in the same area. When they are hunted with arrows or guns, or when threatened by a hawk-eagle, they rush for refuge towards the centre of their home range or drop into lower vegeta-

165

C. a. uellensis.

tion and freeze in the tangles. Loveridge (1933) reports *Colobus polycomos* remaining concealed for half an hour while his party was smoking and hammering their tree. *Colobus polycomos* make a loud goat-like sneeze of alarm. Where they are less persecuted the appearance of man may release threat behaviour from the older males which may be accompanied in *Colobus abyssinicus* by a rapid double pat of the hand on a branch, which alerts all other colobus in the area. The leading male or males in the troop may then investigate, pointing their chins at the disturbance, silently baring their teeth and "chewing" in a ritualized fashion. The patting of branches may be repeated and lead on to the shaking of fronds and breaking of twigs meanwhile "chewing" vigorously, this appears to be displaced feeding behaviour. As excitement mounts, animals may start making alarm calls, nodding their heads, shaking branches and dropping twigs, finally they may defecate over the luckless observer and then flee, usually along well-established arboreal pathways, which seem to pass through thick vegetation whenever possible and thereby conceal their movements. Sometimes a mild disturbance or alarm leads to the formation of small clusters of animals. More often a troop is alerted by the patting of a branch and the response is for all animals to freeze until the danger is past, and there is one account of a colobus drawing branches together round its body.

In Bunyoro, where their woodland habitat has been destroyed by clearing, troops of apparently paralyzed colobus would sit in a tree until it was felled beneath them, others starved to death in isolated dead trees, rather than cross a quarter of a kilometre of open country. Where the habitat is gallery forest or clumps of vegetation in grassland, they descend to the ground more frequently and may flee across country for the shelter of large clumps, if they are approached while in isolated trees or on the ground. One troop of *Colobus polycomos* that I frightened fled to the outermost branches of a tree that overhung a cliff, there they were quite visible and would have been vulnerable to an armed hunter but the behaviour was perhaps geared to lesser predators.

They often appear to be rather phlegmatic and sometimes sleep so heavily in the afternoon that only a shout will wake them. They need to browse intensively for many hours each day to have their fill, the complex sacculated stomach holds one third of their body weight in food and requires long rests to allow digestion of the 2—3 kg of leaves that are eaten each day. The almost bovine demeanour of colobus would seem to derive from their physiological constitution, in any event, their temperament contrasts sharply with that of cercopithecus monkeys. When frightened, they can go through the trees at a fast pace and are agile jumpers. They have been frequently observed to fall head first with arms held out and forward, and even being killed on hitting the ground when the undergrowth failed to take their weight. This is probably due less to clumsiness than to an obscure behaviour pattern, as young animals have been seen to drop from a height deliberately and repeatedly; this "play" may have a connection with escape behaviour, also most colobus "paths" through the forest branches have places where each colobus in a troop will drop or leap from a height when it reaches that point.

On the ground their gait is a bouncy gallop which has been described as being "frog-like" or "squirrel-like" but is in fact peculiar to the colobus. It is probably a clumsy approximation to their arboreal leaps, for in walking their

C. p. sharpei.

gait is very similar to that of other monkeys. It is in springing from branch to branch that a rather different pattern is apparent; they leap with elbows out and land generally feet first with hands and feet close together, sometimes they gain extra momentum for their jump by setting the branch in motion before leaping. They resemble a pole climber when ascending a vertical trunk, clasping with both hands at a time and pulling up. A playful variation on this locomotor pattern may be seen occasionally in *C. abyssinicus*, when an animal being playfully chased may allow itself to swing momentarily by the hands after an upward jump, before drawing the hindlegs up like a trapeze acrobat.

167

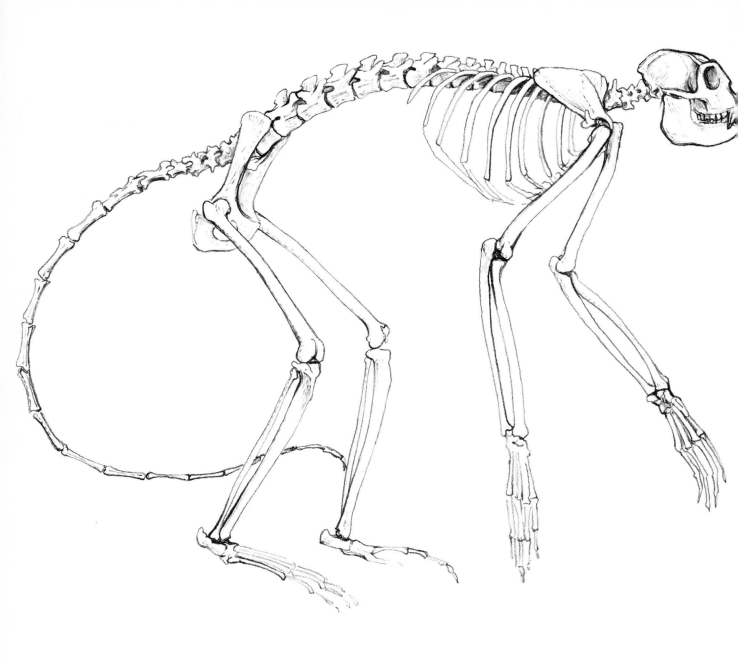

They make a deep throaty croak which has remarkable carrying power. The call is made by all adult members of the troop, particularly the males and it is part of the display generally conducted during three specific periods of the day. The first is in the early morning, sometimes even before dawn and is made while the troop is still in the sleeping trees. As soon as a troop has finished a chorus, neighbouring troops echo it. On a bright day the hullabaloo of colobus can be heard all over the forest as troops answer one another, the animals then become silent as they start the leisurely trip from their habitual sleeping trees to the feeding area. As they arrive at the food trees the troop again bursts into its chorus, animals climb to the higher branches, bounce up and down or leap back and forth from branch to branch in a ritualized fashion meanwhile flouncing their mantle and waving the tail.

168

Neighbouring troops wait for the performance to finish before answering in a similar fashion; when the troops are within sight of each other, the bouncing displays are conducted in conspicuous situations and a response appears to be imperative. Ullrich observed that the display was not performed when the other troop was out of sight. When he watched the interaction of two troops in an area where two territories overlapped and a feeding troop was disturbed, warning cries and threat gave place to a display by males to which the second troop responded, after which one troop withdrew temporarily but the second troop avoided entering their vacated feeding area. Both troops fed for some hours within sight of one another without mingling. The third period when the display is performed is at sunset, usually in the sleeping trees. Often calls may be heard at about 3 a.m. and nocturnal earthquakes also stimulate a very vocal response. The display has resemblances with that of chimpanzees and is sometimes made in similar situations. It would seem to be initiated by the sight or sound of other troops and by the food and sleeping trees which represent foci for territorial behaviour. Following an alarm or threat, the display would seem to be particularly directed at establishing the troops' claim to food trees and to the resting places which generally lie in the heart of their territory. The bouncing display might derive from a small bobbing movement, similar to that seen in many other cercopithecoid monkeys, which appears to be an expression of mild excitement between two individuals. A single bob up and down may be a mild threat between two animals, it often leads to a brief chase in which bouncing on branches is a conspicuous part of the play, vertical dropping and tail pulling are other common elements. The use of branches as catapults, which is peculiarly developed in all colobus species may also have played a part in the development of the bouncing display in the pied colobus.

C. p. adolfi-frederici.

The colobus call has a local reputation as a weather forecast, as they are often silent if rain is imminent and mist and rain inhibit them altogether.

Colobus feed all morning and evening and groom or sleep in the afternoon. They regularly sunbathe in the first rays of the morning sun, or after a storm, using trees specially favoured for this activity, here they loll on the topmost branches in a variety of postures.

Troops vary from two to about fifty individuals. Occasionally old solitary males are seen. In the Southern Highlands, isolated gallery forests of 14 or 15 hectares may support troops of 25 or 30 colobus (*Colobus polycomos sharpei*). Ullrich (1961) studying *Colobus abyssinicus caudatus* estimated a colobus troop's range to be 15 hectares, and he thought they spent three-quarters of their time in a still smaller area of 4 hectares out of which they were led for food, and then only for periods of three or four days at a time. The size of a range varies according to the availability of green fodder and the troop's size, but ranges are by any account small.

The large number of colobus found in appropriate habitats shows that they utilize a vast supply of food very effectively and, in protected forests, the density of colobus monkeys is very high. The troop sleeps, feeds and moves in scattered groups which are generally dispersed over several trees and they seldom clump closely together.

Grooming between two individuals usually takes place during the midday rest, at dawn, or after a storm. A frequently groomed part of the body in

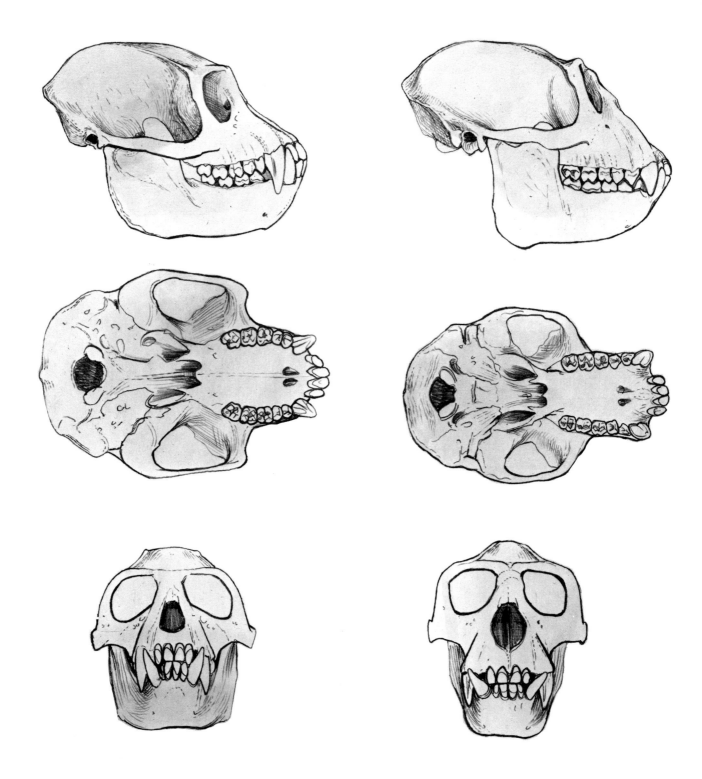

Skulls of *C. polycomos* (*left*) and *C. abyssinicus* (*right*).

C. abyssinicus is the tail, and adult males groom their own tail for long periods. Grooming frequently follows wrestling bouts, which are a common feature of *C. abyssinicus* behaviour; two animals which may be of any age or sex come together and seizing one another by the shoulder, arm or crown playfully spar with open mouths and mock bites, the play is lethargic, both animals remain firmly seated and the bout generally ends in a prolonged hug or grooming. At no time have colobus been reported to fight, and fight scars like those found on male cercopithecus monkeys have not been observed on colobus skins.

No clearly defined dominance hierarchy has yet been observed, but the larger males become conspicuous whenever there is cause for an alarm or a display of threats and adult males often remain in exposed positions while other animals are feeding. Their behaviour may serve to localize enemies and influence the flight reaction of the rest of the troop. In flight the males are usually the last to leave, except that, where they have been shot at, every animal flees or hides with the utmost dispatch.

It is scarcely surprising that an animal that is palatable, decorative, common and saleable (as meat or hide) is hunted enthusiastically where taboos or game laws are not effective. Colobus skins are widely prized for tourists trophies, dance costumes, furnishings, bicycle saddles and hats. Marco Polo

C. a. caudatus.

was the first European to describe the skins, which he found made into capes worn on the shoulders of Mongol Khans in Central Asia. Trade in these skins is very old and over two million skins are supposed to have reached European markets during the latter part of the 19th century. The demand for colobus skins at this time almost certainly accounts for the absence of colobus from the forests of Mengo and Busoga where the animal is a well-known clan totem and the skins were formerly worn at the feudal courts. A very old skin in the possession of a witchdoctor in Kyaggwe is reputed to originate from the Mabira Forest where they certainly no longer exist. When forests are broken up and where there are large human populations the species is unlikely to survive; in the better protected large forest reserves, however, the animal should fare rather better.

The young are born throughout the year; Haddow found no evidence for a peak birth period in Bwamba. This is possibly correlated with a stable food supply throughout the year, but in areas where seasonal changes are marked some pattern to the breeding may appear. The female initiates sexual behaviour, presenting her genitalia with her tail arched over her back. She keeps close to male animals throughout her oestrus and several males may successively mount a female in this condition; pairs seldom form for more than a few hours. Ullrich observed a female present to a male that had been grooming her, they subsequently copulated; he also noted mutual "chewing" to precede a copulation. When the mouth remains closed this action does not constitute a threat and seems to be a sign of excitement without aggressive undertones, as mothers make this gesture at the young.

A curious sexual distinction in *C. abyssinicus* is the female's habit of urinating on her tail, this is not done by the male who therefore retains a bushier, cleaner tail. A comparative study of the sexual behaviour of *Colobus*

Opposite and below
C. p. sharpei.

173

polycomos and *Colobus abyssinicus* would be most interesting as it might throw light on the features that reinforce the reproductive isolation of the sympatric species found in the Congo.

Before birth the mother secludes herself but may be accompanied by a male. Ullrich suggests that her vaginal secretions may attract him, however, the behaviour could have acquired some survival value; a male kept in captivity has been reported to have become very protective and aggressive on the birth of the baby.

When juvenile colobus are captured or confined their cries often attract wild adults of both sexes. Wild females exchange their young quite freely. This behaviour has not been seen in the red colobus.

The young are born almost pure white and are large and well-developed at birth (see drawing). The day following the birth the female rejoins the troop, the baby clings to her belly and is supported by one hand when on the move. After a few weeks the young make their first explorations off the mother's body. They acquire adult colouring within two or three months but grow very slowly and are adults at two years. Much of the weaned animals' time is spent foraging, and play is rarer than in other cercopithecoid monkeys. In a tropical climate they do not appear to be difficult to rear in captivity.

C. a. uellensis at birth.

Baboons

Papio

The genus *Papio* contains two distinct species, *Papio cynocephalus* and *Papio hamandryas*. The former species has four distinct geographic races which have been termed species by some taxonomists. *Papio cynocephalus* is the only species found in East Africa.

Two other baboon-like genera occur in Africa, the mandrills, *Mandrillus*, and the gelada baboon, *Theropithecus*. Huge extinct baboons, *Simopithecus*, resembling the latter genus are known from Olduvai, and a fossil baboon, *Procynocephalus*, is known from the Pliocene of Eurasia. The group therefore seems to have had a wider distribution and probably a greater variety of types in the past.

Adult male *P. c. anubis*.

Baboon (Papio cynocephalus)

Family	Cercopithecidae
Order	Primates

Local names
Nyani (Swahili), Ebbaku (Lukonjo), Galdez (Galla), Chebioyiet (Sebei), Nkerebe (Rutoro), Abula (Kuamba), Nkuka (Lubwisi), Nkobe (Luganda), Abim (Lwo), Echom (Karamajong), Lore (Madi).

Measurements head and body
508—1,143 mm
tail 456—711 mm

weight
22—50 kg (male)
11—30 kg (female)

Baboon

Papio cynocephalus anubis (Olive Baboon), larger darker, more heavily caped, tip of nose projecting. Found in Uganda, West and Central Kenya and around Lake Victoria.

Papio cynocephalus cynocephalus (Yellow Baboon), smaller paler, slender with more turned-up nose. Found in Tanzania, southern and coastal Kenya.

An intermediate form occurs in some localities and the two races interbreed freely in captivity. They appear to be distinct geographic races of a single widely distributed species.

Of the East African primates the baboon is the most conspicuous and the one that comes into the most obvious and immediate contact with man.

Few primates have been more widely studied and there is a vast anecdotal literature. The animal is found in an astonishing range of habitats and eats a wider range of foods than omnivorous man. Since the finding of food and survival under particular conditions determine many facets of the animal's behaviour, it is not surprising that the intelligent, adaptable baboon shows great plasticity and unpredictability in its behaviour. This may account for much that is apparently contradictory in the material published on wild baboons.

▲

Papio c. papioi, western Africa.
Papio c. anubis from Niger to Kenya.
Papio c. cynocephalus from Angola to Somalia.
Papio c. ursinus, South Africa.

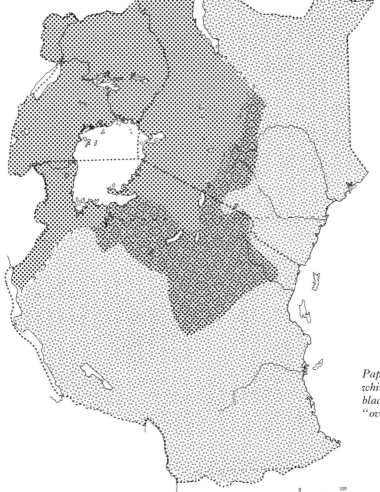

Papio cynocephalus. Range of subspecies:
white dots　　= *P. c. cynocephalus*.
black dots　　= *P. c. anubis*.
"overlap" area = baboons with apparently mixed
　　　　　　　characteristics.

0　　　100　　　200
Miles

Adult male *P. c. cynocephalus*.

The factors determining distribution are not at all clear or well-defined. Rock or tree refuges are a first requirement, but excepting open plains and swamps such shelters are generally available. In Bwamba, an area rich in primates, there is a wide choice of habitats which might be expected to support baboons and which do so elsewhere, yet their habitat in this area is very circumscribed being almost entirely limited to the steep, lightly wooded escarpment of the northern spur of the Ruwenzoris.

In this area Haddow tested some baboons for yellow fever and Semliki Forest virus and only found immunity in animals from Mongiro, where lowland forest grows along the base of the escarpment. Susceptibility to diseases to which other monkeys are immune might be the principal limiting factor in this instance, and the absence of baboons from many forest areas may well be due to hidden factors of this nature, rather than any inherent dislike or

178

disability for the forest habitat as a whole, for in some forests baboons are widespread and successful.

In areas that have been densely settled and cultivated over a long period of time and where secure refuges are hard to come by baboons are exterminated, direct human action and also possibly human diseases may play a part. In most of Buganda and in parts of Ankole and the eastern Province, baboons are rare or absent and, since this was first remarked on in 1925, the situation is one of long standing. Uganda official records show that 20,000 baboons have been poisoned and shot in several limited control areas over a period of thirty years; the unrecorded years and kills, particularly from poisoning campaigns, must swell this figure considerably, yet in many of the control areas baboons are still numerous and only very dense human settlement has effectively eradicated them.

The foods of baboons are largely determined by availability, and the animal's adaptability to what the environment has to offer contributes to its success. In open grassland the seed heads, shoots and storage leaf bases of grasses are the principal food; in forest fruit; in woodland and acacia bush, bark and resin may be important foods for part of the dry season. Vegetable matter normally provides the bulk of the diet; subterranean nuts, bulbs and roots, stems, shoots, pith, bark, flowers, pods, seeds and leaves are found in a variety of situations. At the same time insects, other arthropods, snails, bees nests, reptiles, eggs, nestlings, birds and small animals are encountered and eaten; snakes are avoided but scorpions are eaten after the removal of the tail. In water-logged areas, reeds, water lily roots and stems, frogs, crabs, crocodile eggs, terrapins and fish (scooped out of shallow water) have been recorded. The diggings of orycteropus and pig are examined for left-over insects and roots. Animal foods could not support a baboon troop in most habitats, but where small animals are encountered frequently and easily caught meat may become important, for instance, baboons feeding in grassland have been seen to flush and chase hares with some frequency and with success.

The killing of large animals seems to be a rarer and more fitful occurrence, when the taste is acquired killings are frequent but not usually sustained for a long period. Documented instances may be worth mentioning: a troop near Masindi, Uganda, killed eight goats within a week at the end of February 1957 and later twelve goats were killed in April. During calving peaks many new-born Thompson's and Grant's gazelles are killed by baboons, particularly by the males. A female Grant's gazelle has been seen to chase for three hours an adult baboon that killed her calf and to drive him from one tree to another. In Uganda, a half-grown cob was seized while lying down (part of the face and stomach were eaten). On Mt Meru goats were attacked and their eyes torn out. Vervet monkeys also have been killed and eaten; yet many of these victimized species may live with baboons on the most intimate terms. A single male adult vervet lived for at least two years with baboons at Ishasha and a goat lived with a troop at Bugoye, Ruwenzori, and was seen being "pushed" and "led" by these animals; young baboons have been seen to play with young impala. It is not unusual for both duiker and bushbuck to associate with baboons and one particular bushbuck near Korogwe allowed young baboons to climb on its back. Juvenile baboons and chimpanzees play

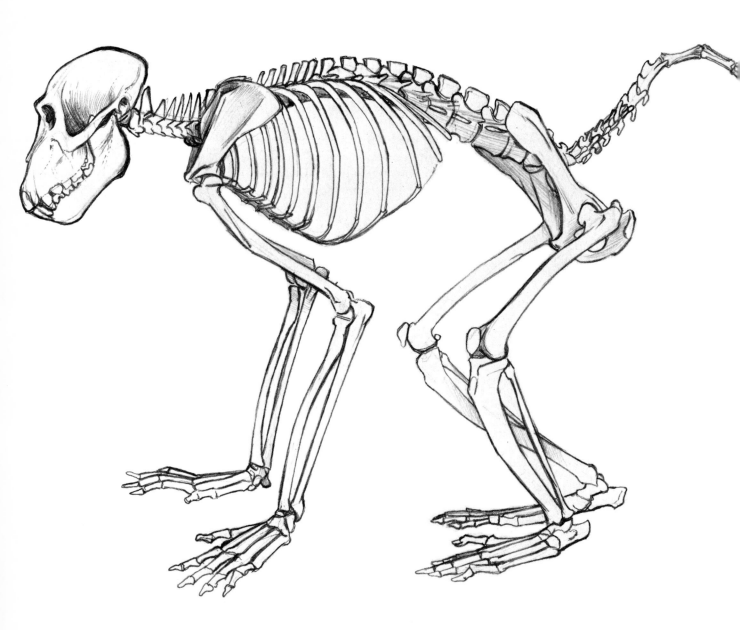

Adult male *P. c. anubis.* ▶

together and baboons have been seen to repeatedly tweak the twitching tail of a resting lion. Where baboons are habituated to people, familiarity may develop between baboons and children or adults, a man in Teso had wild baboons feeding from his hand. On the other hand children are sometimes attacked. Again the attacks may be fitful. In Bunyoro, five children were attacked within a few days of each other, two of which were killed. Near Machame, Kilimanjaro, two children were attacked by a solitary baboon and children protecting crops have been killed in Kinkizi, Kigezi; even an armed hunter has reported being attacked by a baboon.

Captive female baboons frequently show great interest in human babies and other small animals and have been known to carry them away, Darwin (1871) tells how "a female baboon had so capacious a heart that she not only adopted young monkeys of other species, but stole young dogs and cats which she continually carried about . . . . An adopted kitten scratched this affectionate baboon, who certainly had a fine intellect, for she was much astonished at being scratched, and immediately examined the kitten's feet and without more ado bit off the claws". A woman on a farm near Mt Kenya had her baby taken by a wild baboon (and was bitten rescuing it) and a baby was killed at Kissaki, Tanzania, while the mother was at work.

These reports should not give the impression that baboons are habitually aggressive or dangerous, but rather it is remarkable that an animal as obviously capable of inflicting damage, so rarely does so. What induces aggression and particularly sudden bouts of killing remains an interesting problem.

Baboons live within reach of water and drink daily. It is reported that a captive baboon fed on a salty diet was released and after failing to find water in the usual places, went further afield and started digging: water was found when the hole was continued; if this anecdote is true, the baboon's sense of smell may be more developed than is generally supposed.

Eyesight is very sharp in the baboon and is clearly the dominant special sense.

182

The adult male's carnivore-like profile, steady gait, big canines and cape have a distinct leonine appearance. The cape particularly serves to emphasize the teeth of the displaying male, as it describes an enlarging circle round a yawning face in much the same way as a lion's mane. These visual characters, which are extreme in the male, but also perceptible in the female olive baboon, have not been given the importance they deserve, particularly as they make the baboon visually a very "peculiar" primate.

The normal gait of the baboon is a steady dignified pace with frequent relapses into a seated position. They are adept climbers and will pole-climb trees repeatedly when curious, mildly alarmed or in order to threaten; they take to trees when pursued by lions or dogs, but soon learn their vulnerability in this position when faced by armed men.

They have lived 48 years in captivity, but maximum longevity would perhaps be reduced in the wild.

The majority of sounds made by baboons are permutations of the grunt, bark or scream but with changes of frequency and intensity associated with a variety of gestures and postures. Vocal communication seems to be complex and well developed, there is no situation to which the baboon responds with silence, and in this it differs from many other primates. The most frequent noise is a gentle grunting which is generally kept up when foraging, especially in thick cover. This contact call is reminiscent of similar noises made by other highly social animals such as bushpigs and social mongoose (*Mungos* and *Crossarchus*). Younger animals that have become separated chirp or moan loudly; in older animals, young males in particular, a sob or a short gasping bark is associated with rejoining the troop. Between males a grunt may be a threat leading to barking, roaring and fighting. Screeches and gibbering grimaces, which expose the tooth row in an exaggerated grin, indicate fear or frustration. The female may growl or pant-grunt when copulating.

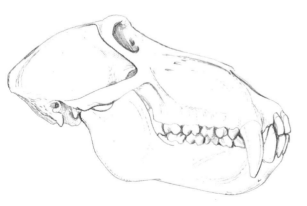

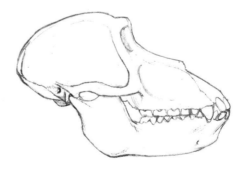

Adult male.   Adult female (to same scale).

183

Barks are used in a variety of circumstances: a single shrill bark by any baboon alarms the troop and an emphatic double bark, made only by the adult males, has a dual role. The larger males can sometimes be "called up" in thick cover by imitating a leopard's grunting when the double bark seems to be employed in a similar fashion to the calls made by mobbing birds, serving to neutralize a predator by establishing its position in relation to the troop and also threatening it at the same time; the troop usually makes off while the males continue to mark the predator's every move. Threats are seldom pushed home, but baboons have been photographed attacking in concert a stuffed leopard, and there are three published accounts of leopards being killed: one in Somalia involved an attack by five adult males.

The size of a baboon troop varies, the smallest being a dozen or less individuals; larger troops may number over 300, perhaps deserving to be called hordes.

Since Zuckerman's classic study (1932) there has been a great accumulation of data from many areas of Africa. It is fortunate that comparisons between two extremes of habitat are possible on the basis of intensive field studies made in East Africa on the olive baboon. The first was made on the open plains and thinly bushed river courses of the Nairobi Park (Irven DeVore, 1963—65), the other was made along the forested banks of the Ishasha River, which marks the Congo—Uganda border (T. Rowell, 1966). Patterns of social organization observed in particular circumstances may turn out to be moulded as much by the environment as by the innate structure of baboon behaviour. After comparing the differences observed in these two habitats the "essential baboon" is perhaps still further from sight, but the great success of this animal adapting to almost every type of vegetation and a very wide range of climatic conditions will be appreciated.

The principal differences observed between "forest" and savanna baboons are summarized below:

	FOREST	SAVANNA
Food	abundant and localized	scattered
Range	approx. 400 hectares	approx. 4,000 hectares
Density	11 per sq km	4 per sq km
Daily movements	approx. 1·5 km	approx. 5 km
Adult male—female sex ratio	1/1	1/3
Birth intervals	12—16 months	18—24 months
Breeding	all year	seasonal peaks
Behaviour	male co-operation without hierarchy	male dominance with hierarchy

On the basis of these contrasts, the forest-dwelling animals might be characterized as lacking certain environmental pressures that are important stimuli in the social life of savanna-living baboons, it would be equally valid to say that the savanna baboons suffer pressures that are not felt in the forest. It is not really possible to assert that any of the features contrasted above is "typical" or "aberrant".

Food supply clearly controls the range, density and daily movements of the troop; in an indirect way it may control other differences. A consistently high nutritional level might encourage continuous breeding, while sharp seasonal contrasts in the availability of food in some savanna habitats encourages breeding at the end of the rains, when animals are in their best condition. Baboon births in these areas tend to coincide with the beginning of the next rains six months later, but it should be stressed that these are birth peaks and not breeding seasons.

The differences in behaviour while real are not as antithetical as they might appear at first and the common ground is worth examining.

The striking characteristics of baboon society is its cohesion which is obvious at all levels. When alarmed, juvenile baboons cluster close together and then may turn to face the cause of disturbance, somewhat emboldened by their companions. Adults of both sexes tend to close up rather than scatter and the mobbing threats and attacks mentioned earlier are conspicuous for their concerted nature; the leadership of a single male is not obvious on these occasions, rather the adult males appear to "egg one another on" in their threats. There is, of course, a clear distinction to be made between behaviour directed outwards towards external danger and the various manifestations of threat inside the troop yet, even within, tension between individuals is often neutralized by one animal standing very close to another as if seeking his shadow. A significant aspect of social organization in savanna baboons is the presence of "cabals" or associations of two or more males that act in concert. Such behaviour could be invoked to illustrate hierarchy or co-operation, depending on which mental model the observation was to be placed in.

It is perhaps necessary for one to be threatened by a group of displaying baboons to appreciate how intimidating the performance is: the bristling capes can create the illusion of very large animals, while deafening, startling roars and barks are reinforced at close quarters by a ghastly display of teeth. That the display is nearly always bluff only emerges from statistics, its function as a deterrent to predators must remain impressively obvious. Capes and canines have, it seems, a manifest advantage as a response to exposure to predators; on the other hand, they do not have any observable function in the sexual life of baboons, yet within a baboon troop the possession of such lethal teeth (which are unnecessary for a largely vegetarian diet) might be thought to complicate social life, particularly when the number of animals may run into hundreds, and fully mature males into scores. These complications created by the bluffing "pseudo-lion" role of the males do not make the baboon the best model for generalized theories on primate social structure for some features of behaviour might be related to the specialized social and protective role of the male baboon.

There is a marked differentiation from an early age in the behaviour of the sexes. Females are ill-equipped by the burdens of pregnancy and nursing infants to act as scouts and protectors, young females are relatively quiet and stay with other females and babies; their tendency to stick together is measurable by the age of four months and is obvious by the end of the second year. By contrast young males are inquisitive and alert, as the most adventurous members of the troop they tend to find themselves on the peripheries, where their vulnerability to predators may account for the greater proportion of females in savanna troops. More important, the exposure of males to danger may give a clue to the selective pressure towards males appearing to be "formidable" animals and ultimately capable of inflicting damage. The evolution of the male's huge muzzle and sexual dimorphism has been discussed earlier.

Intensely fearful of being isolated and equally concerned that no member of the troop be left to straggle, both sexes have strongly centripetal tendencies. Animals of all ages and sex classes will tend to converge on experienced members of the most adventurous class, the adult males. In a relatively small, stable group a single very experienced animal may consistently provide cues for the whole troop, in hordes of over a hundred baboons, centres proliferate and clusters are discernible within the troop, here initiative does not remain the prerogative of an individual. In these widely dispersed troops, smaller groups are interacting on a relatively well-known but constantly changing environment, which of the peripheral fronts or arms of the troop are to draw the others may well be determined by the responses or inhibitions of individual baboons conditioned by a multitude of past happenings; these stimuli are by their very nature hidden from the human observer.

The degree of social consistency maintained in very large troops, or in the smaller groups discernible within them, has not been studied and probably varies a lot, changes in social composition may occur from time to time and under the influence of various events. A large group at Ishasha that occasionally foraged as one troop has been seen splitting up into two known smaller groups to go to neighbouring but separate sleeping places. Some of the males in these groups were interchangeable and there was an exchange of individuals with other troops still further afield.

A protective warning role is played by the male in nearly all the higher primates and, in baboons as in the apes, strutting and a level stare express confidence and threat. Young males and females greet large males by lip-smacking, presenting and cringing: lip-smacking, which is a conciliatory gesture, is used and reciprocated by all classes of baboons. These gestures are also discernible in a less exaggerated form, in apes and in colobus ("chewing" instead of lip-smacking in the latter). Like the chimpanzee a dominant animal may reassure an inferior one by touching or hugging.

In common with *Cercopithecus* and *Cercocebus* but differing from the apes and colobus is the baboon's yawn: this is a displacement activity elicited by many situations of mild excitement or tension. It has been reported as one of the first elements in an action pattern culminating in a full threat display, or actual fighting and is also an expression of fear when it is associated with sidelong glances and a nervous wiping of the nose. Yawning is therefore typical of threat behaviour, containing both escape and attack elements. Males yawn frequently when "harassing" one another. This term has been coined by DeVore to describe an activity that is most frequently directed at copulating pairs; it is first discernible in juveniles getting in the way of the mother's sexual activities, so that it may be an elaborate carry-over of a juvenile pattern into adult life. Harassing is also occasionally seen in disputes over food. The yawn therefore seems to have a broad spectrum of meaning. Functionally, yawning is an effective display mechanism for the most extraordinary and significant feature of the male baboon's anatomy. Primarily directed towards other baboons, the potentially dangerous teeth become components of a ritual in which the face is turned so that a profile view of the muzzle is presented and the use of the teeth as weapons within the society may thereby be inhibited. If this is the mechanism, the advantages for a highly social animal are evident, reducing the possibility of serious fighting. Paradoxically the threatening demeanour of the males becomes a cohesive, attractive force; females and males of all ages tend to run towards the large males even when the threat comes from the very animal they are approaching, the large male himself. These noisy and apparently hysterical incidents, which are not uncommon in caged groups, illustrate that the dual role of the big male as protector and punisher may conflict.

When an inferior animal has been punished or threatened it may redirect its aggression against other baboons. I have seen a frightened adult male slash savagely with his teeth at another baboon as they raced for the same escape path; this may have been an instance of aggression being redirected to another baboon instead of the dangerous predator. If this behaviour were at all common the consequences for social stability and safety in the troop could be serious, but appeasing gestures, lip-smacking, cringing, presenting and running away generally inhibit the large males from doing any damage, while in aggressive encounters between males confident or fearful undertones to the threat pattern generally decide the issue.

The 30—35 day oestrus cycle in baboons is marked by swellings of the perineal skin. During its maximum turgescence, which lasts for about a week, the skin acquires a livid iridescent pink colour. The swelling is discernible for about three weeks. When not actively solicited by the female, male baboons might be attracted visually by this brilliant behind; after smelling the female's perineum or perhaps grooming her, males may mount and copulate. Most copulations however show no visible preliminaries and a male will simply walk over to a female who may be seated and pull her to him.

The new-born are centres of interest for all classes of baboon and the mother is frequently pursued by enthusiastic lip-smacking admirers.

At birth the baby has pink skin and black fur and is carried on the mother's belly until about two months old, whereafter the baby may be seen riding on its mother's rump. Brown colouring appears first at three months and is completed by six months when the baby may be weaned, but females not infrequently continue to give milk for as long as fifteen months. Juveniles may play in groups, sometimes numbering many scores. The young show an instinctive fear of large hawks and rush to cover when they appear, this reaction becomes weaker with age and disappears with adolescence.

Man, leopard and lion are known predators and other carnivores may also eat baboons, but man is today incomparably the most important. If the broken baboon skulls from South African Pleistocene deposits are anything to go by, man's ancestors may also have been efficient killers and eaters of baboons.

Medically the baboon has acquired increasing importance and has recently been shown to be a natural host for bilharzia. As laboratory animals they have been used since 1927, but the trade in baboons has increased greatly in recent years, for the endocrines, nervous system and reproductive physiology are similar to those of man and make the animal valuable in the testing

of drugs. Tubercolosis has not yet been found in baboons, nor the dangerous Virus B (*Herpes simiae*). Baboons have been used extensively for research on atherosclerosis, coronary and other artery diseases. The fact that large numbers can be acquired at a reasonable cost (incidentally reducing agricultural damage) suggests that baboons could be turned into an economic asset instead of a liability as thoroughly destructive pests. It was estimated that a quarter of the total food-crop production of Gambia was destroyed by baboons in 1948, and the damage inflicted in some areas of East Africa may well approximate this proportion. The baboon is therefore classified as vermin and several thousands are killed annually in East Africa.

Male *P. c. cynocephalus*.

Mangabeys

Cercocebus

The name mangabey was wished onto this genus by Buffon, who thought that the first specimen seen in Europe originated from Mangabe in Madagascar; they are, however, exclusively African.

This genus is represented by two species in East Africa, the black mangabey, *Cercocebus albigena*, across the southern areas of Uganda to the Nile, and the crested mangabey, *Cercocebus galeritus*, in the Tana River area. The latter animal is separated from the main population of the species in Central Congo by nearly 1,500 kilometres and considerable geographical barriers.

Mangabeys combine characteristics of both cercopithecus monkeys and baboons in their behaviour. They are like baboons in having ischial callosities and sexual swellings and a rather baboon-like language of grunts. They resemble the cercopithecus monkeys in their smaller size, lack of sexual dimorphism and more arboreal habits.

The more primitive species, the West African *Cercocebus torquatus*, is also largely ground-dwelling. The two East African species represent a relatively successful specialized species, *Cercocebus albigena*, and a generalized species, *Cercocebus galeritus*, with a declining status.

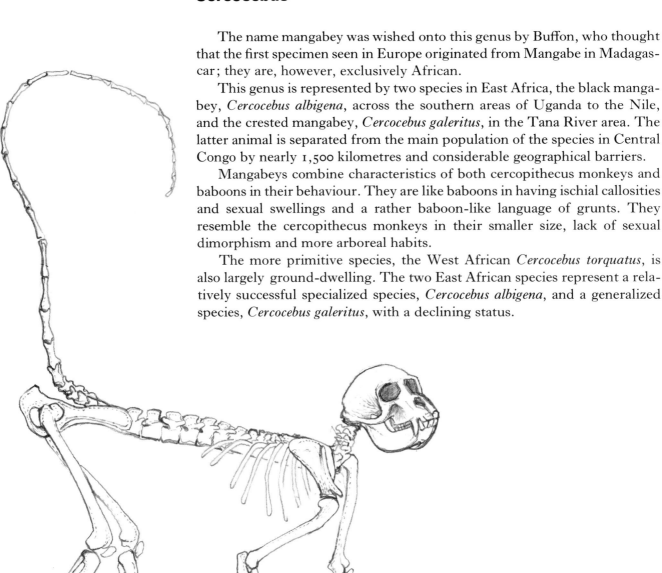

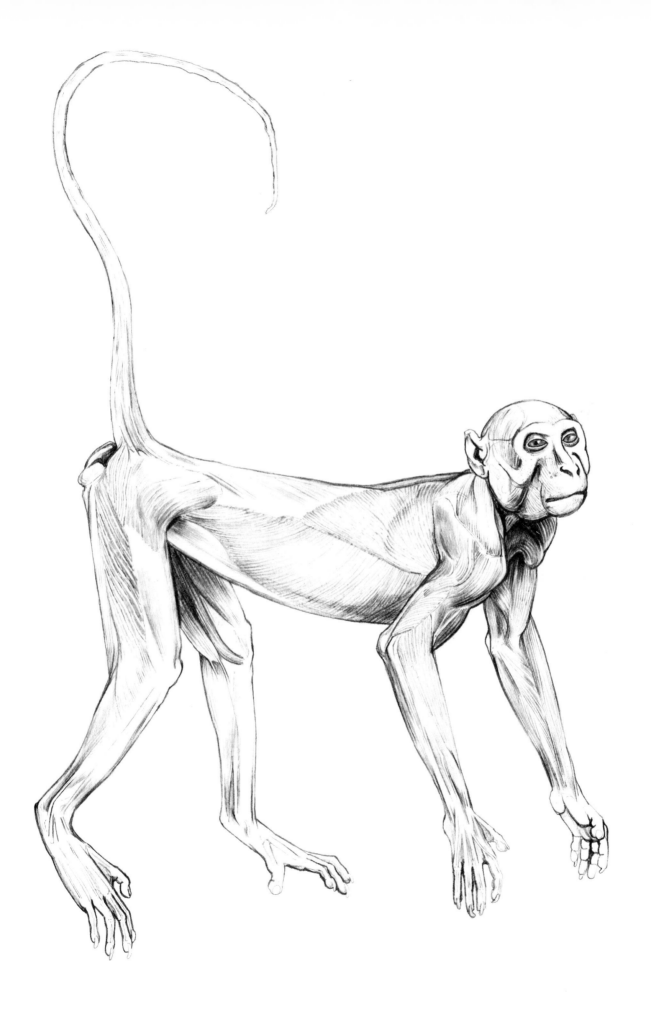

Black Mangabey
(Cercocebus albigena johnstoni)

Family   Cercopithecidae
Order   Primates
Local names

Sserwagabo (Luganda), Ngazi
(Kuamba), Kigazi (Lubwizi).

Measurements
total length

1,365—1,510 mm (males)

1,210—1,425 mm (females)

head and body

540—615 mm (males)

435—580 mm (females)

tail

740—940 mm

weight

7—11 kg

Black Mangabey, Grey-cheeked Mangabey

NOTE: Black mangabey is the designation most generally used in East Africa. The species should not be confused with the Congolese *C. aterrimus* which is also known by this name in some books.

The black mangabey is a large long-limbed monkey with a rather untidy long tail, it is black all over except for grey cheek tufts and a brownish mantle on the shoulders which is particularly marked in the male (see drawing). The face has a characteristic expression and profile due to the sloping chin and deep fossa beneath the eyes.

The black mangabey ranges from West Africa to the Nile, the Congo River seems to be its southern boundary. It has not been found above 1,700 m. In captivity it seems to be sensitive to cold and does not tolerate low temperatures; the wiry coat is certainly ill-adapted as an insulator. It is particularly numerous in swamp forests and is never found far from water. In Mabira and Bugoma (moist secondary forests), where there are few other species of monkeys, the mangabeys range fairly freely through the forest but in Bwamba, where there are a great many primate species and where the forest is also rather drier, they are limited to swamp forest and the immediate environs of water courses. The black mangabey is arboreal and seldom descends to the ground for very long, although they have been seen strung out in a long line crossing a papyrus swamp in Buganda and raiding fields for crops.

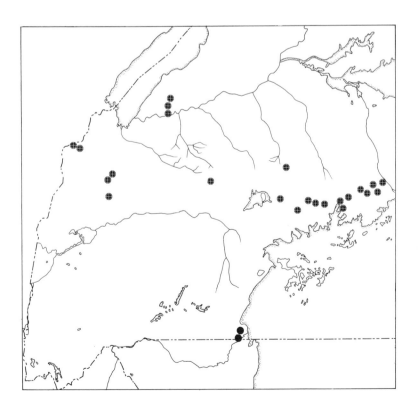

193

Their favourite fruits are from typical swamp forest and forest edge trees. They eat the hard fruits and seeds of *Pycnanthus angolensis*, *Treculia africana*, *Phoenix reclinata*, and in Bwamba, *Elaeis guineensis*. Other fruits favoured are those of *Maesopsis emini*, *Celtis* spp., *Sapium ellipticum* and various fig, *Ficus* species; the shoots of most of these trees and those of other trees are also eaten. Barks are chewed and larvae are extracted from rotten branches by the incisors. The tree-poisoning programme in some Uganda forests has greatly increased the number of dead and dying trees and has made this habit particularly noticeable. The incisors and molar teeth of mangabeys in Bwamba, which feed very largely on *Phoenix reclinata* and oil palm nuts, *Elaeis guineensis*, become worn to the roots even in relatively young individuals, whereas the Mabira mangabeys retain their enormous incisors intact, feeding as they do on a high proportion of buds, leaves, stems and soft fruit. They feed fast and eat a great deal, stuffing their large cheek pouches from dawn until about ten o'clock; there may be some feeding in the middle of the day and then again from 4 p.m. until dusk. In parts of Buganda, where strips of forest abut on cultivation, they may raid "shambas" for maize, they also dig up sweet potatoes, ground nuts and cassava.

Their gait is normally rather staid, the long tail swings up vertically when they stand still with the tip curling over either forwards or back; this is a highly characteristic posture. The tail is mildly prehensile and, in old individuals, becomes naked from habitual twisting along branches; young animals may intertwine their tails with that of their mother.

They have been kept over twenty years in captivity and can probably live that long in the wild (although wear on teeth, where a lot of their food is hard nuts and kernels, probably puts an earlier limit to their age).

Black mangabeys communicate principally by voice and they have a wide range of barks, chuckles and twitters. Males utter a resonating bubbling croak usually ending in a deep chuckle, which may act as a territorial call. Grunting seems to have a similar pacifying and contact function in mangabeys as in baboons, and this is the commonest vocal communication. Threat among mangabeys is accomplished by silent stares, yawns and a menacing posture with little or no noise. An inferior animal shakes its head and smacks

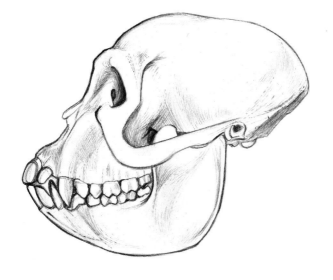

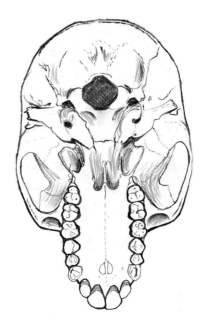

Underside of skull, showing massive incisors in relation to molars.

its lips in appeasement. These postures resemble those of baboons; they are emphatic and are generally only used at close quarters.

The lack of pattern and colour or contrasting skin pigment is accompanied by an absence of the bobbing gestures typical of the cercopithecus monkeys and, like baboons, they are level and deliberate in their movements.

Bands of about 30 are not uncommon but 12—20 animals are a typical group. There are usually as many males as females. According to Chalmers (1968) the adults are rather less tolerant of their own sex than most monkeys, but males are more tolerant to females than is usual, and he has observed a marked pattern in their age and sex associations; pairing of males and females is much commoner than male associations and there is no obvious dominance hierarchy. Grooming is commonest between males and females and between males and infants, while adult males are aggressive to subadults. Although very young babies stay only with their mothers, two infants of over three months, watched by Chalmers, spent 70% of their time with adult males. (He did, however, find very marked differences in a variety of behaviour patterns in a troop from another area.) Later the adolescents gang together as is common in baboons and at five years the animal is mature.

At night and after the morning feed the monkeys form a tightly knit group in a single tree or clump of trees, preferring to sit about 60 ft (18 m) above the ground (this height varies with the locality). They huddle together in pairs or groups of three or four. Between 2 and 4 p.m. the monkeys rest and groom, commonly females do most of the grooming of males, while young animals wrestle and play.

The periphery of each mangabey troop's range is usually within easy hearing distance of neighbouring troops but there is very little interchange of individuals. Chalmers estimated a home range of a twentieth to a tenth of a square mile and an average density of about 200 per sq mile.

When very frightened these monkeys rush for denser vegetation and freeze, but I have seen several large males come down to the lowest branches and shrubs to mob a leopard in the undergrowth. When a pair of crowned hawk-eagles were flying over a mixed group of monkeys in Bwamba, the mangabeys descended into the lower vegetation.

Mangabeys do not show any marked seasonal breeding and in most troops there are usually babies and young of all sizes. The female shows cyclical swellings as in baboons and solicits males during her maximum turgescence. The gestation period is about 174 days. The newborn cling to the mother's breast and use the whole fist, only by about the fifth week is it able to oppose the thumb; this is later than most monkeys but resembles the baboon. Similar also to baboons is the eagerness displayed by females of all ages towards handling newborn babies. While the baby is still very new the mother will show a marked degree of possessiveness but later allows other females to carry and fondle the baby, only responding to its cries of alarm.

Crested Mangabey, Tana River Mangabey
(Cercocebus galeritus)

Family Cercopithecidae
Order Primates

**Measurements
total length**

1,350 mm
head and body
620 mm
tail
750 mm

Crested Mangabey

The crested mangabey is a smaller, slighter animal than the black mangabey and is pale grizzled grey-brown in colour. The hands and feet are dark brown, the underparts yellowish-white and the brown hair on the head is longer than elsewhere. There is a small, white "flash" on the temples. The general impression is of a dirty, nondescript monkey. It differs from cercopithecus monkeys in its absence of colouring and pattern but resembles them in being an omnivorous feeder; it does not have the specialized teeth or diet of the black mangabey.

The Tana River area is the only locality where this species is found in East Africa and the presence of this species in an isolated cul-de-sac, sandwiched between the ocean and semi-desert is significant, for the species formerly had a very extensive range. This is shown by its occurrence over a very wide area of the Congo, with populations north of the river, from the River Dja to Ituri, and others south of the river, on the Kasai and Sankuru Rivers. The species has probably given way to the guenons, having been a successful and widely distributed species prior to their development.

Like the black mangabey it is found near water, but it is perhaps less exclusively arboreal, coming down to feed in the undergrowth and onto the ground. In cultivated areas it raids "shambas". It feeds on shoots, leaves, insects and fruit. Its range of sounds is similar to that of *Cercocebus albigena*.

This mangabey's habitat in the forests along the Tana River flood plains has an ecological resemblance to the Sango Bay forests where *Cercocebus albigena* is a conspicuously successful monkey.

Nothing has been published on the behaviour of this monkey; it is seen in bands which can be quite numerous and is said to be rather tame (which would certainly not help its survival).

Observations on this monkey would be particularly interesting.

C. galeritus distribution.
＊= C. galeritus.

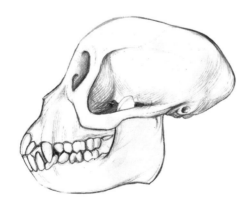

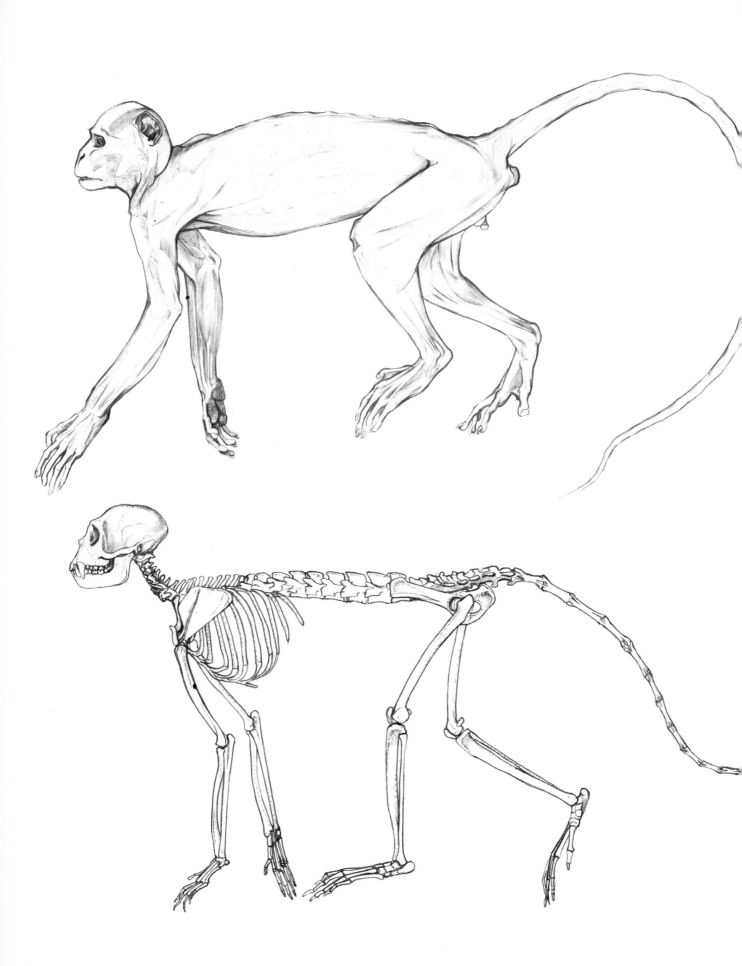

Guenons, Cercopithecus Monkeys

Cercopithecus

Although their external appearance varies enormously, not only from species to species but from race to race, beneath the pelt most species of *Cercopithecus* are anatomically inseparable except by an expert.

Hybridization between many cercopithecus species is common in zoos and is not unknown in the wild, so that the line between a species and a race becomes particularly difficult to draw. It also complicates the study of inter-relationships as there are "hybrid" populations that have been described as "races", and in some circumstances it is difficult to distinguish between a geographic cline and a hybridized swarm.

The anatomical homogeneity of cercopithecus monkeys implies a closeness of relationship which is further confirmed by a detailed examination of the group. A beautifully stepped series of forms is discernible and an evolutionary radiation for the genus as a whole can be postulated, not on the basis of fossils but on the evidence of living forms. Superficial differences in the pelage of populations when correlated with geographic distribution patterns suggest paths of migration and isolation.

Opposite page: *C. neglectus* dissection, *C. l'hoesti skeleto*

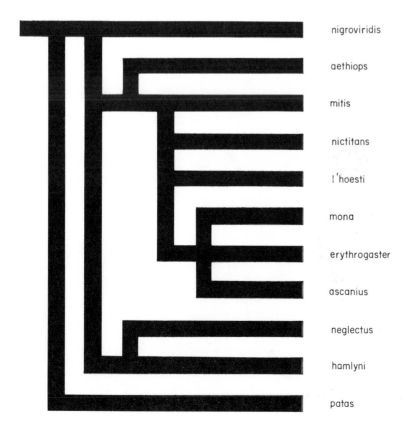

nigroviridis

aethiops

mitis

nictitans

l'hoesti

mona

erythrogaster

ascanius

neglectus

hamlyni

patas

Diagram of hypothetical relationships of some *Cercopithecus* monkey species.

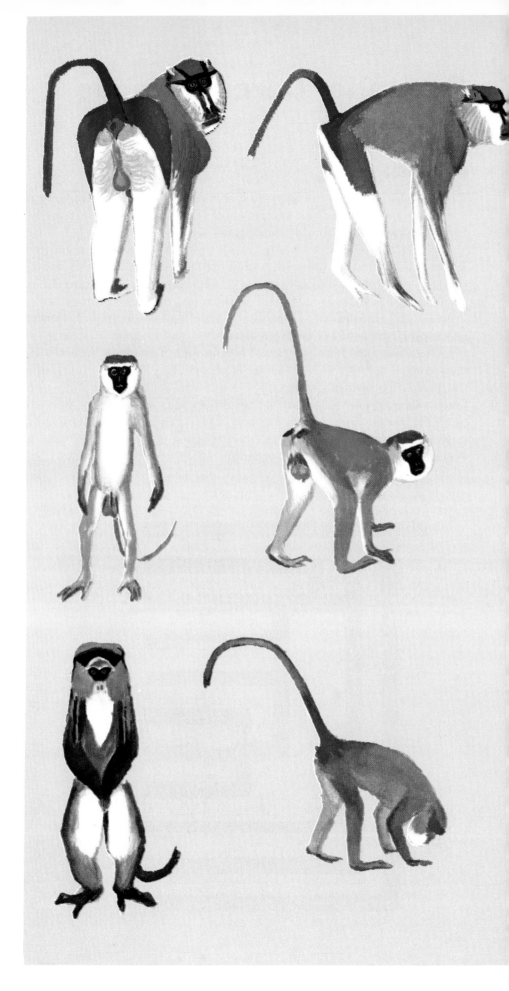

Patterns and postures employed as signals. *Top*, *C. patas*, "bounce" and "presenting" postures; *middle*, *C. aethiops* "red, white and blue" displays; *lower*, *C. neglectus* (*left*) adult threat, and (*right*) juvenile appeasement postures.

The trend found in mangabeys towards the reduction of the muzzle and the minimization of sexual dimorphism continues in the cercopithecus monkeys. Verheyen (1962) has shown that *C. (Allenopithecus) nigroviridis* is closest to the ancestral stock of the group as a whole, and the species is generally regarded as a link between *Cercopithecus* and *Cercocebus*. The skull of *Cercopithecus patas* also shows a close affinity to this species. Thereafter, *Cercopithecus neglectus*, *Cercopithecus aethiops* and *Cercopithecus mitis* also show primitive characters. A somewhat specialized species, *Cercopithecus hamlyni*, probably derives from the same ancestral stock as *Cercopithecus neglectus*.

In the principal radiation deriving from a *C. mitis*-like stock a simplification of the teeth is discernible, the lateral ridges on the molars become less marked and the lower mandible acquires a simpler form.

It is possible that there is a trend towards greater co-ordination in the branches and smaller size, since some species (i.e. *Cercopithecus mona* and *Cercopithecus ascanius*), which appear to be more recently evolved, are fast climbers and are more active than other species. In the main forest blocks these species also appear to be the commonest and most generally successful monkeys.

In forest species the differences in coat and colouring follow a distinct orthoselective trend in which colouring generally becomes brighter, patterns become bolder and more geometric while tonal contrasts increase. Coat colour is also modified in populations living under climatic extremes, but nonetheless the general trend holds true. Striking patterns seem to function as an improved signalling device and the greatest elaboration is reserved for the head and the anal poles. The facial patterns often enhance facial expressions and specifically characteristic movements of the head, while the elaboration of the genital region is associated with "presenting" gestures and certain ritualized movements of the tail that can also be interpreted as signals. A most important aspect of monkey patterns is their species-specific meaning: for instance, the white ruff of *Cercopithecus l'hoesti* is presented as an invitation for grooming and is groomed more frequently than any other part of the body, whereas the head is lowered in threat. The white beard of *Cercopithecus neglectus* on the other hand, is an aggressive feature and is not presented for grooming, the neutral back and the arms being the preferred grooming area. This species of monkey rests and often sleeps in the grooming posture, lying

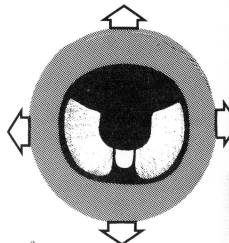

a.

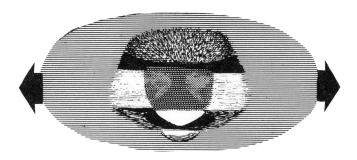

b.

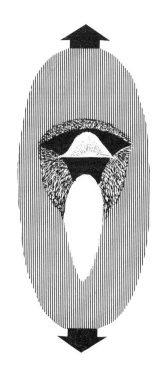

c.

Facial pattern of 3 *Cercopithecus* monkeys indicating principal direction of the head's movement when signalling: a. *C. ascanius*; b. *C. cephus*; c. *C. neglectus*.

203

on its side along a branch. Such postural peculiarities may be a reflection of the significance of signal patterns. There is some evidence that specific signal patterns may have contradictory meanings for other monkey species and could be a factor of interspecific intolerance (see *C. neglectus*).

Complicated patterns of hair growth are associated with the many and diverse "facial designs" exhibited by guenons. For instance, in the case of the spot-nosed *Cercopithecus ascanius* radiation there is progressive elaboration of a parting of hair on the cheeks which culminates with *Cercopithecus ascanius schmidti* in very clearly defined white shapes set within a semi-circle of darker colour. The "T"-shaped pattern in *Cercopithecus neglectus* is emphasized by the banking up of dark colour round the browpatch and the growth of long fur above the ears instead of below as is commonly seen where there are horizontal or circular signals. In the faces of *Cercopithecus neglectus* and *Cercopithecus hamlyni* the verticality of their white signal is emphasized by the cheek fur being pressed close to the sides of the face. The vertical white mark in these species is signalled by an up and down bouncing of the head. By contrast a West African species, *Cercopithecus cephus*, with a horizontal white "moustache" stripe, signals by shaking its head from side to side (see p. 203).

In tracing the course of guenon evolution it is fortunate that the Congolese *Cercopithecus* (*Allenopithecus*) *nigroviridis* has survived. This species probably represents in little changed condition a surviving population of the ancestral type of guenon monkey. There is some elaboration of the hair tracts, which is probably a secondary specific development but there is little facial colouring apart from a paling of the skin of the eyelids and chin which has white hairs. In colouring, pattern and hair texture this species shows some resemblance with races of *Cercocebus galeritus*.

According to Verheyen (1962) the species closest to *C.* (*Allenopithecus*) *nigroviridis* in cranial characters is *Cercopithecus* (*Erythrocebus*) *patas*. In many other characters, however, the latter species has clearly specialized to a great degree and the minimization of sexual dimorphism has been reversed in this ground-dwelling form.

Verheyen (1962) found that *C. neglectus* skulls were relatively primitive, this finding raises the question of how so elaborately patterned a species can be in any sense primitive. To illustrate how a pattern such as that found on the face of *C. neglectus* might have derived from a rather undifferentiated type I have drawn hypothetical monkey faces approximately intermediate in type between *C.* (*Allenopithecus*) *nigroviridis* and *C. hamlyni*, and between *C. hamlyni* and *C. neglectus* (see opposite page).

The main characteristics of pattern formation are a banking up of light or dark coloured areas so that a certain feature or area is emphasized. This is usually correlated with a differential growth of hair which further emphasizes and "forms" the pattern by extending or contracting the fur around the face. The pigmentation of the skin also alters to fit in with the total effect, and special colour, blue, red or pink may develop on the face, on the genital area, or on both.

The simplest example of facial elaboration directed towards emphasizing expression may be seen in *Cercopithecus mitis doggetti*. In this monkey the back is coloured in varying shades of grey or yellowish grey, with grizzling due to

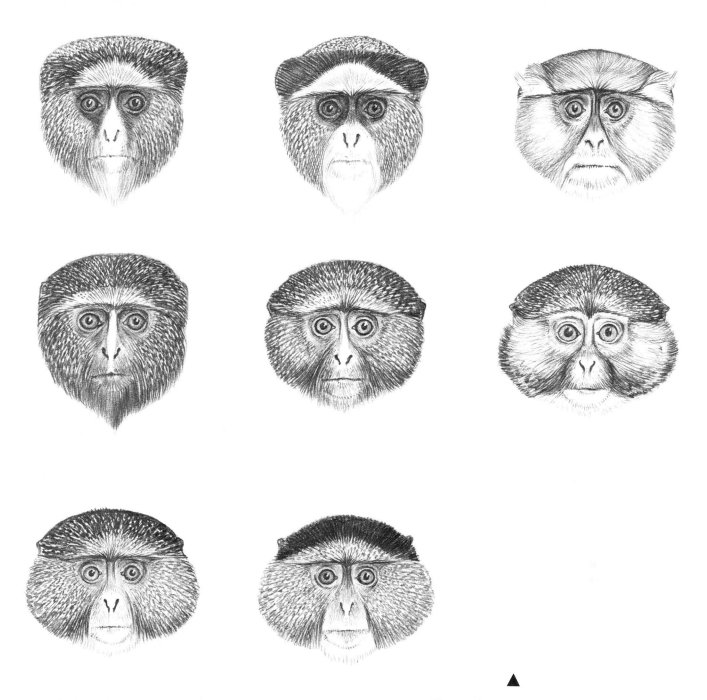

annulation of the individual hairs. This colouring is not very different in *Cercopithecus aethiops*, *C. hamlyni* and *C. neglectus* (in the latter two species the fur is very soft and silky as in *C. nigroviridis* and some races of *Cercocebus galeritus*). In *C. mitis doggetti* the grizzled hair on the top of the head has the tips of the hairs black so that a black cap is formed and in some individuals the ruffling of this hair reveals the grizzled bases of the hairs but in others the hairs are entirely black. Visually the black cap isolates the brow-band which remains grizzled but is of a light colour. This emphasis of the expressive brows may be seen as a first step in the elaboration of facial signals. The further elaboration of pattern in the East African races of *C. mitis* is treated in some detail in the profile of this species (p. 234).

    C. mitis is of particular interest as it appears to be near the stock from which

Faces of *C.* (*Allenopithecus*) (*middle right*), *C. hamlyni* (*middle left*), *C. neglectus* (*top centre*), *C. patecus* (*top right*), *C. mitis* (*bottom centre*), with hypothetical intermediate faces inserted.

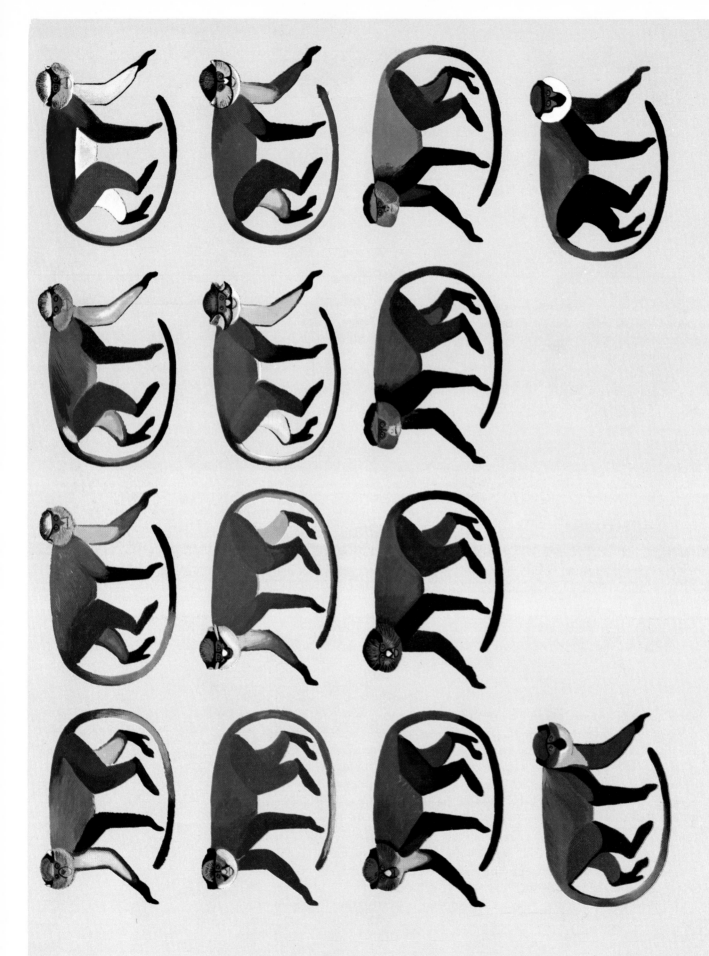

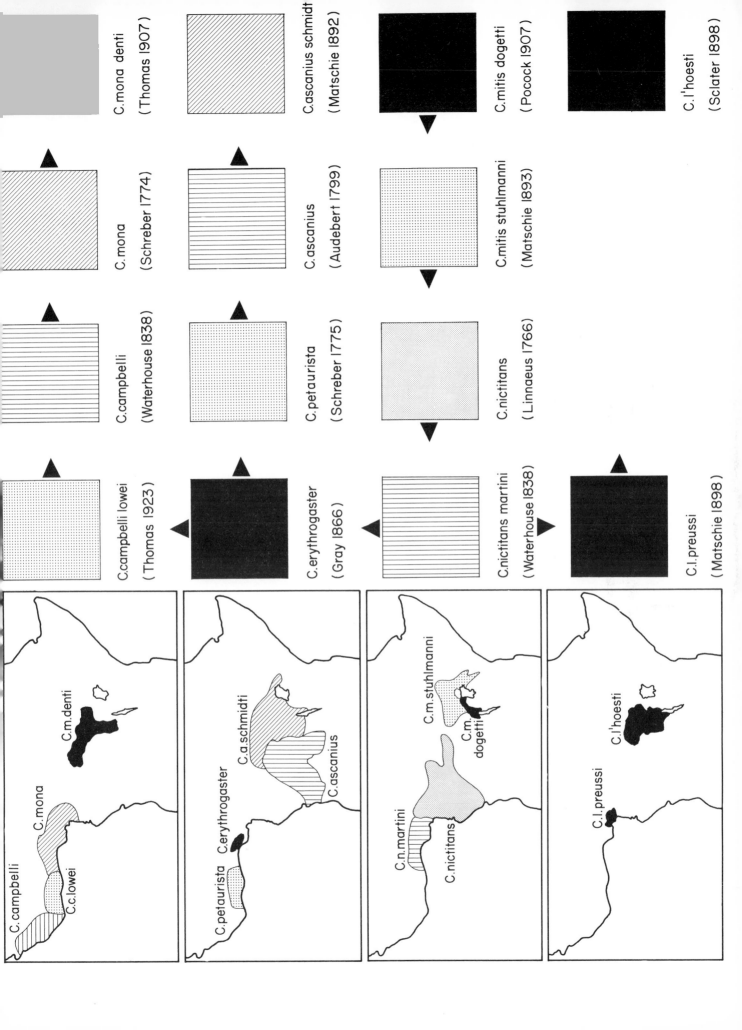

C.mona denti (Thomas 1907)

C.ascanius schmidti (Matschie 1892)

C.mitis dogetti (Pocock 1907)

C.l'hoesti (Sclater 1898)

C.mona (Schreber 1774)

C.ascanius (Audebert 1799)

C.mitis stuhlmanni (Matschie 1893)

C.campbelli (Waterhouse 1838)

C.petaurista (Schreber 1775)

C.nictitans (Linnaeus 1766)

C.campbelli lowei (Thomas 1923)

C.erythrogaster (Gray 1866)

C.nictitans martini (Waterhouse 1838)

C.l.preussi (Matschie 1898)

C.campbelli
C.mona
C.c.lowei
C.m.denti

C.petaurista
C.erythrogaster
C.a.schmidti
C.ascanius

C.n.martini
C.nictitans
C.m.stuhlmanni
C.m.dogetti

C.l.preussi
C.l'hoesti

the main guenon radiation has derived. The race *C. mitis stuhlmanni* which is common in the East Congo extends as far west as the rivers Aruwimi and Uelle. Westwards there is a very similar monkey, *Cercopithecus nictitans nictitans*. In this form the white hairs of the nose have grown more thickly, while the dark skin and hair around the nose have "banked up" to form a striking nose spot. The main behavioural distinction in this species is a frequent bobbing of the head whereby it has earned the French vernacular name of "hocheur" (wagger). It appears that the nose spot in this species takes the place of the browpatch as a signalling device. Three races of this monkey occur in West African forests where they are said to occupy the niche of *C. mitis*. The Nigerian *Cercopithecus nictitans martini* shows two significant differences; it has a distinct white chin and a reddish suffusion in the fur of the cap and back. These trends are more marked in another West African species, *Cercopithecus erythrogaster*, a monkey which shows characters intermediate between several guenon species. Some individuals have white noses and others black, a black band surrounds the crown and another black patch runs from the mouth to the ear; like *C. n. martini* there is a white throat patch but this feature is wider and more clearly defined.

The West African forests are strung along a coast intersected by numerous rivers and cut up by at least one major corridor of savanna in more arid periods. These conditions have been ideal for the past isolation of populations and, in many groups of West African mammals, speciation has occurred which can be attributed directly to these isolating mechanisms (A. H. Booth, 1955, 1958; Moreau, 1969). *C. erythrogaster* may represent a relic population of one of the earliest monkey types to develop an elaborated pattern in West Africa. The species is now very restricted and shows considerable individual variation (which could imply some degree of hybridization with closely related types) but populations of monkeys resembling this species undoubtedly gave rise to more than one line of monkeys. *Cercopithecus campbelli lowei* has resemblances which suggest that it derives from a *C. erythrogaster* stock (without a nose spot) isolated in the Upper Guinea forests; in the extreme west *C. campbelli campbelli* derived from this form through still further isolation, while *Cercopithecus mona* arose later when the forests joined up again and monkeys resembling *C. campbelli* were able to migrate east. Of this radiation the most easterly and perhaps the most recently extended population is *C. mona denti* (see p. 249).

A spot-nosed population of *C. erythrogaster* also became isolated in the Upper Guinea forests giving rise to *Ceropithecus petaurista*, this monkey elaborated the facial pattern while keeping the white nose spot. The most developed type of this highly successful group is *Cercopithecus ascanius schmidti*, which also has the most easterly distribution. This species seems to show a greater adaptability to altitude than *C. mona denti* which may account for its extensive range in East Africa. Other monkey species have evolved in West Africa, but the only remaining type that will concern us here is *Cercopithecus l'hoesti preussi*. This race is restricted to montane forest in the Cameroons and Fernando Po, it has resemblances with both *C. nictitans martini* and with *C. erythrogaster*. It would seem to have derived from some early population intermediate between these types, and its isolation may have involved an adaptation to cooler conditions; it still retains a well-defined nose spot but the hair on it is black instead of white. This race probably resembles the ancestral type of *C. l'hoesti*. The peculiarities of this monkey and its discontinuous distribution are discussed further in the text (see p. 229).

This sketch of the radiation of guenons is necessarily brief, as of the 75 forms of guenons known only a few species are found in East Africa. However, a simplified diagram of the radiation of cercopithecus groups is presented, and the plate on p. 206 represents the series of monkeys discussed above, together with maps of their distribution. This illustrates some of the points that have been raised and details of the East African species will be treated in the text.

One aberrant monkey has had scant mention. *Cercopithecus* (*Erythrocebus*) *patas* is a long-legged open country species which has a rather peculiar distribution, being one of the very few mammals that is restricted to the northern savannas and is not found in equivalent areas in south Africa. The distribution of *Cercopithecus aethiops* raises questions that may have a bearing on that of *C. patas*. This monkey in contrast to *C. patas* is distributed throughout the southern part of the continent. The southern populations of this species are all rather similar, and such differences as there are may be largely a response to ecological conditions. At any rate there is a homogeneity in the southern populations that implies a considerable degree of geneflow. The distinctness of these populations from the northern ones has led "splitters" of the *C. aethiops* group to erect a separate species ("*Cercopithecus pygerythrus*") while the northern vervets have been called "*Cercopithecus sabeus*" and "*Cercopithecus aethiops*" (Dandelot, 1959, 1968). The two populations of *C. aethiops* found in northeast Africa probably represent extreme wings of a pan-African cline. In spite of sometimes living in similar ecological conditions and having some convergence in colouring, different origins are betrayed

Left: *C. a. pygerythrus*.
Centre: *C. a. cynosurus*.
Right: *C. a. tantalus*.

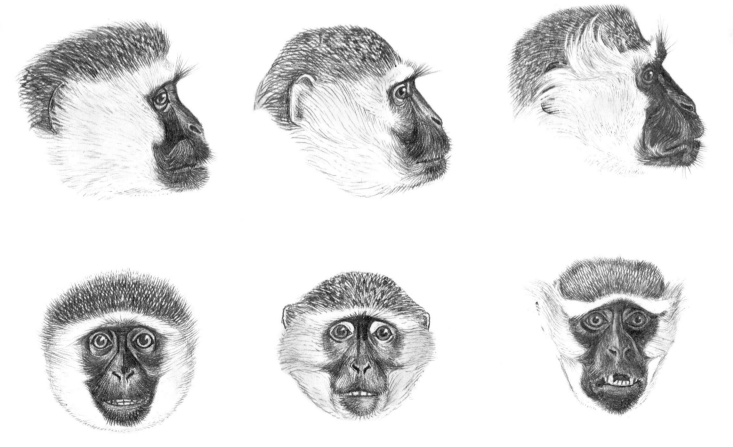

by the arrangement of hair tracts which are characteristic of the southern and northern stocks. The oldest types are probably to be found in the southern population. Their colouring approaches more closely that of a forest guenon, like *Cercopithecus mitis doggetti*; the hands and feet and tail tip are black while the facial whiskers show minimal differentiation. However, there is in Angola a somewhat distinct population, *Cercopithecus aethiops cynosurus*, which shows paling of the extremities and some rearrangement of the facial whiskers. These characters look rather like an anticipation of the characteristics of the northern *C. aethiops* group and, while they might be the result of an ancient period of hybridization, it is more likely that *C. a. cynosurus* represent the stock from which the northern *C. aethiops* were drawn, and that monkeys from this stock moved up into the northern savannas during a very dry period, when there was a savanna connection cutting across what is now the forest of the Lower Congo. The sharply defined differences between northern and southern populations in northeast Africa show that a connection through this area is unlikely, moreover the Ethiopian populations, *Cercopithecus aethiops aethiops*, are almost certainly an eastern extension of the West African type *Cercopithecus aethiops tantalus*.

The relevance of this for *C. patas* is that *C. aethiops* probably developed first in the savannas of the south. It is possible that during the formative period while this species was adapting to savanna the forest belt acted as a barrier to movement. Prior to the speciation of *C. aethiops* in the south, a still more primitive stock of cercopithecus monkeys may have been pursuing a similar course along the northern margins of the forest block. This stock, while it may have adapted further to dry, open conditions, may have occupied an almost identical niche to that of *C. aethiops* right up to the time when this species expanded into the northern savannas. With this invasion the ancestral *Cercopithecus patas* stock might have been pushed further out into the open (see diagrammatic map). The patas monkey would certainly have had to pass through a "stage" virtually identical to that of *C. aethiops*. The pressure to relinquish this rather convenient niche might well have been exerted by competition and the only likely competitor would have been *C. aethiops*. Therefore *C. patas* may not have occupied its present ecological position for long enough to have adapted to the slightly different conditions found in eastern and southern Africa. The species is well established and occurs in large numbers in the Sudanic savanna but is very sparsely and locally distributed over much of its East African range. The distribution pattern suggests that this species is expanding its range in the more open areas but is making slow headway in eastern Africa, which is largely a marginal habitat for this species and also a region where *C. aethiops* is well established over a wider range of vegetation types. The combination of a more wooded habitat and an established competitor therefore may have presented a barrier to the patas monkey's southward expansion.

C. patas has a characteristic that is typical of the guenons as a whole in that the head and the anal poles have acquired an elaborate pattern and colouring. In this the monkey differs from the other open country species the baboon; it also differs in the rarity and quietness of its vocal communications— the two elements appear to be connected. It would seem that as an improved signalling device the possession of striking patterns and colours reduces the need for loud and frequent calling. Both *Cercopithecus patas* and *Cercopithecus neglectus* are strongly patterned; both make movements and assume postures that are exaggerated; both are strikingly quiet monkeys. For the other guenons the term "quiet" is relative, but they also contrast strongly with the rather colourless, monochromatic baboons and mangabeys which are noisy and also deliberate in their movements.

Cercopithecus ascanius, *C. l'hoesti*, *C. mitis* and *C. mona* have very dissimilar coat patterns whereas their louder calls are very alike. As in birds, there are probably advantages in warning calls not being too specific, but there is nonetheless an interesting contrast between long range communication (sound) being relatively undifferentiated, while close range communication (visual signals) are highly specific. There may even be differences between species in the relative importance of olfactory communication (see *C. neglectus*).

The phylogenetic increase of brilliance in the colouring of guenons raises the interesting possibility that a greater sensitivity to colour may also be developing. Reactions of *C. neglectus* monkeys to colour suggest that the emotive response to specific colours may vary greatly in intensity.

Both mangabeys and guenons are restricted to Africa and the evolution of the latter group must be largely a Pleistocene development, occurring after suitable connections with Asia had been broken. Considering the guenon's success in colonizing a wide range of ecological conditions within the continent, their restricted distribution is surely a measure of the recency of their evolution.

Because the radiation is recent a tentative sequence of events can be suggested for some of the East African monkeys on the basis of their distribution patterns. It is convenient to start with a highly developed species, *Cercopithecus ascanius*, which probably arrived in East Africa during the last wet period when the forests were sufficiently continuous to allow the species to reach the Mau. The dating of climatic events is still highly speculative, but this expansion might have occurred about 25,000 years ago. Between about 40,000—50,000 years ago a very arid period is thought to have separated the western and central African forests, at which time the ancestral stock of *C. ascanius*, *C. mona* and *C. l'hoesti* might have been confined to West Africa. During this period *C. neglectus* might have developed in the Central Forest Refuge, expanding its range greatly when the climate changed only to decline when the West African colonists arrived. This very dry period might have been responsible for sealing off the forests of southeastern and southern Africa from further immigration of forest fauna, and the still older *Cercopithecus mitis* stock which became isolated there must have entered these forests at an earlier period. At any rate, no other forest guenons managed to colonize these forests and the differentiation of *C. mitis* along the eastern seaboard and in the southeastern highlands is some measure of the degree of isolation these populations have suffered.

If the suggestions contained in this introduction to the guenons serve to provoke lines of future enquiry they will have served their purpose, for the evolution of the group as a whole has been somewhat neglected. A detailed examination of their anatomy and taxonomy is available in Volume VI of W. C. Osman Hill's monograph on the primates.

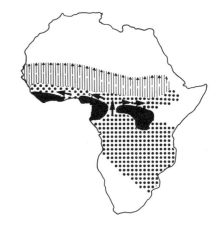

Diagrammatic map of hypothetical expansion of *C. aethiops* through a break between Congo-Gaboon forests and displacement of ancestral populations of *C. patas*.

C. a. pygerythrus.

**Vervet Monkey
(Cercopithecus
aethiops)**

Family	Cercopithecidae
Order	Primates
Local names	

Tumbili (Swahili), Serwagaba
(Luganda), Nkende (Runyoro),
Ongera (Lwo), Edokolet (Ateso),
Kamale (Galla), Suboltit (Sebei),
Girengwa (Kuamba).

**Measurements
head and body**
380—620 mm
tail
420—720 mm
weight
5—9 kg

Vervet Monkey

Cercopithecus aethiops, greenish-grey savanna monkeys with white-fringed dark faces.

Races

Cercopithecus aethiops tantalus (includes *C. a. budgetti*)	Very white frontal band, long cheek whiskers "brushed" up and back, feet pale, tail tip white. Northwest Uganda.
Cercopithecus aethiops pygerythrus (includes: *C. a. johnstoni*, *C. a. centralis*, *C. a. nesiotes*, *C. a. callidus*, *C. a. excubitor*, *C. a. rufoviridis*)	White bonnet all round face, black feet, black tail tip. Tanzania, South Uganda and South Kenya.
Cercopithecus aethiops arenarius	A smaller paler race of *C. a. pygerythrus* type. Arid areas of North Kenya and East Uganda.

NOTE: Island populations, i.e., "*C. a. nesiotes*" on Pemba and "*C. a. excurbitor*" on Manda island, Lamu, might justifiably be regarded as subspecies but they are nonetheless clearly isolates of *C. a. pygery thrus* type.

Vervets with various mixed characteristics of *C. a. tantalus* and *C. a. pygerythrus* are found over most of southern Uganda.

Vervets are found throughout the savanna areas of Africa, but there are three major types which range over the southern, northern and far western savannas respectively; these types contain further subdivisions.

The need to give expression to these greater and lesser distinctions has given rise to taxonomic difficulties. Consequently vervets have been split into several species by some taxonomists, partly in order to accommodate the subdivisions within the trinomial system and partly on a supposed correlation of distribution with vegetation zones (Dandelot, 1959). The vervet is treated here as a single species.

The southern savanna form, *Cercopithecus aethiops pygerythrus*, has at least two distinct but closely related peripheral populations, *C. a cynosurus*, in Angola and *C. a. arenarius* in arid northeast Africa.

The northern savanna type has a distinct easterly population, *Cercopithecus aethiops aethiops*, in Ethiopia and a more westerly population, *C. a. tantalus*. The West African type is *C. a. sabaeus*.

The distribution of these races and a possible dispersal pattern has been discussed earlier (pp. 209—211), while "hybridization" between northern and southern savanna forms in Uganda was discussed on p. 79.

The two drawings illustrating this profile are of a young *C. a. tantalus* male from West Uganda and an adult male of *C. a. pygerythrus* from southwest Tanzania and they allow some comparison of the two most distinct races found in East Africa.

The range of vegetation types inhabited by this species includes all types of savanna, woodland and also riverine, gallery and lake-shore forests. The

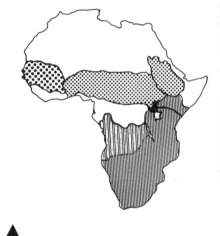

▲

C. aethiops races:
large spots = *C. a. sabaeus*.
small spots = *C. a. tantalus* and *C. a. aethiops*.
stripes = *C. a. pygerythus* and *C. a. arenarius*.
blazer stripes = *C. a. cynosurus*.

species acquires quite extraordinary pre-eminence on some uninhabited islands in Lake Victoria where the density of individuals is very high.

Fruits, buds, seeds, roots, bark, gum and many cultivated crops are eaten by this monkey, which is second only to the baboon as a raider. Insects, small vertebrates and their eggs are also taken occasionally.

Vocal and visual communication among vervets is highly developed, quiet coughing or gargling noises are associated with keeping in touch, or with the young calling for attention and these are often combined with some facial pouting or protruding of the lips. Vervets often show off to other vervets by strutting around the other animal with the tail held aloft, looking over their shoulder and presenting their bright blue scrotum to view. Similarly, if a male vervet faces the object of his attention standing on his hindlegs, the exhibition of his genitalia becomes a prominent display which is often visually reinforced by a rapid bobbing movement of the red penis in a ritualized erection. The performance has been called the "red, white and blue" and the implication of flag waving adds to the appropriateness of this name.

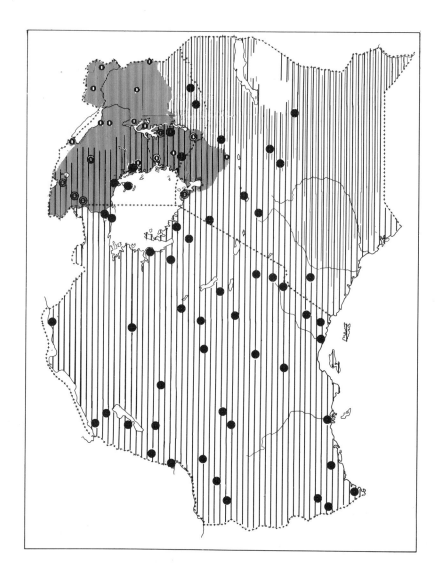

C. aethiops in East Africa:
broad stripes = *C. a. pygerythrus*.
double stripes = *C. a. arenarius*.
fine stripes = *C. a. tantalus*.
double symbol = *"hybrids"*.

▶

214

The female also may exhibit these colours in a miniature form but only while in oestrus. At this time the clitoris becomes prominent and red like the glans of the penis and the surrounding vulva acquires the blue colour of the male scrotum (Wickler, 1967). This author's discussion suggests that the question of mimicry must be raised and discussion of the function of the vervet's genital signals is inseparable from this question. The first possibility to consider is that the female is imitating the male's genital colouring. Struhsaker (1966) considers that hierarchy may facilitate coherence and stability by driving the group into closer formation (as appears to be the case in baboon society) and an implication to be drawn from this study is that the genital display helps stability. Aggression is a predominantly male attribute, but if it were to confer an advantage on the animals' society as a whole, signals with an aggressive connotation might be shared by and benefit both sexes. It is not physically impossible for the female genitalia to mimic the male for the period of oestrus, and in some monkey species females in oestrus may become noticeably more aggressive. It is however difficult to see the advantages for vervets, and they do not in the wild display any behaviour that might support such a theory.

The vervet's reputation for fighting has arisen from observations on caged animals, where they are noticeably more excitable than most forest guenons; they also display a greater propensity for fighting in the wild. However, if excitability and aggressiveness are a specific trait then the prevalence of this trait is inversely related to the ecological and social success of this species, for there is a well-documented and significant inversion between aggressive behaviour and troop numbers. Struhsaker (1966) found twice as many aggressive incidents in a troop of 16 monkeys than in a troop of 27, furthermore Lolui Island vervets, living in an ideal habitat at high densities free from both extra specific competitors and predators, show still less intraspecific aggression, so that supposed social advantages of aggression scarcely fit the facts.

The development of highly coloured genitalia in both sexes and the association of that colouring with the most attractive period of the female's cycle also seems to contradict the theory. The contradiction is evident also in the interest that both sexes and particularly young animals show in the male genitalia; in friendly encounters an animal approaching a male to groom him may muzzle or touch the scrotum and the groomed animal may have erections. In interactions between groups, the adult males stand or strut about displaying their most conspicuous "red, white and blues", but they rarely fight, while the younger males with inconspicuous genital signals are frequently wounded in fiercely aggressive encounters. On this evidence it seems likely that the colouring of the scrotum and penis of the male vervet is a highly exaggerated imitation of the female vulva and clitoris while in oestrus (Wickler, 1967) and that the function is to dampen down aggression by countering threats with an overridingly attractive signal. In this respect the species may resemble the red colobus, *Colobus badius* and the neglectus monkey, *Cercopithecus neglectus*.

Very brightly coloured genitalia in male monkeys have some correlation with ground living, for the mandrill, the patas monkey and the vervet all show a great development of this feature, while other apparently incipient ground dwellers, *Cercopithecus neglectus* and *Cercopithecus l'hoesti*, also have coloured genitalia.

If the greater propensity of vervets to threaten is related to increased danger while on the ground, and if predator selection has played any sort of role in the evolution of vervets, then the Lolui Island population offers an interesting situation, for here the animals live in the highly abnormal state of lacking any form of predator. These monkeys will chatter at humans, bob their head and shoulders, flicker their brows, grimace and stand on their hindlegs; all of which may be observed in vervets elsewhere but with the difference that on Lolui these threats are conducted from a distance of a few feet, instead of from the safety of some distant tree. It seems that an innate threatening behaviour pattern is stimulated by the presence of a large alien creature but the pattern is not tempered by any elements appropriate to a real predator, so that the little animals' confident threats absurdly contradict their patent vulnerability.

When on the defensive vervets conduct threats from a crouching position, often with glances from side to side, these glances sometimes merge into head shaking and they seem to function as a signal for aid, as other vervets may join the defensive animal and threaten or attack its opponent. Like the baboon, these concerted displays or attacks suggest a capacity for co-operation in this species that is not so noticeable in forest guenons. However, fighting is not uncommon, particularly between different troops; one rather poignant incident involved a strange vervet attempting to join a troop; after being twice thrown into a river by some adult males the animal was taken by a crocodile.

Troop numbers range between about 6 and 60 with a mean average of 20 to 30 animals. Sexes are generally evenly balanced with two or three juveniles for every adult. Troop size, density and the size of a territory vary widely and are determined by the habitat, food supply and often by the degree of human tolerance. A sacred grove may provide immunity to a large troop of very bold monkeys in much the same way that macaques and langurs find shelter in temples and holy places in India. In such situations their daily and seasonal movements are generally very restricted. Where it inhabits riverine vegetation it may range up and down a long very narrow strip with occasional or seasonal forays outside; in these circumstances an area measurement of their range is scarcely realistic or helpful.

Lake-shore populations swim readily, and I have seen a fleeing vervet deliberately taking to the water in spite of the presence of crocodiles. An adult vervet has been seen carried away by a martial eagle (*Polemaetus bellicosus*), leopards and pythons are also important predators. Caracals, baboons and possibly servals may kill them occasionally. There are probably greater dangers during the dry season as cover is greatly reduced by drought and fire.

Crop raiding has led to a heavy annual slaughter of vervets by poisoning, trapping and hunting, but in recent years their value as research animals has made live trapping profitable in areas where monkey dealers operate, and this species is in economic terms a most important animal export from Uganda at the present time.

Competition with other monkeys may occur in marginal areas where conditions allow other species to share the habitat with the vervet. Baboons are found over much of the vervets' range and the patas monkey may also be

C. a. tantalus
(Toro Uganda).

found in parts of its northern range. Redtails, *Cercopithecus ascanius*, share
their habitat in parts of Uganda and *Cercopithecus mitis* in Tanzania and
Kenya. Aggressive displays and fighting may be directed not only towards
other vervet troops but also towards redtails and mitis monkeys, but loose,
temporary and amicable associations between species are also not uncommon,
particularly in localities where the monkeys habitually share the same trees
and food resources.

Vervets resemble forest guenons in that males are only slightly larger than
females and they do not show marked sexual dimorphism as patas or baboons
do. Sexual behaviour resembles that of other guenons, the female often

initiating mating by presenting. This is commoner while in oestrus. The gestation period is 140—160 days.

This monkey breeds throughout the year but birth peaks have been observed in a number of localities. Struhsaker found a birth peak at Amboseli in November, and according to a professional trapper in Uganda there is generally a birth peak between July and October around the shores of Lake Kioga; however, births appear to be dispersed in one year and more concentrated in another—an investigation of the factors responsible for irregular patterns of breeding would be interesting.

218

In captivity young vervets have been observed to grow perceptibly faster in their first few months than do other guenon species and it is possible that this is related to the species' greater vulnerability. The young suckle from both nipples at once; they may be carried or groomed by other adult and juvenile females but, in contrast to baboons, adult males are not generally tolerated near the infants. The animal is sexually mature at two or two and a half years.

Redtail Monkey, White-nosed Monkey (Cercopithecus ascanius schmidti)

Family	Cercopithecidae
Order	Primates

Local names

Nakabugo (Luganda), Nkunga (Lunyoro, Rutoro, Lukiga, Lubwizi), Nkembo (Lusoga), Nkende (Lukonjo, Kibondo), Ikondo (Luhya, Ragoli).

Measurements
total length
1,080—1,380 mm (males)
930—1,185 mm (females)
head and body
480—515 mm (males)
340—485 mm (females)
weight
3—6 kg (males)
1·8—4 kg (females)

Redtail Monkey,
White-nosed Monkey

The redtail monkey is the most easterly representative of a group of forest monkeys characterized by nose spots or red tails; all of them are relatively small, exceptionally active and fast and make bird-like calls.

In the discussion of guenon evolution (pp. 204—208) it was pointed out that *Cercopithecus ascanius schmidti* is an extreme form, having the brightest colouring of its group (see colour plate p. 206) and the most geometric facial pattern (see drawing opposite); this latter feature has been brought about in phylogeny by progressive elaboration of a hair parting on the cheeks.

This particular race is found from the Central African Republic eastwards to the Mau Forest and south to the Mahari Mountains. It has probably expanded its range in East Africa relatively recently. Other races occupy most of the forested areas south of the Congo River, while closely related species, *Cercopithecus petaurista*, *Cercopithecus erythrotis* and *Cercopithecus cephus*, replace *C. ascaniis* in West Africa where they have colonized the entire forest zone. Eleven races are contained within this complex of species with no significant overlap of range between forms.

They seem to occupy with success a special niche in which their visually oriented communicative system, smaller size and great activity confer on them an advantage over heavier slower types. They have evolved an emphatic visual signal system and, perhaps correlated with this development, they make fewer and less loud noises than other guenons, depending instead on very bird-like chirps and twitters.

C. a. schmidti appears to be the most numerous, most densely populous and most generally distributed monkey in the greater part of southern and western Uganda, particularly in the recently formed secondary forests where it seems hardly any other guenon species can compete with it as a colonist.

Redtails also inhabit montane, lowland, riverine and swamp forests and also lake-shore thickets. Within the forest, redtails favour areas of secondary growth, river courses and the margins, they are only attracted into the top canopy by fruit or flowers and generally prefer the middle layers. Where the forest is expanding, colonizing and secondary trees provide ample food and cover, but where the forest is in retreat before cultivation, "shambas" may be raided regularly. Crops taken are banana, millet, maize, bean, pumpkin, pineapple and passion fruit, they also dig up roots. The bulk of redtail diet in the forest is made up of the leaves, shoots or fruits of many species of trees and shrubs and the flowers of *Markhamia* and *Spathodea*. In ironwood forests, redtails presumably subsist very largely on the leaves, shoots and fruit of *Cynometra*. Comprehensive lists of food plants are given in Haddow's monograph on this species (1952), and he mentions ants, gum and grasshoppers are also eaten.

The period of greatest activity is during the early morning and late evening. On Buvuma Island these monkeys were reported to have learnt to make nocturnal raids and it would be interesting to know whether this was an

Cercopithecus ascanius:
black = *C. a. schmidti*.
flicks = *C. a. whitesidei*.
stripes = *C. a. katangae*.
spots = *C. a. ascanius*.

extension of the evening or early morning activity. There is a break in feeding during the middle of the day, otherwise the animals generally continue to eat over a lengthy period. An animal eats about half to one kilo at a meal, and Haddow records one juvenile as containing its own body-weight in food. When raiding "shambas" they feed in a less leisurely manner and the cheek pouches may be filled with food. Haddow never found the commonest food-stuffs (leaves and shoots) in the food pouches, but frequently found fruits, aromatic kernels of *Pycnanthus* and the greatly relished *Markhamia* flowers. It has been suggested that these foods are retained as much for the flavour as for rapid consumption, particularly as the stones and kernels of the fruit are later spat out. Captive redtails like to keep all sorts of coloured odds and ends, sweets and chewing gum in their pouches. The monkeys feed quietly, occasionally stopping to peer around (this is more obvious while raiding).

This monkey is very much in evidence in many Uganda forests where its chirps and shaking of the vegetation are generally the first signs of its presence. An explosive warning cough made by male redtails has some resemblance

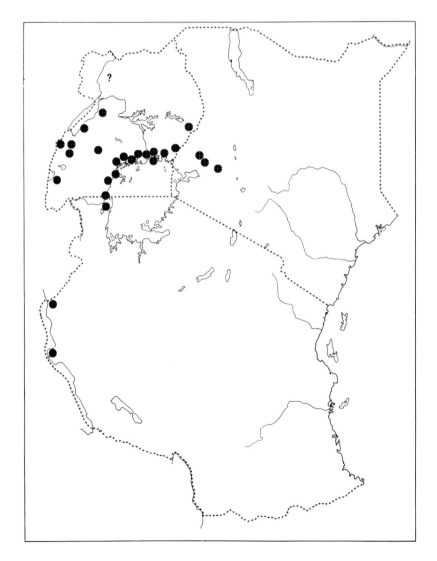

with that of *Cercopithecus mitis* but lacks its carrying power, otherwise the vocal repertoire is rather dissimilar. No other loud noise is made and the chrips and twitters mentioned are elicited by any mild excitement or alarm. When undisturbed, very faint calls may be heard and females call to their young with high-pitched whistles.

When suddenly alarmed, adult animals sometimes hurl themselves into tangled growth, obtaining a maximal disturbance of the vegetation and creating thereby a visual and auditory warning as effective as any call or postural signal. Sometimes redtails remain silent and hide in dense vegetation; this is most noticeable in areas where they are frequently hunted. Where dogs are not used and the undergrowth is thick, redtails may descend to the ground and flee.

Very large numbers of redtails, even numbering hundreds have been seen together, but usually these aggregations centre on some very local source of food. Furthermore the animals tend to disperse into smaller units with the decline of the fruit or the harvesting of crops. The size of the grouping can therefore be correlated with food supply and this suggests a very plastic social organization.

The character of redtail territorialism remains undefined, the willingness of small groups to coalesce into larger groups and, perhaps, these in turn also to join still larger groups suggests that the species is well-adapted to disperse and reform on a sliding scale of social units in response to fluctuating opportunities offered by the environment.

Haddow and Buxton have found changes in the composition of social units in the course of a single day, the smallest groups being "family parties" of four to five individuals always containing a single male. These groupings are thought to represent a distinct sleeping pattern in this species, but larger groups of twenty to thirty redtails may also sleep in two or three trees wherein separate sleeping parties are not discernible. The trees used for sleeping are seldom occupied for more than a night or two in succession, but a limited number of trees within the home range may be used, so that a constant cycling or changing of sleeping localities is evident.

A group of redtails can usually be predicted to be present within a limited locality and the suggested range for a troop has been estimated at about 130 hectares.

Solitary individuals are a common occurrence in redtails. Haddow found that the majority of these were males of all ages and classes and suggested that "the breeding male in the company of his females day after day may gradually fail in strength, energy and sexual potency and may then be replaced by another, possibly a solitary one". A strong sexual drive is not very evident in redtails, Haddow and Buxton had failed to see copulation even after several years in the field, so that the male redtails' sexual initiative seems unlikely to act as a social cement. The male may in some sense be the active agent of the group, but it is also possible that the females and young attach themselves to males and that in the amalgamation of groups the females may realign or change their attachments. When dispersal follows, some males may find themselves without followers and remain solitary until some social accretion allows them to become once more the centre of a group; perhaps males may also become accustomed to a solitary existence. The majority of solitary females, on the other hand, were found to be either very old or in an advanced state of pregnancy, and as captive mothers with newly-born young are severely harassed by the constant handling and interference of other monkeys, it is possible that in the wild some females may withdraw themselves as birth approaches. Notwithstanding this I have twice recorded females with very young and newborn babies running with the troop. I collected one solitary male in Bwamba probably forced to live on its own because it was unable to keep up with normal monkeys. This young adult animal was disabled from really rapid movement through trees by an abnormally humped condition of the back. On dissection this appeared to be the result of a broken back which had subsequently healed and fused two vertebrae together at an angle. It seems extraordinary that a monkey should in the first place break its back and then survive.

When disturbed a troop will often make off rapidly while one male, generally hidden in thick cover, continues to herald every movement made by the intruders by making the explosive cough mentioned earlier. This call was being employed with typical threat postures by a male redtail which, with several mangabeys, was mobbing a leopard from very close quarters and presumably in full view of the animal itself—so that hiding from people may be a response to man's capacity for killing with arrows and guns.

While performing his warning role the male may become separated from the troop by some distance, Haddow suggests that this behaviour may show the "presence of a master male in one aspect of behaviour" and it would be interesting to gain more information on the nature of this behaviour polarizing as it does on a single individual, for indications of a hierarchy are not perceptible in other aspects of redtail behaviour. No fights have ever been observed and scars are rare on males. All sexes and ages of redtails bob the head with raised hind-quarters, lash their tail and make staccato chirps, this threat display is commonly directed at people, dogs, cattle and sometimes other monkey species; however threats have not been seen between redtails themselves except very rarely in captivity. An isolated component of this display, raised hindquarters with lashing tail, may sometimes be seen when wild juveniles play but does not seem common in adults. A young male reared "en famille" by Miss H. Lock lashed its tail above the head when strangers arrived in the compound or after fits of activity, mild annoyance or frustration. Whenever it was smacked, or when an object was taken away from it the same animal would, when young, walk backwards, screaming and lashing its tail. When older (subadult) it would walk backwards towards strange people or animals (and occasionally towards male members of the family) and when close it would reach between its legs with a hand; this odd behaviour appears to be very similar to that displayed by *Cercopithecus neglectus*. Like many captive monkeys this animal appeared to make distinctions between men and women and he reserved an excited head-bobbing exclusively for Miss Lock and young women of similar complexion. The sexual distinction may in this instance be more apparent that real, as imprinting of his foster parent may have influenced this display.

In contrast to the up and down bob which is frequently associated with excitement in wild redtails and is probably elicited by a variety of stimuli, there is a wagging of the head, this gesture may be similar to that found in

Cercopithecus aethiops, as it is typical of "lost contact" and may derive from a side to side searching movement. The circular form of the redtails' facial pattern with its brilliant white spot is appropriate for versatile all-direction flagging or signalling and may be contrasted with the predominantly vertical pattern and movement of *C. neglectus*.

In common with other cercopithecus monkeys the female redtail has no external signs of her cycle except for menstrual bleeding; she also exhibits very little overt sexual behaviour.

The exact gestation period has not been determined but probably resembles that of *C. mitis*, about 120—130 days. There appears to be a birth-peak between May and September so that mating would tend to coincide with the driest period of the year, between December and April. The redtail seems to have a high threshold of sexual arousal and if, as Haddow suggests, mating is primarily nocturnal, it is possible that wet conditions might discourage sexual activity during the rains. Also generally optimum feeding conditions for the redtail might be expected for the six or seven months preceding December.

The newly-born infant is covered in rather woolly grey fur, a nose spot is perceptible, but the side-whiskers are not apparent and the tail is dark. After some three weeks the coat starts to differentiate, rufous hairs appear on the underside of the tail and by about three months the animal is a smaller, paler and duller replica of the adult. The juvenile's tail is often corkscrewed around long twigs and also seems to be a useful counterbalance to the rather clumsy movement of young animals.

Redtails often associate with *C. aethiops*, *C. mitis*, *Colobus polycomos* and *Cercocebus albigena* when feeding. A wild, fertile *C. mitis stuhlmanni-C. ascanius schmidti* hybrid is known (P. A. Blake, 1968). As the hybrid animal was with a troop of blue monkeys it is probable that the father was a redtail. The most intimate associations appear to be with *Cercopithecus mona denti* and mixed bands appear to intermingle and feed together. When disturbed, such mixed parties may still remain together, but I observed one mixed group become thoroughly alarmed when pursued by hunters, whereupon the two species followed different escape tactics: the redtails remained in groups but the mona dispersed very much more rapidly, scattering as individuals or pairs.

The possibility that this species has influenced the distribution of *C. neglectus* is discussed elsewhere and the suggestion is made that the redtail represents a recently evolved and very successful species of monkey. It is therefore interesting to find that in the seasonal swamp forest of Sango Bay, in contrast to its abundance in comparable forests elsewhere, the redtail, though present, is relatively scarce while the silver monkey, *Cercopithecus mitis doggetti*, the mangabey, *Cercocebus albigena*, and the colobus, *Colobus polycomos*, are abundant. This forest shows signs of faunal and floral isolation and it is possible that the redtail is a relatively new arrival which has not yet effectively established itself there. It is also possible that extensive annual flooding or some other local condition may support a disease that is inimical to this monkey, or one to which it is not as fully adapted as are the other longer established primate species.

The hair of this monkey has frequently been found in pellets of crowned hawk-eagles, *Stephanoaetus coronatus*, and in the dung of leopard. R. White

(personal communication) observed a leopard make a kill in Uganda. The animal apparently waited until some redtails were in a relatively isolated tree, it then rushed up the trunk, panicking the monkeys into dropping out of the branches and then leapt out of the tree in time to bat with its paw at a falling monkey, hitting it before it reached the ground. In the Mabira Forest where redtails are numerous and only two other monkey species occur, several pairs of crowned hawk-eagles are known to nest in a relatively limited area of the forest and it is probable that redtails are their principal diet.

Haddow (1952) remarks "in many areas, notably Bwamba county, monkeys are still eaten by a large section of the populace and in that area monkey meat is still sold at a higher price than beef, mutton, goat flesh or pork. The writer has tried the flesh of various species of cercopithecus and colobus and has found it excellent".

The redtail is most important in the epidemiology of yellow fever. The majority of redtail monkeys in areas of lowland forest show evidence of having been infected with yellow fever, while other monkey species are also reservoirs of yellow fever. The habit of crop raiding suggests that the redtail is the principal agent in spreading this monkey disease to man, the vector being a mosquito, *Aedes simpsoni*, commonly found around settlement. Haddow (1952) gives a detailed account of this monkey's relation to yellow fever and other virus diseases. It is possible that the redtail is also a reservoir of filariasis (Loa loa), through tabanid flies of the genus *Chrysops* which bite both monkeys and man.

This monkey has been used as an experimental animal in the study of virus diseases at the East African Virus Research Institute and may in future prove to be a useful laboratory animal. At present the species is treated as vermin.

L'Hoest's Monkey, Mountain Monkey
(Cercopithecus l'hoesti)

Family Cercopithecidae
Order Primates
Local names
Nyaluasa (Lukonjo), Engende
(Lutoro), Embeya (Lukiga).

Measurements
total length
1,700 mm
head and body
545—700 mm (males)
460—515 mm (females)
tail
480—800 mm

L'Hoest's Monkey, Mountain Monkey

The mountain monkey is a very dark, richly coloured animal with a pure white chin ruff framing the face. The body is black with a grizzled mahogany back, the lips and nose are black on a pale grey face. The broad nose has a black heart-shaped "spot" of dense hair on it, which is visually obliterated by being incorporated within a black facial triangle. The males are larger than the females and have bright mauve testes.

This race is an eastern population of the Cameroon mountain monkey, *Cercopithecus l'hoesti preussi*. This monkey probably developed from an ancestral stock resembling *Cercopithecus erythrogaster* and *Cercopithecus nictitans martini* during a period when the conditions that are now restricted to the Cameroon mountains were more general. Its montane distribution resembles that of a number of other mammals that are today restricted to the Cameroons and the East African mountains. In common with these species its eastward spread might have been connected with a generally cooler period.

C. *l'hoesti* distribution:
black = C. *l'hoesti*.
stipple = C. *nictitans nictitans*.

Mountain mammals are often replaced by other closely related species at lower altitudes and their present isolation on widely separated massifs can only be explained by a very great extension of montane habitats or cooler conditions in the past. During these periods they were probably ascendant over the lowland species which are adapted to warmer temperatures and lower altitudes. The montane species' subsequent elimination from the intervening lowland is presumably due to rising temperatures tipping the balance in favour of the lowland competitor. This pattern seems to fit the distribution of several mountain species. The pattern is particularly interesting for C. *l'hoesti*, because while both races are montane the eastern population, *Cercopithecus l'hoesti l'hoesti*, also occurs in lowland forest as far west as Kisangani, while *Cercopithecus l'hoesti preussi* is strictly confined to montane forests in the Cameroons. The closely related C. *nictitans* inhabits the intervening area, and the most easterly point at which this species is known stops short of Kisangani. This species presumably occupies a niche analogous to that of C. *l'hoesti* which may therefore be excluded throughout the lowland range of C. *nictitans*. The presence of *Cercopithecus mitis stuhlmanni* in the eastern Congo is perhaps the principal reason why C. *nictitans* extends no further east. This species, while not greatly different from C. *nictitans* is very distinct from the highly elaborated C. *l'hoesti l'hoesti* which has differentiated more than the western race; consequently the question of competition with a lowland relative does not occur. The species has therefore held its own in lowland forest, where it is locally common over a wide area, while its most conspicuous success is in the mountains.

Most of the East African populations occur in the montane areas of western Uganda, but the species is also found in Kibale, Kalinzu and Kayonza which are medium altitude forests. In these areas the species prefers the thick regenerating growth in felled compartments. Haddow (1952) reports that parties regularly come down from the montane forests of the

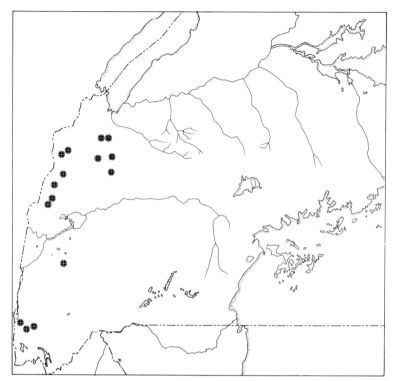

West Uganda. Distribution of *C. l'hoesti*.

Ruwenzori to raid fields in the foothills, descending over 600 m and even return twice in the course of a single day. The bamboo forest zone seems to define the upper altitudinal limits of the range. Ground feeding is common and the species will venture a long way across open country, wooded savanna, heathland and cultivated fields. They eat the fruit of *Podocarpus*, *Pygeum*, *Hagenia*, *Polyscias fulva*, *Myrianthus arboreus*, *Maesa lanceolata* and other montane trees, also the leaves of various trees, shrubs and herbs. In the undergrowth they eat the fruit of *Rubus*, *Aframomum*, *Eugenia* and *Cucurbitaceae*; the sporangium of *Pteridium aquilinum*, bracken shoots and mushrooms are common foods. *Albizzia* resin is relished, insects and lichen are also eaten. Bananas, peas and cassava attract the mountain monkey to the fields and they have acquired a taste for the fruit of the Australian black wattle.

The dense, tangled undergrowth which grows beneath the broken canopy of the mountain forests determines much of the monkey's behaviour in this habitat, and in lowland forest it seems to prefer comparable areas of regeneration or secondary growth where the canopy is broken or non-existent; this preference may be due to less competition from other monkeys, as many of its food plants are in the undergrowth much of its time may be spent on the ground or in the lower growth. They associate with other species and I have seen a mixed band of *C. l'hoesti*, *Colobus abyssinicus* and *Cercopithecus ascanius schmidti*. They are rather quiet monkeys but make low chirping noises not unlike those of the redtail; two individuals meeting make a low mutter, described by Møller as short "puffs". A sharp bark of alarm similar to that of other forest monkeys is also made.

230

The only field study made on this species was conducted by Møller in the Kayonza Forest and most of the information given here derives from his observations. He found solitary males were not infrequently met with in the vicinity of larger groups, parties were generally not large and it was reckoned that 17 individuals constitute an average troop. Møller estimated the range at 7—10 sq km, this probably varies a lot regionally and their lengthy raiding trips suggest that they are more mobile and adventurous than the more strictly arboreal monkeys. One, or more rarely two adult males were found in a troop. Of East African monkeys this species seems to bear the closest resemblance in behaviour and social structure with the redtail, *Cercopithecus ascanius*.

The night is usually spent in one of a limited number of favourite sleeping trees, which are often situated in higher locations, while daily foraging may take the monkeys down to the valleys for much of the day. Males although occasionally aggressive are usually very tolerant and the leading male in a troop will allow young animals to take food away from him. They are sociable and spend much time grooming, an activity in which the neck, chin, cheeks and armpits are the most attractive parts of the body. As these parts are hidden in the aggressive posture, the white signal patch on the throat would appear to have an "opposite meaning" for *C. l'hoesti* to that of the white beard of *Cercopithecus neglectus*, which has an aggressive connotation and in which species the belly and chin are deliberately exposed in an aggressive display. Grooming partners generally spend as much time grooming as being groomed, but adult males are groomed more than any other age or sex class. The species

is very alert to danger and individuals constantly climb vantage points to peer around, this behaviour may be a response to the greater risks involved in ground dwelling. Adult males take up the rear when there is an alarm and they tend to stay on the periphery of the group.

Predators might include leopard and golden cat, as these carnivores are common in their habitat. The handsome skin of the mountain monkey is sought after as a shoulder bag by the Bakonjo and the Bakiga, and the former organize elaborate hunts with packs of dogs. The difficult terrain does not make these hunts an easy matter and the species is certainly not threatened by this practice, but it does perhaps explain why the animal is generally very wary.

232

Details of the reproduction and sexual life of this species are imperfectly known. There is some evidence that a limited birth peak may occur in February. The newly born baby is brown and acquires the adult colours and contrasts gradually over the first two or three months. It clings to the mother's belly or rides on her back and often intertwines its tail with that of the mother.

C. m. moloneyi

**Mitis Monkey,
Blue Monkey,
Sykes Monkey
(Cercopithecus mitis
includes
Cercopithecus mitis
albogularis)**

Family   Cercopithecidae
Order   Primates
Local names
Kima (Swahili), Sengwa (Kikami),
Nkima (Luganda, Runyoro),
Cheptjegayandet (Sebei), Nko
(Kuamba).

Measurements
(vary with subspecies)
total length
1,070—1,520 mm
head and body
440—670 mm
tail
630—850 mm
weight
6—12 kg

Mitis Monkey, Blue Monkey, Sykes Monkey

Races

Cercopithecus mitis stuhlmanni	Uganda and W. Kenya
Cercopithecus mitis kandti	Kigezi Volcanoes
Cercopithecus mitis schoutedeni	Kigezi Volcanoes
Cercopithecus mitis doggetti	S.W. Uganda and N.W. Tanzania
Cercopithecus mitis moloneyi	Southern Highlands
Cercopithecus mitis monoides	S.E. Tanzania
Cercopithecus mitis albogularis	Zanzibar Island
Cercopithecus mitis kibonotensis	N.E. Tanzania
Cercopithecus mitis kolbi	E. Kenya
Cercopithecus mitis albotorquatus	N. Kenya Coast

Range of *C. mitis*.

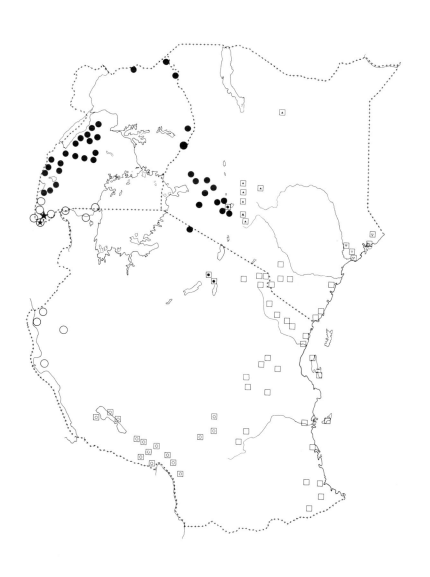

black star = *C. mitis kandti*
white circle = *C. m. doggetti.*
black circle = *C. m. stuhlmanni.*
white circle in
    white square = *C. m. moloneyi.*
white square = *C. m. albogularis,*
        kibonotensis and
        monoides.
black triangle in
    white square = *C. m. kolbi.*
white triangle in
    white square = *C. m. albotorquatus.*
double symbols = *"hybrids".*

This species is very widely distributed and its many geographical races show marked differences in coat colour, size, whisker arrangement and length of the fur. It is the only forest guenon with a wide distribution in southern and eastern Africa, climatic changes having isolated many of the forests of this region and closed them to those cercopithecus species which have developed more recently. Lengthy isolation has therefore been one factor in encouraging the diversity of this group and some taxonomists have split the eastern population into a separate species, i.e. *Cercopithecus (mitis) albogularis* (Dandelot, 1968), *Cercopithecus albogularis* (Sykes, 1831; Hill, 1966).

The broken topography of East Africa causes physical isolation and ecological diversity which has its reflection in the appearance of some populations. For instance, there is a correlation between darker, redder, thicker coats and higher, wetter, more humid forests, while monkeys in dry, hot areas tend to have thinner, paler coats with less red. In addition to this diversity, climatic changes have brought formerly isolated populations together again and interbreeding between races of this species can be demonstrated in several localities (differences in size and colouring do not seem to be a deterrent to interbreeding within the species). Indeed a wild hybrid between this species and the redtail monkey *Cercopithecus ascanius* is also known and hybrids with other cercopithecus species are common in captivity.

The voice, anatomy and general disposition of *Cercopithecus mitis* are recognizably the same throughout its range, which includes forests of many types growing at every altitude; coastal mangrove swamps and high altitude bamboo forests represent extremes.

All these variables and the relatively greater age of the species in relation to the other forest guenons have complicated the picture of *C. mitis* dispersal. However, its distribution pattern is similar to that of other forest mammals in East Africa.

The species as a whole is typified by the race *Cercopithecus mitis doggetti*. This race is found in lowland and montane forests in the area between the lakes Edward, Kivu, Tanganyika and Victoria. In view of its central geographic position and a radiating pattern of *C. mitis* populations all showing more similarities with this form than with one another, it is reasonable to suggest that this form might bear the closest resemblance to an ancestral type. *Cercopithecus mitis doggetti* has a grizzled grey back varying from ash to tawny grey, the arms and feet are an intense shiny black and the legs are dark grey with some grizzling on the upper thighs. Long grizzled whiskers on the cheeks encircle the face and greatly enlarge its visual impact. A black cap creates a sharp contrast with the pale, grizzled brow patch. Grizzling occurs in many individuals under the black tips of the hairs on the cap suggesting that the black has developed in order to emphasize the movements of the brow and facial expression. The brow patch is an important signalling device for communication among guenons and this feature will be discussed further in relation to other races of this species.

Cercopithecus mitis stuhlmanni has an almost melanistic darkening of the coat but otherwise bears a close resemblance to *C. m. doggetti*. In common with some other forest mammals, this race until recently had a continuous distribution from Uelle in the Congo to the eastern rift wall, and the stabilizing effect of gene-flow can be perceived in an absence of variability in what

236

Left : C. m. stuhlmanni.
Right : C. m. albotorquatu

Left : C. m. doggetti.
Right : C. m. kolbi.

Left : C. m. schoutedeni.
Right : C. m. monoides.

Left : C. m. kandti.
Right : C. m. moloneyi.

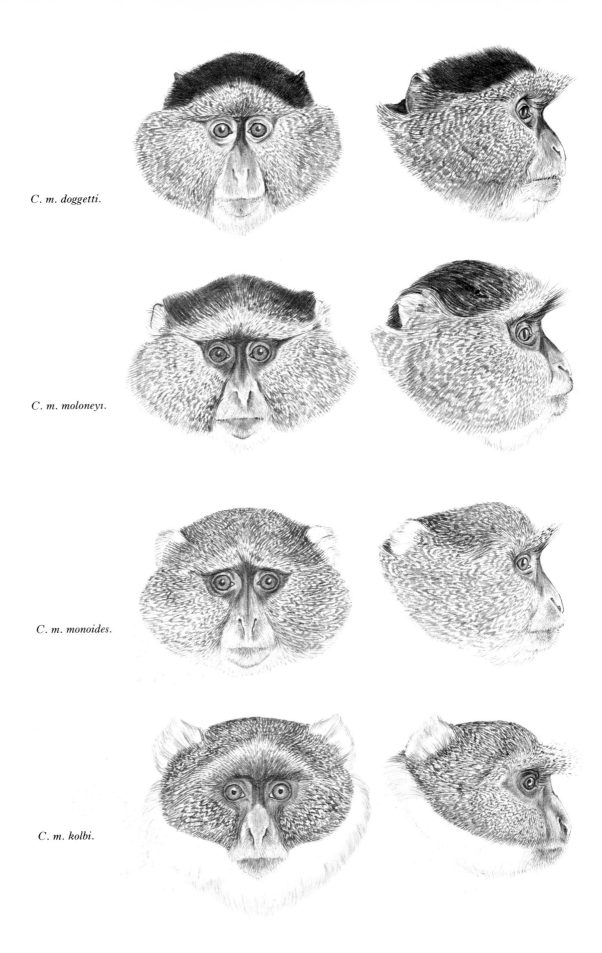

C. m. doggetti.

C. m. moloneyi.

C. m. monoides.

C. m. kolbi.

are now isolated populations. Some *C. mitis* monkeys in the forests south of Lake Edward are somewhat intermediate in colouring between *C. m. doggetti* and *C. m. stuhlmanni*. Whether this is due to a clinal gradient or interbreeding between formerly isolated races is difficult to say, but confused field identifications of monkeys from this area (Stott, 1960) have led to the belief that the races are sympatric in this area (Hill, 1966). This is not the case.

To the south, Lake Tanganyika divides populations, *Cercopithecus mitis opisthosticus*, a race with a pale, steel-grey back is found to the west and does not occur in East Africa. Down the eastern coast of the lake, 400 km of dry woodland and savanna separate *Cercopithecus mitis moloneyi* from *C. m. doggetti*. A long-standing break between these two populations is evidenced by an absence of intermediate forms. The chief differences are an extension of grizzling on the hindlegs, the cap and the shoulders. The red colouring typical of *C. m. moloneyi* may be correlated with the high humidity and rainfall of the Southern Highlands' montane forests; as monkeys from a population in the drier, lower altitude forest at Mafwamera have olive backs. In *C. m. moloneyi*, grizzling on the cap dissolves the sharp contrast between the black cap and brow patch found in *C. m. doggetti* and *C. m. stuhlmanni*, instead the lower part of the brow patch has become almost white, while the skin colour on the face has darkened (see opposite). In this way a strong contrast above the eyes is preserved. *C. m. moloneyi* merges with *Cercopithecus mitis monoides* and *Cercopithecus mitis kibonotensis* in a cline in which tonal differences between the brow and cap disappear. Correlated with this development is an increasing emphasis on the white collar and ear tufts and the appearance of a distinct brow bonnet, an ingenious arrangement of very long brow hairs that alter their direction with small changes of facial expression. These long sparse hairs reflect light and therefore look paler than the surrounding hair, thus there is some optical compensation for the absence of a pigment contrast. A random decline of features which serve to communicate expression has led to an equally random development of alternative features which change the total effect drastically, but which serve the same purpose. Thus, *C. mitis* populations along the whole eastern seaboard of Africa have distinct collars or ear tufts and bonnets which give these monkeys a distinctive and, by human standards, rather surprised expression.

The greatest colour contrast is achieved by the monkeys occupying the forests on Mt Kenya, *Cercopithecus mitis kolbi*. These have long, bright ear tufts and a broad snow-white collar against black shoulders. Correlation between wet mountain forests and colouring is again apparent in their dark, red backs; these become lighter in colour with decreasing altitude. Along the North Kenya coast the back is a pale yellow colour and the hindlegs and shoulders are dove-grey, *Cercopithecus mitis albotorquatus*. On Mt Kilimanjaro and on Mt Meru, the mountain populations of *C. m. kibonotensis* also show darker and redder coats than those of the surrounding lowlands, although to nothing like the same degree as the Mt Kenya *C. m. kolbi*, or the Southern Highlands *C. m. moloneyi*.

These eastern populations of *C. mitis* are separated from *C. m. stuhlmanni* by the width of the eastern rift, an ecological barrier that would be readily crossed by the monkeys on either side were the niche on the opposite side not already occupied. That individuals do cross and interbreed is shown by hybrids from two localities. Individuals showing grizzled caps and reddish tints (*C. m. kolbi* characters) have been collected in *C. m. stuhlmanni* populations on the Mau. Intermediate animals, in which either bonnets or brow bands appear and showing a variety of mixed features are known from the rift wall above Lake Manyara.

Such crosses are unstable and the genotype of the odd wanderer seems to be quickly absorbed into and lost in the dominant group, but some of these mixed types do show a strong resemblance to a race found in Ethiopia, *Cercopithecus mitis boutourlinii*. During the later Pleistocene, when forests expanded into northeastern Africa, the forests were principally colonized by mammals from the Central Forest Refuge in the eastern Congo, but the Ethiopian populations might be the result of *C. mitis* stocks from two directions colonizing these northern forests with subsequent widespread interbreeding, which eventually stabilized into a type exhibiting the general colouring of the western *C. m. stuhlmanni* with the bonnet and white throat of the eastern races.

The complications created by interbreeding are nowhere more obvious than in Kigezi and the area around the Bufumbira Volcanoes; here three races have been described, *C. m. doggetti*, *Cercopithecus mitis schoutedeni* and *Cercopithecus mitis kandti*. *C. m. kandti* is restricted to high altitude and bamboo forests in the volcano area and it exhibits the reddish colouring that is so often found in *C. mitis* from montane habitats. Also found in the same habitat and possibly in the same troops are *C. m. doggetti* with pale grey backs. The third "race", *C. m. schoutedeni*, would seem to have been described from animals that are intermediate in almost every particular, and which show every shade of gradation between *C. m. kandti* and *C. m. doggetti*. The situation could represent interbreeding occurring between formerly isolated races, although it is difficult to see what the isolating mechanisms could be. *C. m. doggetti* tolerate the cold bamboo forests as well as *C. m. kandti*, so that ecological separation does not seem likely and it is difficult to posit imaginary physical barriers in an area where none are apparent. An investigation of this problem would be most interesting.

Although *C. mitis* will live under many climatic conditions it is very dependent on shade, captives show signs of discomfort if exposed to hot sun and wild populations in well-developed forest habitat will relinquish, as the sun becomes stronger, a canopy food supply that was attractive in the early morning, and the monkeys will retire into the undergrowth and shade, even if food is very much scarcer there. Perhaps this factor contributes to exclude *C. mitis* from drier habitats where the smaller *Cercopithecus aethiops* is the dominant monkey.

The diet of these monkeys alters according to the physical conditions and forest types in which they are found and from season to season, so that their feeding habits and social organization show a corresponding flexibility. In lowland forest, where a rich and consistent food supply is available, over 60% of their food is fruit, with young leaves, shoots, flowers and occasionally insects accounting for the rest. The high altitude races feed largely on herbs, leaves, bamboo shoots and berries. One individual *C. m. schoutedeni* collected in Kigezi had a stomach full of mushrooms. *C. m. doggetti* living in swamps and riverine forests frequently feed on the soft base stems of papyrus, and the

C. m. albotorquatus.

C. m. stuhlmanni.

population in Sango Bay forest also eat moss and lichen. They are fond of termites, grasshoppers, ants and they extract grubs from bark with the teeth and fingers. They have also been seen eating "cuckoo spit". In Sebei and in Kenya, *C. mitis* occasionally take maize, bananas and other crops, they are known to thieve eggs and also take young birds and ducklings but, on the whole, this species seldom raids cultivation. *C. m. kolbi* has become a menace to forest nurseries on Mt Kenya by eating the growing shoots and bark of exotic softwoods.

C. m. kolbi.

The way in which food is dispersed in the habitat influence many aspects of *C. mitis* behaviour, for instance, some populations forage on the ground or in low bushes. *C. m. kandti* are sometimes found in troops of thirty or more animals scattered through thick bamboo forest or undergrowth, seeking food mostly at ground level; the monkeys keep up a constant twittering and chirping that is reminiscent of the social mongoose, *Mungos*, which also forage noisily in thick cover. By contrast *C. m. stuhlmanni* in forest are hardly ever seen on the ground; small parties of four to six animals will feed quietly on their own and the members of a troop may only come together at a favourite fruiting tree or other very localized source of food. This is also true for

C. m. doggetti in similar habitats and, in general, social life seems to be very flexible.

The principal mode of communication appears to be vocal, consisting of eight or ten quite distinct noises, most of which are rather low-pitched and some rather bird-like. A very loud nasal cough is uttered by the adult males when alarmed and they also make a more rarely heard harsh croaking which may be associated with territorial behaviour. A very deep soft ventriloqual "boom" has been heard in wild *C. mitis* from both sides of the eastern rift and has been heard from a large captive male (Forbes Watson personal communication). Booming calls are made by several cercopithecus species but their significance is not known.

The troops appear to be very stable in membership and no behaviour suggestive of a hierarchy has been observed. P. A. Blake (personal communication) saw displays when two troops converged on a single fruiting tree. The mature males bounded to and fro with loud croaks, bouncing heavily on branches and shaking the foliage. Other members of the troop noisily threatened and chased one another but no serious wounding has been observed. However, I have seen a large male fighting a smaller individual and nearly force it to fall out of the tree. Staring and movements of the brow are the commonest expression of threat, these are reinforced by lowering the head with a more violent movement of the brows and open mouth; the teeth are only exposed in extreme stress and fear. Yawning and eyebrow flickering with sudden "uneasy" movements are a frequent reaction to being watched; this is probably a displacement activity with threat-flight impulses conflicting. Male *C. m. kolbi* sometimes sit facing an observer with their legs well parted to expose the pale blue-grey genitals.

Grooming is commoner in the larger troops and is invited with a variety of postures, the commonest of which is a half crouch with the tail over the back and the head turned away. This is also the gesture with which the female solicits copulation, but token mounting and the overtly sexual nature of "presenting" is absent, although this is presumably its derivation. The range of a troop and its density must vary very widely with the resources of their habitat and the size of the troop. A. Blake estimated that a small troop of about fifteen *C. m. stuhlmanni* ranged over one tenth to one sixteenth of a sq km with a density of about 200—300 individuals per sq km. This species of monkey was numerous in his study area, which was very rich in food plants, so that these figures probably approach the optimum.

A small troop of *C. m. doggetti* isolated in riverine forest in South Ankole was first encountered and given protection by Mrs T. Nuti in 1957, the progress of this troop was followed over a period of ten years and the diagram overleaf summarizes the natural growth of population from 15—64 individuals over this period. The graph only shows an approximation of the sex structure and does not include nine deaths, one of which was thought to be of old age; three other adults apparently died of disease and the remaining juvenile mortalities were mostly caused by falls into the river. A large male which was thought to be old in 1957 was still alive in 1968. This troop showed a great attachment to their home range and never wandered out of it. All monkeys in the troop followed well-defined arboreal "paths", which continued to be used for many years.

243

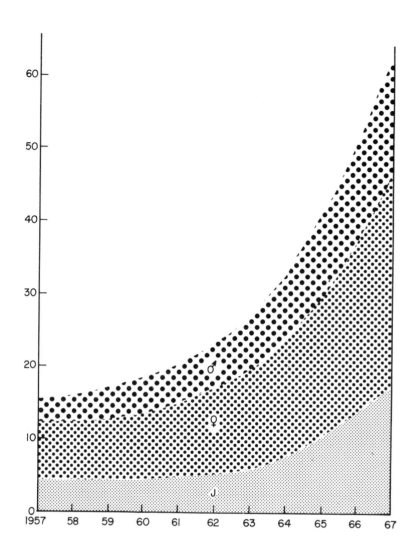

Diagram of population growth in *C. mitis doggetti* over 10 years. ▶

The crowned hawk-eagle, *Stephanoaetus*, kills *C. mitis* and it has been reported from Mt Elgon that an attack by these birds in the tree tops was utilized by a leopard that waited for the monkeys to drop out of the trees within its reach.

Through most of its range in East Africa, *C. mitis* shares the forest with colobus but competition is limited to leaves and fruit favoured by both species. When feeding on a common food supply the troops often share a tree quite amicably, but they may perform a display or make perfunctory threats on first meeting. It is difficult to assign a rank to either species, as each seems to show an equal tolerance and lack of fear towards the other, but they seldom remain together for very long. On forest margins and in riverine forest or thicket, particularly in Kenya and Tanzania, interactions with *C. aethiops* are common, threats may be exchanged and some chasing may occur, but the species not infrequently feed quietly together without animosity. *C. m. doggetti* mixes with black mangabey, *Cercocebus albigena*, in

244

Sango Bay, and with *Cercopithecus l'hoesti* in Kayonza Forest. In many forests *C. mitis* mixes with redtail monkeys, *Cercopithecus ascanius*, and this association of species is often sustained for much longer periods than that with colobus or other cercopithecus monkeys.

Two redtails have been seen in a troop of *C. m. stuhlmanni* and a solitary redtail male and a *C. m. stuhlmanni* female have been found together; a fertile wild hybrid between these species is also known (A. Blake, 1968). Occasionally the great apes seem to be an attraction for *C. mitis*. Solitary individuals of *C. m. stuhlmanni* have been frequently seen with troops of chimpanzees in Budongo. Notwithstanding this, on one occasion a young *C. m. stuhlmanni* was seen to be killed and eaten by chimpanzees (Suzuki, personal communication). Schaller (1963) frequently saw *C. m. kandti* with gorillas.

C. m. moloneyi.

In view of the easy association of *C. mitis* with other species in most parts of their range, other factors than competition must be sought to explain the distribution of *C. m. stuhlmanni* in Bwamba where it is only found in montane forest and a small area of lowland forest at the point where the spur of the Ruwenzori rises sharply out of the forest. In this small locality, Haddow found *C. m. stuhlmanni* immune to yellow fever and Semliki Forest virus but not immune to other local monkey viruses. On the mountain slopes away from the lowland forest, *C. m. stuhlmanni* was non-immune to the viruses tested. Several unique viruses have been discovered in Bwamba and the lowland forest contains amongst its extraordinarily rich fauna numerous very locally distributed tabanids and other biting insects that might be potential hosts and vectors for diseases likewise limited to the lowland forest of the Central Refuge. The absence of *C. m. stuhlmanni* and the baboon from this area of lowland forest could therefore be due to pathological factors and might be an interesting problem capable of experimental investigation.

C. m. stuhlmanni is unlikely to be implicated in the transmission of yellow fever to humans, as the species very seldom raids "shambas" and so does not come into close contact with settled areas. With the exception of Mt Kenya the species is nowhere a serious danger to human health, agriculture or forestry.

The species will not be in danger while trade in their skins continues to be prohibited. Prior to 1925 large quantities of their skins were exported and uncontrolled slaughter probably accounts for the early elimination of this species from most of Buganda and Busoga.

Breeding certainly continues throughout the year in several localities but peaks in breeding activity have been suggested for a captive population (Rowell, personal communication). A. Blake saw females that were presumably in oestrus solicit mating and groom the male after copulation.

Gestation is 120—130 days, one young is born at a time and the infant scarcely leaves its mother for the first months. Its plain grey coat begins to acquire grizzling before it is two months old, at which time it also begins to explore its surroundings. Later still the young animal may be left for brief periods with other females and juveniles while the mother feeds alone. As the young monkey becomes more independent the mother will leave it, cross gaps by jumping or running over slender saplings and sit waiting at the other side for it to cross on its own. The performance is repeated frequently at gaps and can be interpreted as learning behaviour by which agility is developed,

and the arboreal "paths" of the home range are learned. Juveniles play during the midday rest period and indulge in mock biting, wrestling, pouncing and tail pulling. In common with other forest guenons the young grow at a slower rate than in *C. aethiops*.

C. m. doggetti.

Mona Monkey, Dent's Monkey
(Cercopithecus mona denti)

Family   Cercopithecidae
Order   Primates
Local names

Mbenge (Lubwizi and Kuamba).

Measurements
total length
1,100—1,400 mm
head and body
400—700 mm
tail
700—900 mm
weight
3—6 kg

Mona Monkey, Dent's Monkey

Dent's mona monkey is immediately recognizable in the forest by the brilliant contrast of the white belly with the dark back; this contrast is reinforced along the flank by a line of silvery, almost luminous white. The inner surface of the arms also contrasts strongly with the rich black outer surface. The face has dark hair and skin forming a band between the dirty-white brow band and the yellowish cheeks and white lips. The ears have long white tufts. The back is a rich grizzled burnt umber colour, becoming yellowish on the outer surface of the hindlegs, the feet are black and the tail is grey with a black tip, there is no difference in colouring between sexes and, although the males are larger, the largest females exceed the dimensions of smaller males.

Mona monkeys are a heterogeneous group mainly found in West Africa, where they are reported to be numerous and highly successful. In the light of their success in West Africa it is difficult to understand why *Cercopithecus mona denti* the most easterly representative of the group has not spread further east than Bwamba. Mona monkeys are only found in low-lying forests and it is possible that this species has been unable to invade the higher altitude forests of East Africa, or failed to compete with well-established monkeys already there. If the subspecies *Cercopithecus mona denti* represents a recently evolved type it is also possible that the continuity of the East African forests had been interrupted before it had become really established or adapted to local conditions.

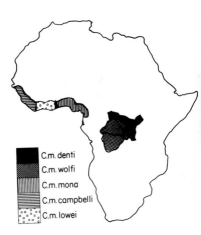

C.m. denti
C.m. wolfi
C.m. mona
C.m. campbelli
C.m. lowei

It is fairly common in the high-canopy mixed forest in Bwamba where it feeds on shoots, leaves and fruit. In Bwamba the mona is often mixed with troops of redtails, *Cercopithecus ascanius*; small parties of up to a dozen mona monkeys accompanying much larger troops of redtails have been seen on many occasions by an experienced Mwamba informant. My own observations confirm this, and the association may be common rather than exceptional.

Solitary mona monkeys have also been reported from Bwamba; one was reported to have followed a redtail group closely for several days, but was only seen to feed in their food tree after they had left it at about 9 a.m. There are many resemblances between these two species, but an interesting difference in behaviour emerged on one occasion when I observed a mixed troop that had been feeding together being chased by a party of hunters. At first the monkeys fled through the tree tops as one band, but soon the mona split off dispersing in twos and threes in different directions. They seemed to be very much faster than the redtails and are superbly co-ordinated arboreal animals, racing through the canopy at a great speed and seeming scarcely to touch the branches in their passage. Their capacity and willingness to disperse is rather exceptional and might imply a greater ability to reassemble.

The mona may freeze when there are people under them and they will stay still in the face of considerable disturbance. Even in a light canopy they are extremely difficult to see, as they crouch among knots of foliage in the

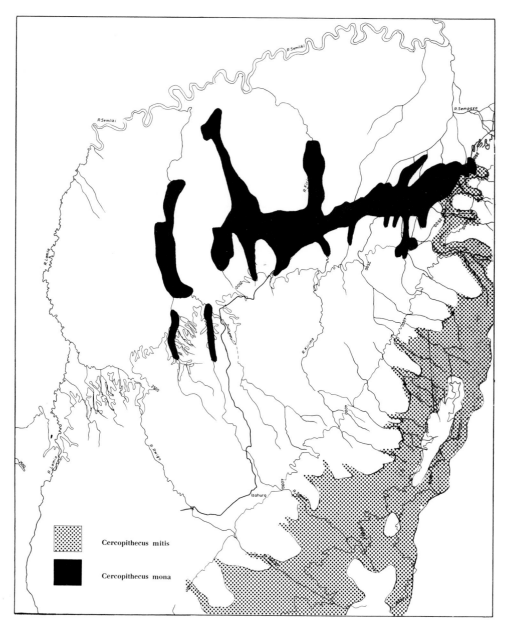

Distribution of *C. mona denti* in Baramba (see vegetation map on p. 46).

roof of the forest. They raid crops together with redtails and persistently enter the newly cleared "shambas" that are replacing their habitat.

Haddow found the liver parasite, *Hepatocystes kochi*, in mona monkeys. His tests showed immunity to yellow fever and Semliki Forest virus. As crop raiders they are implicated together with the redtail in the epidemiology of yellow fever.

Mona monkeys in zoos have hybridized with several other species including *Cercopithecus neglectus*, *Cercopithecus mitis* and *Cercopithecus aethiops*. No field study has been made to date on this species.

Brazza's Monkey, Neglectus Monkey
(Cercopithecus neglectus)

Family Cercopithecidae
Order Primates
Local names
Kalasinga (Swahili), Adu (Kuamba),
Ebubusi (Itesot), Enyuru
(Karamajong)

Measurements
head and body
465—595 mm (males)
400—495 mm (females)
tail
630—850 mm (males)
530—630 mm (females)
weight
5·8—7·8 kg (males)
4·5—4·98 kg (females)

Brazza's Monkey, Neglectus Monkey

The impression given by this monkey in the wild is of a compact greeny-grey animal with white markings, which are very distinctive and form beautiful geometric permutations according to the posture of the animal; the white beard and bright orangeish brow-patch are very conspicuous in adults.

Cercopithecus neglectus monkeys have an extensive range in the northern Congo basin, extending nearly as far as Cameroon; they also occur patchily to 11° south of the equator in the Congo. They are found in several localities in southern Ethiopia, while in East Africa they occur in a number of small pockets in western Uganda and in the vicinity of Mt Elgon.

C. neglectus is not a West African species but the presence of the related *Cercopithecus diana* in that region suggests that at an early date a branch deriving from a common ancestral stock reached West Africa where, like many other forest mammals it developed in isolation from Central and East Africa.

If the development of forest monkeys approximates to the sequence and form suggested for them in this book, the greatest proliferation of cercopithecus species and races occurred in West Africa and it is probable that during a long but relatively recent period West African and Central African forests were separate and that the only species with highly elaborated colour patterns in the central and East African forests was *C. neglectus*. Assuming that pattern conferred an advantage, it was probably highly successful, and this is borne out by the present distribution of *C. neglectus*. The lack of any variation in one species of a genus that is otherwise highly variable in external appearance implies that this species must have had a very widespread and virtually continuous distribution in the recent past. These monkeys must have been very numerous to maintain a degree of gene-flow capable of sustaining the present homogeneity of the species over so vast an area of Central and East Africa.

The species is not limited to swamp forest throughout its range, it has a high tolerance for altitude, ranging up to 2,100 m on Elgon and it is found in closed forests, in bamboo and along the water courses of dry montane forest. The species almost certainly ranged through all forest types in the past.

Its former pre-eminence in Central and East Africa may, therefore, be contrasted with its present sorry status as a "swamp" monkey.

Apart from the very recent decline of forests which has, of course, fragmented its range, what other factors can have brought about so great a change in the status of this monkey? An examination of its distribution in East Africa suggests that the species was "too late" to enter the southern forests, which may have become isolated before *Cercopithecus neglectus* developed. It also shows that, apart from the few pockets where it is very strictly limited to swamp forest, it now occupies those outlying areas of the northern forests where the redtail, *Cercopithecus ascanius*, is either absent or rare, and it is possible that these two monkey species are inimical to one another.

C. neglectus: Distribution.

253

The redtail may be a relatively recent arrival deriving from an essentially West African stock, if so it is a very successful colonist and occurs in most Uganda forests where it presumably competes with formerly dominant monkey species. However, its co-existence and co-dominance with *Cercopithecus mitis stuhlmanni*, *Cercocebus albigena* and *Colobus abyssinicus* in many forests makes it unlikely that straightforward competition for resources is involved.

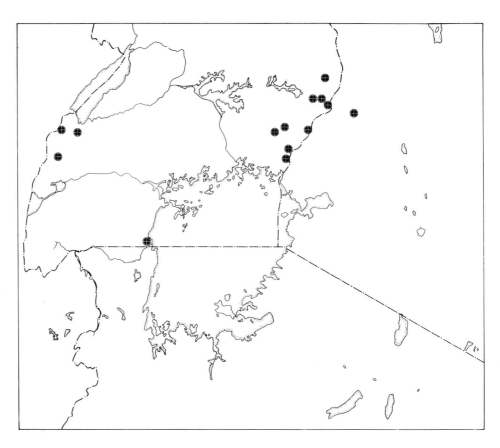

Distribution Uganda—West Kenya. ▶

I have not seen *C. neglectus* with *C. ascanius* in the wild, and Haddow (1952) omitted *C. neglectus* from a long list of monkeys associating with redtails. *C. neglectus* have, moreover, been widely reported as being "ill at ease", "hesitant" or "suspicious" when put in the same cage as *Cercopithecus petaurista* and *C. ascanius*. Hill (1966) says "it does not agree well in captivity with the livelier species . . . for it resents teasing".

In Bwamba knowledgeable informants have also noticed that *C. neglectus* and *C. ascanius* do not associate. A pigmy hunter witnessed a fierce fight between adult males of the two species in which they actually fell out of the trees and continued fighting on the ground.

Redtail troops are much larger than those of *C. neglectus*, the densities of individuals are high, the monkeys are more active and inquisitive. It seems from incidents between captives that the appearance of *C. neglectus* in particular excites the redtails' curiosity. The highly developed signal patterns of these two species have the same emotive components of red, blue, black and white, but they are structured differently and the context of their signals

254

differs fundamentally. The continued harassment of *C. neglectus* by *C. ascanius* —as observed in captivity—could, therefore, under wild conditions, conceivably constitute an obstacle to their co-existence. A comparative study of these two species, their communicative systems and interactions would therefore be of the greatest interest and might throw some light on what could be a special form of interspecific intolerance.

In Busia and Bukedi, *C. neglectus* mix with *Cercopithecus aethiops* and settle down in captivity with this species. In Bwamba, they are frequently associated with *Cercocebus albigena*, a monkey with which they share a preference for palm nuts, *Elaeis guineensis* and *Phoenix reclinata*, and the fruits of the swamp tree, *Pseudospondias microcarpa*.

C. neglectus frequently feeds on the ground, eating the leaves and berries of various plants, including *Aframomum* and *Ficus urceolaria*, a low, creeping shrub with scabrous, latexy leaves, which it "works" with the fingers and makes into a plug before eating. R. White (personal communication) found this species crossing a quarter of a mile of open country to reach fruiting figs, and reported occasional "shamba" raids for maize and green cotton bolls. It also eats grasshoppers, lizards and roots, these are also worked with the fingers and gheckos are broken apart meticulously, viscera and eyes pulled out and rejected, while the flesh is shredded off the bone and some of the bone discarded. The gum of *Albizzia gummifera* is a favoured food.

This animal is a peculiar monkey in several respects and Verheyen (1962) concluded that, on cranial characters, *C. neglectus* is closely related to *Cercopithecus nigroviridis*, the most primitive type of cercopithecus monkey. This conclusion is also supported by an examination of the infant coat of *C. neglectus* which is very similar to that of young *C. nigroviridis*. The species retains very soft hair which is characteristic of *C. nigroviridis* and some of the more cercopithecoid mangabeys.

One phenomenon in *C. neglectus* that led to great taxonomic confusion in the past is the striking difference between the infant, juvenile and adult coats. The different intermediate stages were even given different specific names. The newly born *C. neglectus* has a uniform soft-brown coat with an indistinct white chin. As it grows the first feature to appear is a rich chestnut colour to the brow which may extend to the cheeks and nape; a similar colour appears simultaneously round the rump and root of the tail. This reddish suffusion on the head and the anal pole is retained in the next stage when the white beard and delimitation of the brow patch by dark borders starts, and the femoral stripe begins to differentiate. Thereafter the process is one of steadily increasing contrast and definition until the adult colours are achieved; the last colour to go is the brown suffusion.

There would be nothing particularly extraordinary in this, were it not for the fact that the stages have no correlation with the size, and almost no relation to the age of the animal, so that a juvenile still with milk dentition may have acquired a nearly adult colouring, while adult-sized animals retain juvenile coats. Dobroruka (1966) notes a zoo animal only losing the "juvenile colouring" in its sixth year of life. This phenomenon has been noticed over a long period of time and in large numbers of captives in East Africa, and it may be correlated with behaviour in the wild and in captivity. The first correlation to be made is that of disposition; an adult male *C. neglectus* with full colouring is a very confident and aggressive monkey, prone to attack other monkeys in captivity and to make bold displays and exhibit protective behaviour in the wild. By contrast, female *C. neglectus* and males with colouring that has not reached its full development are more timid and mild mannered, this applies in captivity as well as in the wild, where troop members are markedly more retiring than the adult males. The female *C. neglectus* is smaller than the male and even the largest females do not approach an adult male in size. The only other sexual difference is a highly significant one, for the female has a rich brownish-red perineum. All males without the completely adult colour are, by retaining the brown colouring, probably mimicking the female's one distinctive characteristic. A newly captured juvenile male, put with an adult male, displayed a very marked ritual, he dropped onto the elbows and hid his face, presenting the raised hindquarters and tail. This was sometimes followed by a backward walk in the same position, looking between the legs and waving the rump and tail, then he lay on his side, presenting his back to the adult male.

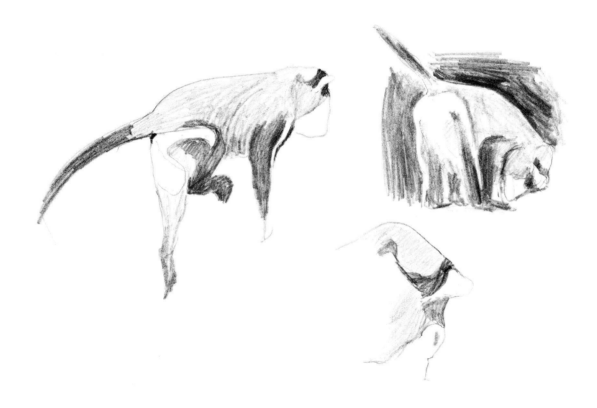

This behaviour is not dissimilar to that of other monkeys in which presenting is a common appeasing gesture, but the gesture is much more stereotyped and consistent in this species. It has the visual effect of presenting a patch of orange-brown colour, which is emphasized in younger animals by a ritualized movement of the brown tail and rump, while hiding the white beard and face. I have seen a very similar ritual in an adult captive male, although no trace of brown colouring remained on the rump and the animal appeared to display this behaviour as an appeasement ceremony towards me. He initially "presented" with raised rump and the feet widely parted after a bout of threats and snatching; the rump was presented several times after the threatening and on each occasion the animal looked over its shoulder. Later, when I confined the animal in a large aviary which I entered he was uneasy about my closeness and after he had been followed about a few times, he initiated the following display; he suddenly backed towards me, meanwhile looking between his legs and, having wrapped his tail round my arm, reached out with the hindlegs to take the weight of his hindquarters on my hand, he was then standing on his head. This performance, which was repeated several times, sometimes led to the monkey losing his balance and falling over, when he was willing to be groomed. The gesture was obviously more elaborate than a simple "presenting" and it is possible that the animal was attempting to "force" an appeasing gesture, as has been reported by Simonds (1965) for macaques.

In order to test the effect of colour on an adult male *C. neglectus*, the same animal was presented with cloths of different colours, which were waved at him just within reach of his cage. Marked differences were noted in his responses; orange-brown was touched eagerly with the fingers, which were immediately smelled intently. This colour elicited great interest and so did yellow which was touched repeatedly, even when these were yellow patches printed on a patterned dress worn by a visitor to his cage, the hands were smelled after touching the colour. In great contrast was his response to bright blue, particularly when a patch of this colour was presented together with a piece of black and white colobus fur, or with a piece of *Cercopithecus mitis*

stuhlmanni fur, the latter combinations elicited violent chattering, while small pieces of blue cloth provoked less extreme agitation but were snatched at violently nonetheless, the hands were also immediately smelled after these snatches. Other colours elicited no response, including black and white and the blue monkey skin alone and he showed no reaction to grey's. Orange-brown therefore excites interest but not aggression and this would seem to be because of its female association, while bright blue is the colour of the male scrotum and also of the mature male's facial skin and this elicits very strong aggression. These tests suggest why presenting is stereotyped in this species and they also illustrate other aspects of the social behaviour of this monkey.

Any monkey, including the adult male, may solicit grooming, but this is always accomplished by what appears to be an almost furtively shy approach, in which the animal tries to advance its neutral coloured back. A male will raise his beard up vertically and look upside down over his back and through his eyebrows as he approaches crab-wise to be groomed, but will never make a direct frontal approach, unless it is with the head lowered between the arms so that the beard and white lining to his thighs are hidden from view. Animals will lie down with their backs towards the monkey beside them or, in solitary captives, towards the human groomer. The full length lying posture elicits grooming and was observed in the juvenile mentioned earlier as an appeasement gesture. This throws an interesting light on the prostrate sleeping and resting posture of the species; it differs from that of other monkeys, which more usually sit upright. The hiding posture is also flat, on the ground, or

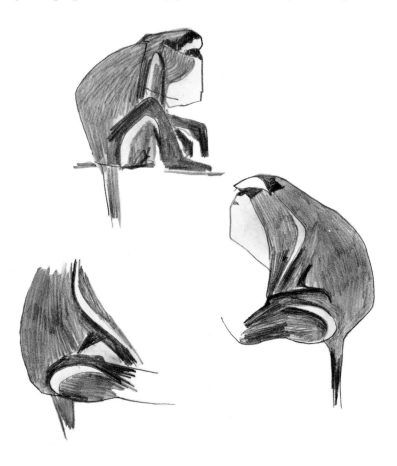

along a branch. The red patch on the head is probably retained to counter-balance the aggressive connotation of the beard, and the strange contortions of posture are calculated to present only this pacifying feature in social situations. Threats, instead, display a maximum of white and threatening *C. neglectus* males, whether on the ground or in the trees, will rise on the hindlegs and violently bob their hindquarters and beard up and down. This movement emphasizes verticality and it is the avoidance of any element contradicting an emphatically vertical signal that explains why *C. neglectus* and *Cercopithecus hamlyni* have closely adpressed hair at the sides of their faces, instead of the typical ruff of broad cheek whiskers so characteristic of other cercopithecus monkeys.

A second, milder form of threat creates a triangle. Here the legs are widely splayed and the broad spread of the white thighs is capped by the beard; sometimes the blue and red genitalia are prominently displayed in this posture like the vervet males. The genitalia are sometimes made more visible by jerky erections, but this is rarer than in *C. aethiops*. Unexpected but indeterminate alarms, provoked by sudden movements or noises may be responded to by a rapid rising and spreading of the thighs to display the white. The display of white thighs and genitalia may not always be accompanied by obvious signs of aggression. The pigmies, who are keen observers and know this monkey well, claim an excited troop sometimes gathers round another animal displaying in this way, while another animal circumcises it! What sort of behaviour could have given rise to this legend?

There is a rarely seen posture, the meaning of which has not been determined. This is a sudden rapid movement in which the monkey hangs upside down for a moment, with all limbs clutching the branch as its swings or swoops under it.

The onset of mature colouring in captive *C. neglectus* males is synchronous with a marked change in disposition from timidity to confidence and aggressive behaviour; it can probably be assumed that both colouring and behaviour are under linked hormone control. While faulty nutrition and stress in captivity may contribute to the delay of adult colouring, it is significant that of two *C. neglectus* males in the Prague zoo observed by Dobroruka (1966) one had adult colouring and the other not, both were presumably acquired and kept under the same conditions, so that the six years delay of adult colouring in one individual is less likely to have been controlled by physical conditions than by psychological factors deriving from his relationship with the other male; disappointingly no data on this are published. However, observations on other captives and wild troops suggest that the control of colouring might be psychological. Troops generally contain 15—35 individuals, but smaller "family" groups are not unusual and solitary males are also common. Troops have been observed to contain only one male with adult colouring, and it would therefore seem that a young male on growing up will either continue to live in the troop, behaving and looking rather like a female, or acquire adult colouring whereupon he will probably be driven out. In larger troops he may take the nucleus of a new troop with him, as the power of a single male to lead a very large troop must be limited and the absence of large bands may reflect this limitation, while the very small groups that are encountered may have recently formed by splitting off from larger ones. Similarly, old males

past their prime are probably driven out by younger ones. Solitary males are frequently met with, and one collected in Bwamba was an old, thin animal with worn teeth and suffering from a severe infestation of mango fly maggots and other parasites. This individual was found "freezing" with the head held flat to conceal the white beard, as he crouched at the base of a palm frond. Freezing is a common means of avoiding danger, but troops will often flee through the trees unless hunted with bows and arrows in which case they come to the ground. Hunters in Bwamba, using both bows and arrows and dogs force the monkeys to descend to the ground, where they can be caught by the dogs.

The adult males have a slow, rather cat-like swagger on the ground but they are fast and adept runners and have been reported as being able swimmers; certainly their range is never limited by rivers, not even by the Congo, which is a barrier for many other species.

260

They are very quiet monkeys. In captivity, a short nasal grunt is associated with the anticipation of food. The young chitter, try to bite and make a squealing cry when hurt. Adult males are distinguished by being more vocal than other members of the troop, they make a chattering croak of alarm, which sounds rather like the wooden rattle of a football fan, and may be rendered gutturally as "chi-chi-chi-aarh", followed by single barking croaks. They also make a hooting "ow" or "awoo" and the latter noise is often associated with the decamping of a human observer and may be a release of tension noise, typical of many monkeys.

Sexual behaviour and breeding have been observed in captivity, where the species breeds readily. It is popular in zoos and as a pet in East Africa, as it is appealingly gentle when young. They breed throughout the year. Hybridization between a female *C. neglectus* and a male *Cercopithecus mona* has been achieved in captivity.

These isolated observations and suggestions serve to illustrate how much remains to be learned about this most interesting species.

Patas Monkey
(Cercopithecus (Erythrocebus) patas)

Family   Cercopithecidae
Order   Primates
Local names

Ayom (Lwo), Elwala (Karamajong),
Engabwor (Ateso), Akahinda
(Runyoro), Naggawo (Luganda).

Measurements
head and body
605—875 mm (males)
485—770 mm (females)
tail
540—740 mm
weight
10—25 kg (males)
7—14 kg (females)

Patas Monkey

The patas is a highly distinctive monkey which has adapted to a life in the open where its long limbs and slender build have allowed it to survive in an environment otherwise hostile to monkeys.

Young animals and females are coloured rather like dry grass with shades of fawn, russet and grey. By contrast the adult males are very much larger than the females and have brilliant reds with strong tonal contrasts on the hindquarters and face. These very striking physical differences between sexes are matched by differences in all aspects of behaviour and social life.

The distribution pattern of the patas coincides broadly with open wooded steppes and savannas south of the Sahara, it is rather patchily distributed from Senegal to Somalia. A local population on the Serengeti lacks some elements of the head pattern; this peripheral population may therefore have been isolated for some time.

The opening up of land and the opportunities for raiding crops may favour the patas and in many cultivated areas it is said to be commoner than in unsettled areas; but this is difficult to determine as denser settlement may simply increase the frequency with which the animals are seen or reported.

They are found in vegetation types ranging between open grassland to dry woodland and are commonest in thinly bushed *Combretum* and *Acacia* savanna. The pods, seeds and leaves of these trees are important foods. Many varieties of fruits and berries, grass seeds and storage leaf bases are eaten, insects, small vertebrates and fungi are also taken. They are reported to take eggs and young from birds nests. They damage growing cotton by eating the young bolls and dig up newly planted wheat; they are also fond of millet, bananas, groundnuts, cassava and maize. While feeding they pluck up the food with one hand, meanwhile keeping within sight of one another, but remaining well spaced apart. They do not squabble over food but when a local source of choice food is encountered senior members of the troop chase away the others.

Patas are difficult to watch near cultivated areas where persecution has made them very wary, but in common with other animals, they have become more trusting in the National Parks, nonetheless observation is generally only practicable while grass is not too long and thick.

They are notoriously quiet monkeys, an occasional burring alarm note or bark uttered by the male is the only sound that is at all commonly heard. At close quarters, however, the females and young have a restrained but audible communication of chirrups and hoots. A very characteristic noise is a low moan; in juveniles this is a want call. Captive females make this call frequently when they are in oestrus and males when anticipating food. Other sounds are low whistles and a repeated guttural cough similar to other monkey alarm or annoyance calls.

Daily activity generally starts with the animals leaving their sleeping trees and grooming one another before starting to feed. Like other monkeys, a localized source of food, such as a fruiting tamarind, may keep them in one small locality for much of the day, but when food is dispersed the troop may be on the move almost all day. During the heat of the afternoon, from about 1 to 4 p.m., all the animals in the troop congregate together in a shady tree. This is followed by an evening feeding period. At dusk the monkeys disperse in neighbouring trees singly or in pairs. On several occasions Hall (1966) saw an adult male reconnoitre the entire neighbourhood of the sleeping trees before settling down. There are several reports of patas being active at night; like the redtail, *Cercopithecus ascanius*, this may be a habit developed to circumvent human vigilance as the reports concern nocturnal raids on crops.

Patas monkey running, analysis from film and dissection.

In open country patas customarily move with a long lope. When alerted they may stand on the hindlegs in a very characteristic and frequently-seen stance, and sometimes run along or hop on their hindlegs while carrying a heavy object such as a maize cob or root in the hands. The adults are decidedly clumsy in the trees and easily lose their footing. When running they have been clocked at 55 km per hour. The hands and feet are peculiarly elongated and this general lengthening of the limbs is part of the patas adaptation to fast movement on the ground.

264

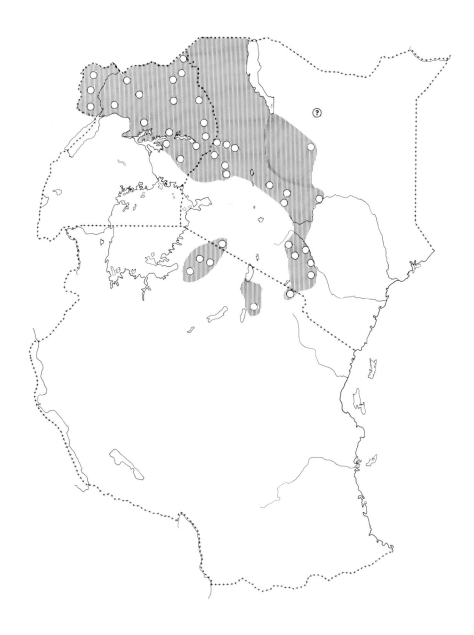

In East Africa, patas troops vary in size from 5 to 50 individuals, this may be influenced by habitat, and troops in *Isoberlinia* woodland are said to be larger than the ones in grassland. Over a hundred patas together at a time have been reported from West Africa so that the social pattern found in East Africa is unlikely to hold throughout the range. In the smaller groups there is usually only one adult male, who attacks any strange male, indeed all members of the troop generally display aggression towards other troops or strange individuals. Solitary adult males are also seen. There are often several adult males in larger troops, the organization of these social groups has yet to be investigated, as the only observations to date have concerned small one-male units.

Male patas monkeys are conspicuous members of the troop, sitting alert in low trees or on top of termitaries; they keep watch almost continually and even while feeding will rise on their hindlegs to peer around from time to time. They tend to be aloof from the games and squabbles of the troop and their orientation is mostly out into the environment as they scout the terrain

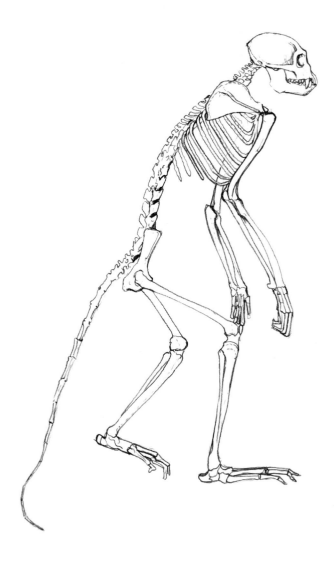

about the troop. The acute sight and alertness of the male patas was appreciated by a professional hunter who employed one to find lions and leopards for his clients to shoot. The animal would give its burring noise on sighting any carnivore, but could see distances that often required binoculars to determine the source of its alarm (it was rewarded with beer). After the death of this animal, two different females were also taken out, but significantly they were not particularly alert and proved useless. In this the patas contrasts very strongly with other monkey species, especially baboons where much of the adult male's attention is taken up with his immediate surroundings and the affairs of his troop, while juveniles and young males on the periphery, are the most likely members of the troop to encounter danger and give the first alarm.

Conspicuous and making no effort to conceal himself the male yawns nervously while on watch, his head bobs as he yawns and stretches the face upwards before closing the mouth. The gesture is clearly an involuntary release for tension, but the patas male has for his size the longest canines of any African monkey and the bobbing yawn may well advertise this to potential predators which are likely to sight the male before other members of the troop.

As males are often at some distance from the troop and may be absent altogether for some hours at a time, most social life is therefore thought to centre on the adult females. In the wild and in captivity female patas tend to display more aggression than the females of most monkey species. In both captive and wild groups, priority for choice food items is asserted by a dominant female and a peck order is discernible; this order is maintained by threat and chasing but fights or wounds are rare. Captive female patas monkeys often make very clear sexual distinctions between humans and will frequently threaten or bite women but seldom threaten men, they usually seek to groom their owner or any well-known man. If they are smacked by a man they generally cringe and lip smack in appeasement, while any gesture of aggression by women or children is responded to in kind.

Captive male patas make a remarkable ritual of bouncing up and down on the same spot. The main effect is to dance the brightly coloured posterior up and down, attracting attention to the strong contrasts on both the head and genital poles. Another variation is to see-saw in a rocking bounce, which may then be followed by a fast run in which the animals hurls itself against some tree, rock, sapling or the cage wall bouncing off in another direction. In solitary captive males, the whole performance or parts of it are frequently associated with the anticipation of food or the sight of a keeper, and may be accompanied by the "want" moan which was mentioned earlier as characteristic of juvenile behaviour. The ritual clearly releases tension and also seems to be associated with "wanting" or anticipation, but in spite of the apparent absence of an overt aggressive connotation, it is possible that for captives the keeper may release conflicting tensions in which there is an element of threat. The display has some resemblance with the "red, white and blue" of vervets but with the important difference that no trace of male genital colouring is found on the female. The dance also resembles the interspecific threatening bounce of other cercopithecus species, *Cercopithecus neglectus*, *Cercopithecus mitis*. There is also the possibility that jumping up and down in long grass, associated with interspecific encounters, became displaced and ritualized. There is also a hint of the presenting gesture, for while bouncing, the tail is held erect and sometimes to one side suggesting that appeasement of aggression is a component of this ritual. Adult males, sometimes turn and look over the shoulder after this display as if to see its effect; this is typical of presenting behaviour. A wild male has been observed to bounce off a sapling after fleeing from some members of the troop that threatened him while copulating; this is the only component of the display seen in the wild so far. Perhaps an element of play chasing may be added to the possible origins of this unique and complicated display. Further observation of wild patas is needed as very little is known of many aspects of its social life and nothing of the significance of this "dance", which patently advertises the spectacular visual qualities of the adult male.

The area covered by a patas troop will vary with the season, the topography and the food resources of the home range, but may be up to 80 sq km. They sleep in a different locality every night, but are thought to keep to one area for long periods of time while other troops also range consistently in adjacent areas. All members of the troop display animosity when two troops meet.

The habitat of patas may be shared by vervets and baboons, but the riverine vegetation and rock faces on which these species rely for shelter, are avoided by patas which flee from danger by scattering and running away on the ground rather than climbing trees. Young animals may crouch in the long grass and I have seen an adult enter a culvert tunnel when approached by a vehicle. It is possible that orycteropus holes may provide shelter. Hall noticed no change in behaviour when a martial eagle flew over them. Leopard, cheetah, hyaena, lion and hunting dogs are the most likely predators apart from man, and all these species elicit alarm calls. A baboon troop has been seen to make way at a water hole for an adult male patas and wait until he had drunk and departed before returning to the water. On the other hand, a troop of patas have also been seen to withdraw before baboons. When hunted with dogs, the males fight fiercely and may succeed in driving off the dogs. They are not infrequently killed in tribal hunts but are generally adept at avoiding punishment for their lightning raids on cultivation. They approach the fields quietly, but are easily driven off if seen and are not considered the serious menace that baboons and vervets are. Limited numbers and a rather local distribution in East Africa make their economic exploitation unlikely, although a few animals are exported to zoos. The large populations in West Africa are being exploited for commercial medical purposes.

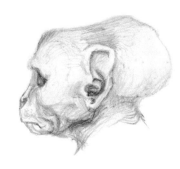

The breeding of this monkey tends to follow a distinct seasonal pattern, gestation is thought to be somewhat less than six months, so that a mating season would occur during the later rains, as practically all young are born between January and May, during the latter part of the dry season and early rains.

Females have a thirty day sexual cycle of which bleeding is the physical sign. During oestrus females show very marked behavioural changes. In captivity whether in the presence of a male or no, the female creeps about on bent legs and wheezes in an asthmatic manner, meanwhile blowing out her cheek pouches with air, occasionally she makes the "want" moan. This behaviour distresses the uninitiated observer as it looks like the last stages of some painful disease. In the wild, the female solicits the male in this crouched posture. The male's bouncing display can therefore be seen to be the antithesis of the female's oestrus display.

Patas females are unlike most other monkeys in being intensely possessive of their young and in not allowing other animals to touch them. As a result orphans may not be cared for by foster mothers as has been observed in *Cercopithecus aethiops*, *Cercopithecus mitis* and other species. The protective role of the male towards all members of the troop is well displayed in an incident where an adult male was observed to accompany a small juvenile. When their troop was approached all the animals scattered except for the male while the juvenile crouched in the tall grass; only when it was picked up did the male run away. The young animal was found to be very thin and covered in ticks (which is most unusual for patas), so that it can be assumed that it was an orphan (J. Brown, personal communication).

New-born babies are dark brown and are able to see and climb about on their mother by the second day. At two weeks mother and baby may be walking about separately when undisturbed, but on the least alarm the baby will

cling to the mother. Male and female juveniles behave differently—the most noticeable difference being the appearance of the "bouncing display" in juvenile males at about five or six months. In a captive this was elicited by strange animals, such as dogs or cats, and by mock threats from the human foster parent; the display may contain conflicting aggressive and fearful impulses and seems to be related to displays observed in *C. neglectus* and *C. ascanius* in that it may be followed by a backward walk towards the strange animal or foster parent. Bouncing off bushes is also typical of young males when they are playfully chased. They also bounce back and forth off posts or walls when deliberately confined in a room or cage, at which time it seems to be an expression of frustration.

Females become adult at three and a half years, while the evidence so far indicates that a male may reach his full size at five years of age. Observation suggests that differential rates of growth may occur in both patas and neglectus monkey males, and it is just possible that a hormonal mechanism could delay development in the males of this species. If the female represents a physiological norm for the species, it is not impossible that the extraordinary features and great size of a mature male may need a behaviourally controlled biochemical trigger for full development.

Adult female patas, above opposite; adult male patas below and opposite.

The Prosimians

LORISOIDEA **Lorisidae** **Perodicticus potto**
 Galagidae **Galago crassicaudatus**
  Galago demidovii
  Galago zanzibaricus
  Galago senegalensis
  Galago inustus

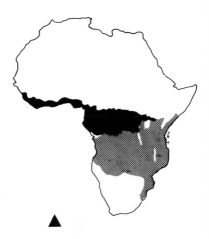

black = Perodictus potto.
stipple = *Galago crassicaudatus.*

Diagram of incisors, tongue and
sublingua in a prosimian
(Perodicticus).

Prosimians have been called both progressive insectivores and primitive primates. However, the modern African prosimians have proceeded so far along an evolutionary tangent that their unique specializations are as striking as their affinities with the ancestral primate stock, or their nearness to insectivores.

The numerous Palaeocene prosimians were already highly divergent in their adaptations, and fossils from Kenya suggest that the divergence between *Galagidae* and *Lorisidae* was already marked by the Miocene. The presence of lorises but not galagos in Asia is an interesting problem, for the Miocene fossils suggest that it should have been possible for galagos to reach Asia. In this connection, it would be worth investigating what factors might be responsible for the lack of overlap in distribution between *Galago crassicaudatus* and *Perodicticus potto*, for although the smaller galagos may be found in the same tree as a potto, mutual intolerance between the larger species could be a factor of some significance historically. The biochemical resemblances between *P. potto* and *G. crassicaudatus* are greater than those between *G. crassicaudatus* and another galago species *Galago senegalensis* (Walker, personal communication), so that the relationship between the lorises and galagos is altogether problematic.

The East African species appear to be fairly numerous in their appropriate habitats and the group may be said to be specialized and successful. All are arboreal and nocturnal, with large eyes, a moist nose and cleft upper lip, and have cryptic woolly coats coloured in a wide range of greys and browns. Their limbs are mobile, with long spatulate fingered hands and feet with opposable thumbs. All species carry a specialized grooming claw on the second digit of the foot. This claw is augmented by a scraper or comb of specialized teeth in the lower jaw, which is an adaptation confined to modern prosimians. The tooth structure is not found in any of over a hundred Cenozoic prosimian species. Correlated with these teeth is a "second tongue" or sublingua, with denticles to fit the space between the teeth. This barrage of combing devices is apparently needed to cope with the seasonal moulting of the fur. Seasonal moulting or the occasional appearance of large quantities of fur in the dung or stomach has been observed in all East African prosimians. The biochemistry of the skin reveals that the hair follicles are rich in glycogen and phosphorylase and that they are largely quiescent at any particular time (Yasuda *et al.*, 1959), indicating that the group have special biochemical adaptations to cope with the periods of general shedding and regrowth of the hair.

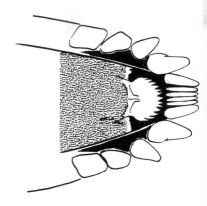

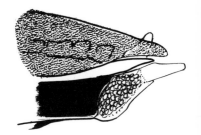

◀ *G. demidovii, top ; G. zanzibaricus, 2nd top ; G. senegalensis, 2nd bottom ; G. inustus, bottom.*

In a captive greater galago, *Galago crassicaudatus*, the shedding of the coat and the growth of a new one was coincidental with the addition of a multivitamin concentrate to its diet. In a female *Galago demidovii* the moult coincided with the weaning of her offspring and in *Perodicticus potto* there is some evidence that the shedding of hair includes the neck vibrissae, which are probably involved in the sexual rituals of this species. In birds, moulting is mechanically functional, as it is necessary for worn-out feathers to be replaced. Lorisoid fur too may have a special function, perhaps as a "scent dispenser", and the condition of the fur in *P. potto* at least may have a sexual significance. It is possible that the enzymes and glandular cells, with which the skin of all African prosimians is richly endowed, may produce secretions that scent the fur with special odours of social significance for the species. A change in the condition of the naked areas of skin in some prosimians has been observed; a bright yellow pigment appears on the naked areas of the face, ears, hands, feet and on the genitalia of particular individuals. This phenomenon is most marked in *P. potto* and *G. demidovii*. On present evidence this pigment might be associated with a mechanism for appeasing or inhibiting intraspecific aggression. Montagna and Yun (1962) found strong acid phosphatase activity in the face and feet of the potto, while alkaline phosphatase was characteristic of most other areas of the body; this is also the case with galagos. The occurrence of locally restricted pigment must be connected with the different chemistry of the naked areas of the skin.

Hand (*right*) and foot of galago (*Galago inustus*) showing the grooming claw.

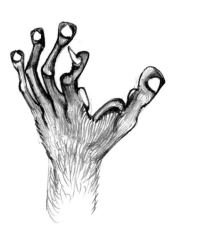

Prosimians have cycles of sexual activity; the periods when the females are in oestrus or the males have active testes seem to be followed by long periods of anoestrus or sexual quiescence. Thus there tend to be annual or biannual breeding peaks and the degree of sociability seems to fluctuate widely.

Their young are born with open eyes, are well furred and grow rapidly, particularly galagos.

The principal foods of East African prosimians are insects, molluscs, fruits and tree gum. The availability of these foods fluctuates with the seasons but there are, nonetheless, interesting differences of emphasis from species to species. *Galago demidovii* is primarily insectivorous, *Perodicticus potto* is very dependent on gum whereas *Galago crassicaudatus* is more frugivorous than other species. All species feed on gum and saps exuding from certain trees that are common in their respective habitats. This diet may well be an adaptation that has assisted their success as a group, for insects and fruit can be well nigh unprocurable in some seasons.

A typical feature of lorisoid behaviour is the use of urine for marking; the potto expels it drop by drop while on the move, but galagos water their hands and rub urine on the soles of the hands and feet. All species maintain a network of self-scented pathways. In captivity scent is renewed whenever hygienically-minded keepers clean their cages, and scents are clearly of the greatest importance in every aspect of their life history.

For nocturnal animals scent trails have an obvious advantage and the evolution of scent marking presumably served originally as an orientation device for the individual animal. This function can be seen in many primitive mammals, but prosimians seem to have developed elaborate social behaviour in which scent marking has acquired a communicative function.

Lorisidae

The family comprises two African genera, *Perodicticus potto* and *Arctocebus calabarensis* in West Africa and two Oriental genera *Loris* and *Nycticebus*. They are all forest animals.

The skull of the potto is more robust than that of most galagos and the eyes have a more upward orientation, so that stereoscopic vision is achieved without reduction of the snout. In the potto's case this is correlated with a generally more prominent and vulnerable position for the neck and shoulders which are a specialized area, the peculiarities of which are discussed in the subsequent profile.

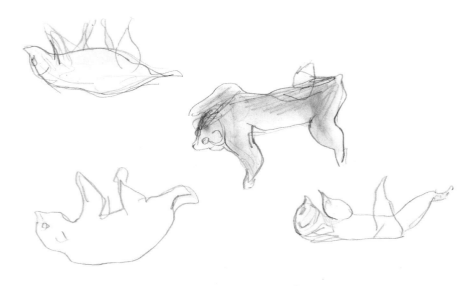

Galagidae

Galagos are jumpers with long hindlegs and bushy tails which serve as balance. They have long ears which resemble those of bats, being ribbed and capable of folding back against the head with specialized muscles to operate them.

The three forms show a considerable range of size on a roughly similar body plan. The relative size of the eyes and ears and the proportions of the limbs vary but within fairly narrow limits. When unhurried *Galago demidovii* has a dormouse-like little scurry, while the greater galago, *Galago crassi-caudatus*, has a ponderous cat-like prowl, however all species can and do leap when they wish to move fast.

The skulls of the smaller species are delicate and are balanced above the vertebrae. The eyes are large and rely upon movement of the head as they are fixed in the orbits.

The evolution of some galago species can be suggested on the evidence of living forms. *G. demidovii* may represent a primitive type, as teeth indistinguishable from those of this species have been found in Miocene deposits in Uganda and in western Kenya, this galago, *Progalago minor*, could be a chrono-species of *G. demidovii*. The savanna-living *Galago senegalensis* was well-established by the earliest Pleistocene, it is therefore most interesting to find a form of galago intermediate between typical *G. demidovii* and typical *G. senegalensis* in *Galago zanzibaricus* living in some of the relict forests of East Africa. Isolation over a very great period of time is necessary for speciation, and the eastern forests may have provided both the isolation and also the environmental pressure to adapt to drier conditions. It is worth noting that this galago is not the only instance of an isolated species with a related form in the main forest block.

The jump of a galago (*G. senegalensis*). From dissection and from Hall Cragg (1966–68).

Furthermore there appears to have been a re-invasion of the forest by a population of *G. senegalensis*; this forest population has become specifically distinct, i.e. *Galago inustus*. The spread and subsequent isolation of this type into the West African forests led to the development of a highly distinctive species, *Galago elegantulus*.

Galago crassicaudatus appears to be more distantly related to the other galago species and may have been a distinct branch since the Miocene.

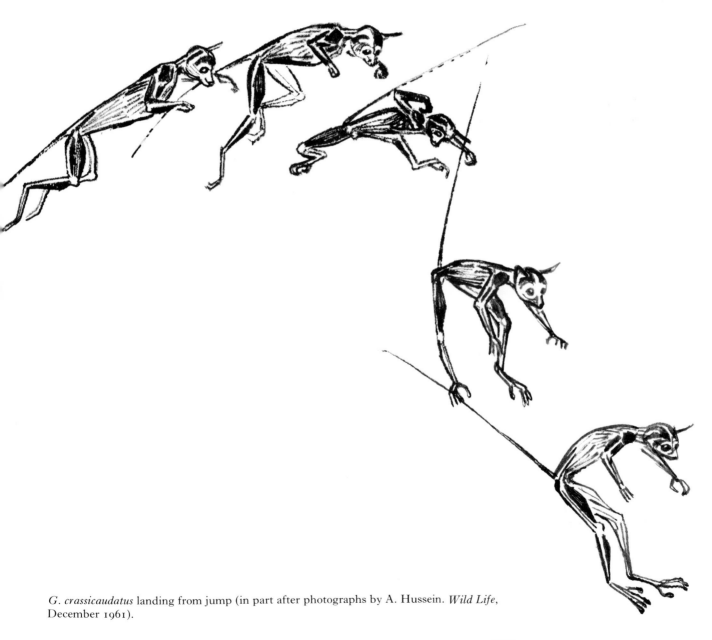

G. crassicaudatus landing from jump (in part after photographs by A. Hussein. *Wild Life*, December 1961).

Potto

The nocturnal potto is a woolly, slow-climbing, arboreal animal. The body and legs, disguised by thick fur or a hunched position are both in fact long and slender. The head is round with small ears and golden-brown protuberant eyes. The shallow lower jaw is overhung by a robust short muzzle. The grasping hands and feet have flat nails and strongly opposable, spatulate thumbs. The hands have a rudimentary knob for an index finger and on the hind feet the small second toe carries a grooming claw.

The head is broad and there are very characteristic bony rims round the eyes and ridges for the attachment of the temporal muscles; the skull is thick and almost without sutures. The facial region is short and blunt with vertical, peg-like upper incisors, canines and premolars. The incisors are small, the canines and first premolars are long and thick. The lower incisors and canines are procumbent, forming the typical "lemuroid comb".

Potto
(Perodicticus potto)

Family	Lorisidae
Order	Primates
Local names	

Orunaku (Runyoro), Kikami (Luganda), Shakami (Tiriki), Kabende (Kuamba).

Measurements
total length
370—455 mm
head and body
320—395 mm
tail
50—100 mm
weight
1—1·5 kg

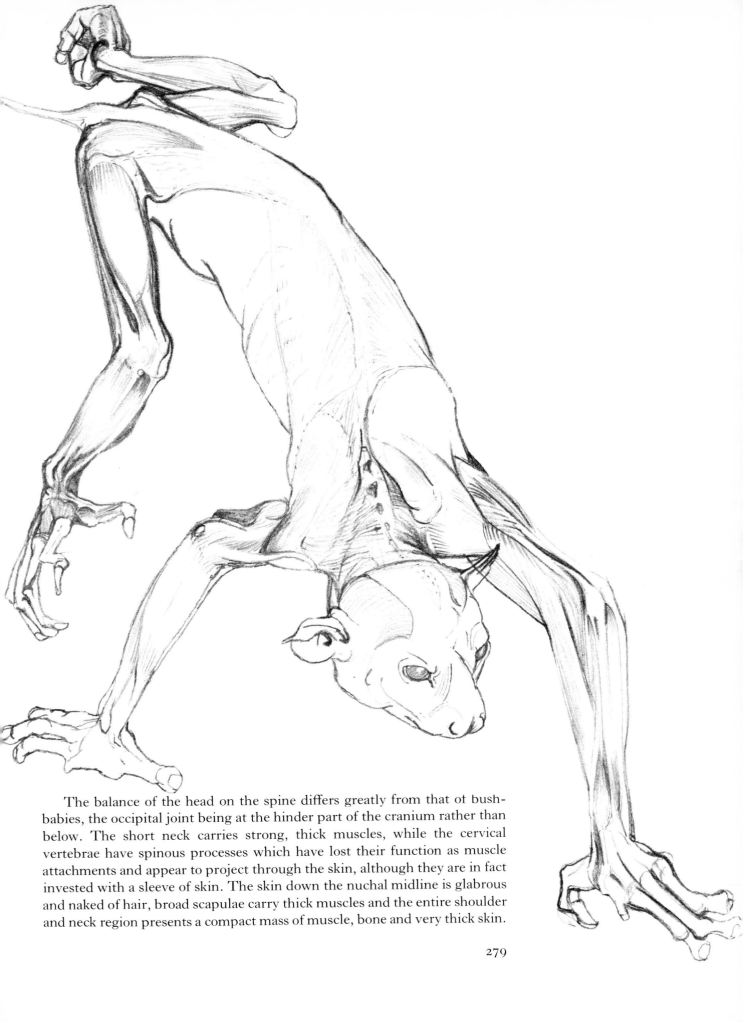

The balance of the head on the spine differs greatly from that of bush-babies, the occipital joint being at the hinder part of the cranium rather than below. The short neck carries strong, thick muscles, while the cervical vertebrae have spinous processes which have lost their function as muscle attachments and appear to project through the skin, although they are in fact invested with a sleeve of skin. The skin down the nuchal midline is glabrous and naked of hair, broad scapulae carry thick muscles and the entire shoulder and neck region presents a compact mass of muscle, bone and very thick skin.

The long thorax has a greater number of ribs than galagos and an exceptionally large number of vertebrae (the exact number varies), the spines of the vertebrae are very low; this may increase general flexibility and facilitate the lateral spinal movement so typical of the potto.

The arms and legs have powerful muscles. The wrists are small and highly flexible. A peculiarity the potto shares with other slow lorises, the South American sloth and the pangolin is the presence of a multibranched arterial and venous system in the limbs called *rete mirabile*. This system has been investigated in some detail (Suckling *et al.*, 1968). Functionally the vascular bundles appear to allow circulation to be maintained over a long period of time in limbs which remain flexed and motionless; they are therefore a fatigue-reducing mechanism. The grip of the potto is very powerful and has become a symbol of tenacity in African legends and proverbs.

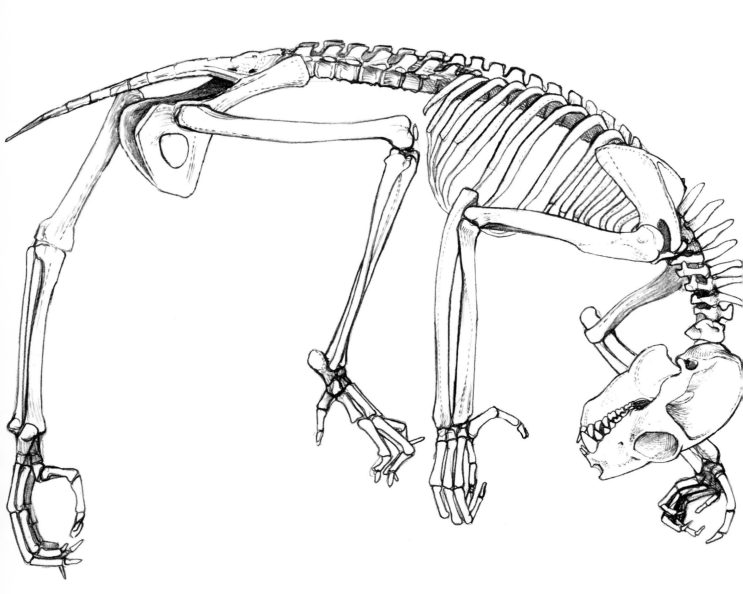

There are various colour phases ranging from pale bay to dark mahogany, silvery-grey guard hairs on the back with a black shoulder "mantle" characterize young adults. Long vibrissae are present on the upper surface of the neck in some individuals.

The potto is found from Sierra Leone, across the forest zone, to the Mau Forest in western Kenya. In East Africa, it is found in almost all the moist forest areas of Uganda and West Kenya and in the lower montane forests of Elgon and Ruwenzori. They are commonest in secondary forest and in clearings along the forest margins, where they may wander out into isolated trees in savanna or cultivation. They also occur in swamp forest but are not common in climax forest with a high closed canopy.

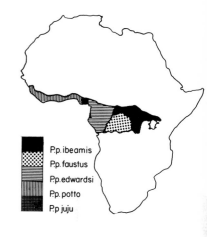

P.p. ibeamis
P.p. faustus
P.p. edwardsi
P.p. potto
P.p juju

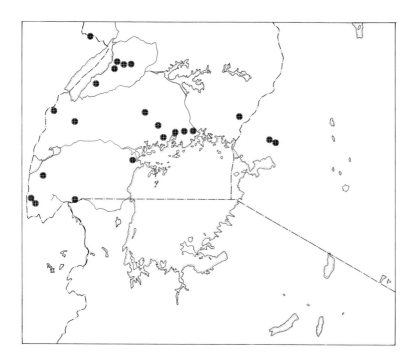

They are occasionally seen in the daytime when their presence may be betrayed by mobbing birds, but they are more reliably found by the reflection of their eyes in the beam of a torch. They are quite numerous in suitable habitats. Most of the trees on which pottos have been seen or collected are typical of the forest edge or of regeneration areas. These trees can in most instances be shown to be sources of food for the potto.

Pottos will eat a wide variety of foods: insects, fruits, leaves, moss, lichen, eggs, molluscs, fungi and tree resins have been recorded and birds have been taken by captives. Of insects, caterpillars, grubs and beetles are the most commonly eaten types, while figs and the fruit of *Musanga*, *Parkia*, *Avoun*, *Uapaca*, *Myrianthus*, *Conopharyngia* and *Parinari* have been noted. Leaves of *Urera* and the gum of *Albizzia* and *Sterculia* are recorded by Rahm (1960) who emphasizes the important part played by gum in the diet of the potto. Trees in which pottos have frequently been seen or collected are *Musanga cecropoides, Parkia, Sapium ellipticum* and *Phoenix reclinata*, all these species exude saps or resins and the last named species used to be a source of toddy for the Baganda. On the evidence of at least 19 stomach contents, about 60% of the wild potto's diet is resin, about 30% is insects and the remainder is fruit and other foods, only two animals had not eaten gum. The sample is small, regionally biased and only represents seven months of the year, nonetheless it suggests a pattern to the potto's feeding habits and the data are presented in graphic form in the figure below. Resin appears to be the principal food for the greater part of the year, but a sharp increase in the consumption of insects and fruit can be observed in the March—April sample. Dominique (1966) found variation in the foods eaten by Gaboon pottos which also ate large quantities of resin. In his sample of 8 stomachs animal food constituted 20%. The rains are a period of greater abundance and observations on pottos confined to open runs in Uganda have revealed an increase in appetite and activity at this time. Furthermore, thick fat deposits have been found on male pottos between May and September.

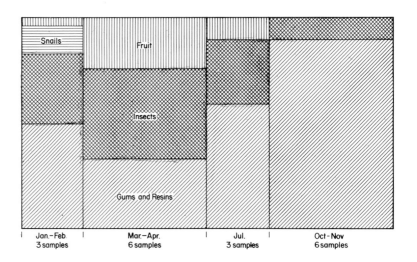

Snails
Fruit
Insects
Gums and Resins

Jan.–Feb.   Mar.–Apr.   Jul.   Oct–Nov
3 samples   6 samples   3 samples   6 samples

Resin is principally carbohydrate with aromatic flavours and some vitamins, it is a favourite food for many primates and squirrels, but the importance of this food in the life of the potto may be the cause of unusually slow movements of the alimentary canal. Honigman (1936) found that particles of straw took between 34 and 190 hours to pass through the system. For comparison a vervet monkey took between 1 and 68 hours (the slowness is relative, however, for the two-toed sloth may take almost seven weeks to excrete its meal). Mastication of food is prolonged and the identification of insect species is often difficult, but Cetonine scarabaid beetles, *Diplognatha* spp., *Neptunides Smaragdesthes*, *Dicranorrhina* and *Chelorrhina* are often identifiable in the stomach contents, where they are usually found together with resin. These beetles are very common on tree exudates and are presumably caught at the sites; anyone who has attempted to catch insects with their fingers while they are feeding on sap will know that caution and steadiness are required. The potto's movements are well-suited to this task, and sleeping birds or cicadas

may also be vulnerable to its approach. Nonetheless, most of the potto's foods are immobile and it is unlikely that the potto's slowness is an adaptation to assist its stealthy approach of living prey, since the grubs, beetles, caterpillars and snails commonly eaten are scarcely able to escape. Pottos have been caught searching for beetle larvae in the decaying thatch of houses.

Inconspicuous movements, "freezing" or "humping" when disturbed or alarmed have obvious advantages against predators and the potto's light grip and well-knit shoulders would probably defeat most animals trying to dislodge it. Furthermore the potto is remarkably silent, depriving itself thereby of an important aid to social life. Baudenon (1949) reported a "long sad monochord" with considerable carrying power, but the call must be rare as other observers have not noticed it. Silence and caution seem to be concomitant with a largely solitary life although their sociability in captivity has led Rahm (1960) to call the species a "contact animal".

In spite of the pottos social limitations, Dominique found wild animals able to detect conspecifics at up to 60 m distance. He discovered that both sexes have territories that are defined by the use of urine marks. The females range over 3 to 6 hectares while the males have a larger range which overlaps that of one or more females. Each potto has a preferred home area and sleeping sites within the territory (often where food is commonly abundant). The whole area is regularly patrolled and urine is constantly being re-deposited; the borders of male territories (where they may overlap another male territory) are marked almost nightly. Males generally avoid each other but at least one fight between wild males has been observed. Females are less aggressive but also avoid one another. Urine marks are subject to constant scrutiny and are clearly an important means of communication for an animal in which smell is the dominant sense. Marks probably betray the sex, the condition and the individual identity of the depositor.

Dominique found male–female contacts occurring at any time of the year in Gabon. In East Africa there might be changes in the intensity or sexual significance of these contacts which could be related to greater seasonal definition.

The main stimulus to social activity would probably be a rise of sex hormones. Sexual activity seems to be seasonal in the potto and there is some evidence of behavioural and physiological changes that may be correlated with the rise of androgens in the male which presumably takes a more active part in seeking a mate. Male pottos with large testes tend to have fat deposits, are more aggressive and much more active. They also become very smelly; this is due to numerous large apocrine glands in the skin of the male scrotum and in the skin covering the swellings that occur on either side of the vulva in the female. These glands are surrounded by numerous nerves which are very rich in acetylcholinesterase (Montagna and Ellis, 1959). As this enzyme assists impulses to cross synapses, the genital skin is clearly extraordinarily sensitive. These apocrine glands and nerves are associated with skin that has become highly specialized and is arranged in a mosaic pattern, with pigmented clefts demarkating octagonal areas of colourless skin. This specialized skin is found on the genital area of both sexes, which has led to the area above the female's vulval swellings to being called a "pseudo scrotum".

The pseudo-scrotal form of the female potto's genitalia is due to the

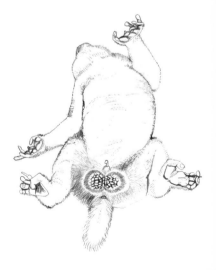

presence of two large glands containing a thick white secretion with a strong (and to the human nose, foul) smell. The glands are only possessed by the female and would seem to play some special role in the sexual behaviour of pottoes. Dominique (1966) found that small quantities of this secretion threw male pottos into great excitement. He also observed that the anal glands were active in 6 female pottos with both open and closed vaginas, so that glandular activity was not related to oestrus. The glands may facilitate the meeting of sexes but it would appear that this is not restricted to the rut. The mosaic patterned area of specialized skin which occurs on the genitals of both sexes also occurs very rarely in aberrant individuals on the sides of the neck.

The nuchal area resembles the genital region in being richly innervated (Montagna and Yun, 1966). The glabrous skin surrounding the spines is highly sensitive and observations on captive pottos show that the two areas are complementary to one another. Pairs sharing a box while in breeding condition will sit hunched over one another in such a way that the spines of one animal come into direct contact with the erogenous areas of the genitalia of the other. On stimulation by the spines the apocrine glands produce a secretion which is a very pungent musk (so strong that it can be smelled by a human observer in the forest). The ritual transfers the secretion onto the shoulders of the stimulator and the typical sitting position of the potto places the genitals in an ideal position for this behaviour which might be described as glandular masturbation (see figure p. 286).

In the male, the most copious flow of strong-smelling musk seems to be correlated with the enlargement of the testes, at which time the scrotal skin has sometimes been found to be red in colour and the animal exceptionally active. Observations suggest that this behaviour is related to the male's rut and may be restricted to the breeding season. Glandular activity may therefore function as a stimulus to seasonal, socio-sexual behaviour helping to overcome highly ritualized defensive reactions in an otherwise solitary animal. The elaboration of the peculiar neck and skin may therefore be connected with the defensive caution of the potto and the need for a social bond. The evolution of such an odd "bonding device" in only one genus of the widely distributed *Lorisidae* is interesting. The Oriental lorises have neither specialized skin nor spines, but the rare and localized golden potto, *Arctocebus*, of West Africa shows rudimentary specialization of the scrotal skin which could be regarded as anticipatory of the potto's condition. A potto-like hunched posture is also characteristic of this species, but the nuchal spines do not project beyond the surface of the neck. It is interesting that *Perodicticus* is the more successful and widely distributed of the two genera.

The marking of one animal by another is a common means of cementing a social bond, for instance, individuals of the aggressive but highly social striped mongoose, *Mungos mungo*, mark one another with their anal glands. Like the potto they have a stimulus to restrict their marking to a particular part of the body, but in their case it is the visually differentiated area of the striped rump (see Vol. III), while the potto's stimulus is tactile and olfactory. The stimulation of a glandular area by sharp points is also not unique, for male antelopes with pre-orbital glands have periods of intense glandular activity correlated with the rut, during which time they mark every twig or thorn of their territory. In the antelope this may function as a territorial mark but in both cases the activity is clearly self-rewarding (except that the "twig" appears to be also rewarded in the potto's case). A young female, hand-reared in isolation from other pottos "invited" neck tickling from humans by presenting the nuchal area. The same animal also simulated the reciprocatory act by rubbing her genital area vigorously on her keeper's wrist watch, this was a ritualized rhythmic activity in which the pelvis was swung forward under the body.

Experimental scratching of the neck region in various pottos elicits different responses, from a rhythmic swaying trance to lunging and biting. For one individual potto stimulation of the neck seemed to act as a form of reassurance; this adult captive female, when presented with food, consistently refused to eat until her neck was gently scratched (A. Walker, personal communication).

Reactions to the stimulation of the nuchal area vary widely according to the disposition of the animal. This could be partly connected with sexual condition, for the activity and aggressiveness of pottos varies very widely and not only male pottos with large testes but also female vary in their tractability.

There is also an ambiguity about the nuchal region which is reflected in the contradictory statements of those who have written about pottos; proposals for an aggressive, defensive or sensory function to the neck have been made, all supported by sound behavioural evidence. The erogenous function of the nuchal region is not incompatible with its being exposed to injury. The sensitive lips, tongue and gums are next to the biting teeth of many mammals, which have occasion to use them daily in attack or defence. The pottos shoulders are well-adapted to use as a shield or a battering ram area (see figure below), being a well-knit complex of bone muscle, thick skin and connective tissue and it is this area which must take the main impact of an attacking predator or other potto.

The defensive behaviour mentioned earlier involves an initially immobile crouched position, on being approached closely the animal raises its shoulders, this is followed by an extension of its limbs, or a raising of the body and butting with the neck. In extreme irritation the potto makes a snapping lunge, which uncoils the long back anchored by the hands and feet (see drawing), allowing a snake-like strike with gaping mouth which is startling for its reach and suddenness. With further agitation it intersperses lunges and growls with bat-like twitters and attempts to escape. In the wild, this may be achieved by dropping out of the tree and making for the nearest thicket.

More significant are the reactions of pottos to one another, Cowgill (1964) describes two male pottos sitting neck to neck for long periods without further reaction. If pottos are indeed "contact" animals, this hieratic sitting apart without moving could be interpreted as a mild form of threat, since a closer approach seems to be inhibited. This must remain as a speculative interpretation, as the behaviour has not yet been seen in the wild and the sexual condition of the animals was not stated. Threat behaviour of a higher intensity has been observed and described by Blackwell and Menzies (1968): "Our two adult animals were introduced for the first time, they approached along a branch and came to a halt about one length apart . . . the back was arched, the head held down between the forelimbs and the neck fur was

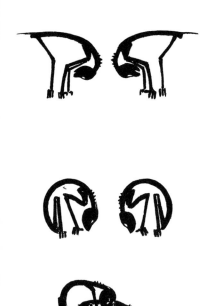

parted to reveal the tubercles. Suddenly they swung towards one another, limbs remaining stiff and pivoting only at the upper and lower joints. Their nuchal regions met in a heavy blow. This was repeated several times until both animals were bleeding about the region of the tubercles. . . . The whole combat was completely unprimate-like and resembled more the fight of two goats or antelopes.'' In this instance the incident occurred between a male and a female that subsequently mated together, so that the male was clearly in a reproductively active condition. Aggressiveness may therefore have been linked with the animal's sexual condition. One is reminded again of ungulates, for captive males have frequently been observed to damage females while in rut, and aggression in all species tends to be more easily provoked under the stress of captivity. This observation can be contrasted with that of Rahm (1960); one of his male pottos growled distinctly when he saw another male and took up the defensive position, whereas the same animal confronted with a female approached her uttering barely perceptible sounds and licked her. Another male potto, retreating after a fight with a second male, approached his mate and "hugged" her most persistently.

In East Africa and the eastern Congo, pottos have a very marked birth season between November and February (nine birth records), but there may be some extension of this season or there might be a second birth season, as there are five births recorded between April and June. There is some evidence that the male potto is solitary and sexually inactive for many months of the year and it is possible that the female may also have long periods of anoestrus. The triggers for the physiological changes initiating sexual behaviour are unknown, but the onset of the rains appears to be correlated with changes in the potto's metabolism. Animals collected during the later rains tend to be plump, with fat deposits between the muscles and thick layers of fat up to one cm thick in the groin and armpits; in this condition the male's testes tend to be large and active. Seasonal differences in diet have already been noted. Captives kept in open runs have also exhibited an increase in appetite and activity during the early rains (Walker, personal communication). By contrast, male pottos collected in the later dry season (births period) tend to be lean and thin with small testes.

The observation that the potto's physique waxes and wanes is not new, although the interpretation given in Bosman (1704) is in the finest bestiary tradition. "Some writers affirm that when the creature has climbed upon a tree, he doth not leave it until it hath eaten up not only the fruit but the leaves entirely; and then descends fat and in very good case in order to get up into another tree; but before his slow pace can compass it he becomes as poor and lean as tis possible to imagine and if the tree be high or the way anything distant, and he meets nothing on his journey, he invariably dies of hunger betwixt one tree and the other.''

The conspicuous thinness of male pottos in the dry season may be the result of exhaustion due to intense activity while the animal was in rut. The seasonality of breeding might therefore be determined by the need for the male to recover his condition. Such a pattern is not unusual in mammals, for instance, the males of several ungulate species cease eating altogether or drastically reduce their feeding during the rut (Fraser, 1968).

A marked seasonal rhythm may also be associated with colour phases and

moulting. Evidence for intermittent moulting is the occasional presence of large quantities of fur in the dung and stomach. Young adults develop a silvery back and a black shoulder mantle which later gives way to dun brown, mahogany or bay; the length of the fur also changes. The presence of quiescent hair follicles implies that a periodic shedding of the hair is a physiologically determined and normal pattern in all lorisoids. The long tactile vibrissae that grow on the neck of pottos are also capable of being shed, for instance one potto put in a sack lost its vibrissae simply by being in rough contact with the hessian. The vibrissae are found on both sexes but are often most developed on males. Their development is probably associated with the erogenous nature of the nuchal area and the scenting of the neck in "glandular masturbation". Their moulting may coincide with the decline of social and sexual life.

Pottos observed in captivity mate in the usual mammalian way and they also mate belly to belly (Sanderson, 1940). The gestation is thought to be in the region of six months. It is not known whether the female in the wild segregates herself at birth, but this is common in other lorisoids.

One or very occasionally two young are born, white or pale cream with blue eyes and a tuft of longer fur on the forehead. Captive mother pottos sometimes ignore a dropped infant, suggesting that they may need the stimulus of clinging, but one mother held on to her dead baby in her box until it decayed.

A young potto is carried by its mother on the belly and has its first adult foods by licking the edges of its mother's mouth while hanging underneath her, it takes its first solid food at five weeks. A captive mother used to "park" its infant on the wire ceiling of its run while it went off in search of food; the infant took the initiative in both leaving and re-mounting the mother (A. Walker, personal communication).

A hand-reared potto made a clicking noise whenever it was insecure and responded positively towards imitation clicking. It has a powerful clinging reflex for any woolly material; at the age of about five months it played rather like an aberrant kitten, rolling over, making short rushes and swinging by the hindlegs. From an early age there were long periods of grooming with teeth and tongue. At six weeks this animal exhibited the first signs of the nuchal ritual, humping the neck at its keeper, then when the neck was scratched it would hold on with a tight grip and thrust the body back, up and forward in a repeated and almost trance-like rhythm, with much flexing and stretching of the neck. If the scratching was discontinued it evinced signs of distress, and at the age of nine months would hump feverishly and then make mock lunges with the mouth open; the humping and lunging appeared to be a means of inviting neck tickling.

At six months pottos have adult colouring, are independent and behave very much as adults, they are fully mature at nine months.

Dominique found male juveniles abandon the mother's territory of their own accord and go through a roaming stage. Juvenile females instead appear to be accommodated by the mother altering or displacing her own territory. Once again the active role of the male contrasts strongly with the passive one of the female.

The potto's relationship with the palm civet, *Nandinia*, is interesting and could bear some investigation as this animal is often cited as its principal predator, an assertion for which there is no evidence. Pottos and palm civets have some convergent characteristics when one compares them with their respective relatives, the galagos and the genets. Each species has small ears, a thick, cryptic woolly coat and a pungent scent gland with which they mark other individuals; both eat largely immobile or inactive foods and are more deliberate in their movements than their relatives.

Owls, pythons, genets and perhaps monitors are the potto's most likely predators. They are occasionally electrocuted on high tension cables or run over on roads.

Postscript : Since going to press P. Charles-Dominique has published a These d'Etat on the natural history of the potto.

Greater Galago
(Galago crassicaudatus)

Family Prosimii
Order Primates
Local names
Komba (Swahili),
Kulo adadi (Galla).

Measurements
head and body
270—465 mm
tail
325—520 mm
weight
1,034—1,241 g

Greater Galago

Galago crassicaudatus is separable into a small brown northern type and a larger grey southern type.

Galago crassicaudatus crassicaudatus	Southern Africa
(*G. c. argentatus* and *G. c. montieri*)	H. and B. 285—465 mm
	T. 340—520 mm
Galago crassicaudatus agisymbanus	Northeastern Africa
(*G. c. lasiotis* and *G. c. kikuyuensis*)	H. and B. 270—330 mm
	T. 325—390 mm

Galago crassicaudatus panganiensis is intermediate in colouring between these two races and also occupies a geographic position between the two.

The numerous races described in the older typological classification tend to obscure the well-defined groupings indicated above. The subspecies synonymized above within the two major races were originally described on

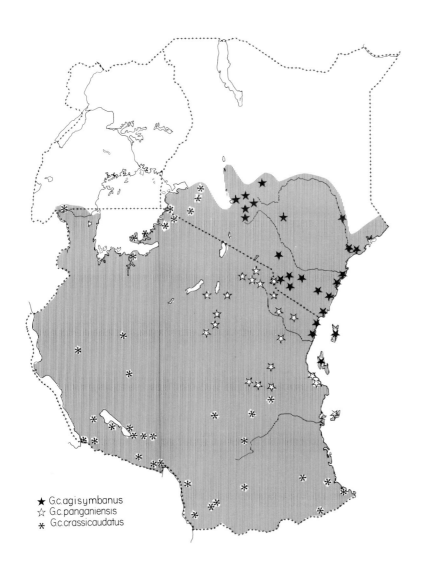

★ G.c.agisymbanus
☆ G.c.panganiensis
* G.c.crassicaudatus

the basis of geographically ill-defined differences in the tonality of the animals' extremities, i.e.: hands, feet, muzzle and tail tip and in washes of silver, fawn or brown in the colour of the fur. Some of these differences can be correlated with climate. For instance, galagoes from the Kenya Highlands east of the rift tend to be a darker brown while animals from more arid areas near the coast tend to be small and pale (sometimes with white tail tips). In the Highlands west of the rift *G. c. crassicaudatus* has a local tendency to melanism; the majority of (but not all) the specimens from the Mau and Kisii Highlands being black. The populations of *G. c. crassicaudatus* flanking Lake Victoria and the southern edges of the Congo forests are larger than those from the dry savannas in southern and Central Tanzania.

The greater galago has most of the specialized features characteristic of the galagos but, in keeping with its greater size and weight, it has limbs of heavier build and a stouter skull. Compared to the other galagos the face is broader and longer and the eyes are smaller in relation to the head. The southern populations often have a very long, silky coat. As in other prosimians the coat has been observed to moult, at which time the hair is vigorously combed out with the incisors and swallowed.

This species ranges across the southern part of the continent from Angola and Southwest Africa to the East Coast from Natal to Somalia. Over this extensive range it inhabits a variety of relatively moist vegetation types. It has been able to adapt itself to changing conditions and is found in plantations of exotic trees, eucalyptus, cypress, mango and coffee as well as in suburban gardens. The breadth of its range is misleading however, for over large areas it is confined to riverine forest, isolated thicket, bamboo and some montane forests. It is particularly numerous along the coastal plain in thicket and forest. on the lower slopes of many East African mountain blocks (the Uluguru Mountains, Kilimanjaro and the Kikuyu Highlands) and along the southern and eastern shores of Lake Victoria. It occurs in areas with well-defined seasons and is confined to south of the equator excepting the coastal strip and wadis of Somalia. The factors restricting this galago to its present geographical range are difficult to understand as there are large areas of apparently suitable country in Uganda from which it is absent, this is odd in an animal that is both adaptable and successful in other parts of its range. There is hardly any overlap of range with the potto, *Perodicticus*, that occupies a comparable ecological niche; this could have some significance in relation to recent vegetation changes or obscure factors involving disease or competition.

The diet resembles that of other galagos, being composed of fruit, seeds, resin, insects and occasional vertebrates. This species appears to rely to a greater extent on fruit than other galagos and in those areas where they live among mango or avocado pear trees they seem to feed almost entirely on these fruits while in season. In northern Somalia this species is reported to subsist largely on the seeds and gum of a species of riverine acacia which in that area is the only large tree available to it. An interesting regional peculiarity appears in Pemba and Mafia Islands where the teeth of quite young animals become worn and broken, presumably from grit and sand in some locally favoured food.

In common with other African prosimians it has a fondness for saps and resins and some observers have claimed that it subsists largely on tree exudates, however, this is more likely to be a seasonal food even if it is a critical one for its survival in areas like Somalia. They eat many other fruits, seeds, berries, leaves, petals and a wide variety of small animals (including lizards and chameleons) birds and their eggs, and they have been reported to kill sleeping game birds and poultry of which only the head and brain were eaten. They are caught on the coast by leaving palm toddy or "pombe" (local beer) near a favourite tree or along its customary path. After drinking it the animal is found unconscious on the ground. These drinks resemble in smell and probably in taste rancid sap or over-ripe fruit which galagos eat in the wild.

Insects and living prey are caught with a grab of both hands or a leaping pounce. They drink both by lapping and licking their wet hands. An interesting difference appeared in the diet of a mother and her two half-grown young from southwest Tanzania. The mother had fed on fruit, seeds and gum with a few insects, the young had fed principally on giant land snails and one had also eaten insects, but neither had taken fruit or gum. As the animals were together the different choices suggest a tendency towards an invertebrate diet in the young and that a reliance on or preference for fruit and resin in adult life may be a secondary development.

In this species urine is transferred onto the hands by the back feet, and marking is particularly conspicuous when the animal is alarmed; captives in new situations are uneasy until their immediate surroundings have been thoroughly marked. They usually walk along branches with a gliding motion and with the tail following the line of the body. On the ground they are slow and generally walk with the tail held erect. They are adept jumpers in spite of their size.

For several months of the year these animals advertise their presence with a loud croaking wail repeated at frequent intervals during the early part of the night. This monotonous call may be made 50—130 times in an hour; like that of other bushbabies it appears to be seasonal and connected with territorial behaviour and the cries are interrupted by chattering whenever another galago approaches the caller. They owe the name "bushbaby" to this call for its resemblance with the cry of a child. Along the Kenya coast these calls have been noticed to be most obvious from May to October, which would

appear to be the period before, during and after the mating season and before the young are born. There is also a shrill cry that seems to act as an alarm, and the young make a curious high-frequency clicking sounding very like a grasshopper, which is associated with discomfort or hunger. Adults make a similar click when they are distressed, this has presumably derived from the infant call.

The greater galago is generally strictly nocturnal but is very occasionally seen about on overcast days. Its nocturnal activity has been investigated to determine how suitable a host the species is to mosquitoes. Different patterns of activity have been found which suggest that the sex of the individual animal and probably also the season influence the pattern. Haddow and Ellice (1964) found that one male had frequent long periods of complete immobility between short bouts of activity continued throughout the night; females showed a biphasic concentration of sustained activity in the evening and in the early morning, with a long lull during the middle of the night. All animals are most active in the early evening.

These galagos resemble *Galago senegalensis* in being seasonally gregarious. On present evidence this appears to revolve round an annual breeding season, but most data derive from populations living in areas with a single rainy season. This may not hold for the northern populations, indeed captives have bred twice a year and this suggests that it is climatic and environmental conditions that limit breeding to a single season. Twelve animals have been seen at a time and as many as nine have been recorded sleeping in one nest. These are large collections of leaves and twigs which are generally hidden in very thick tangles and may be used for years, large bird nests are also reported to be used. As the females generally remain alone with their young for two months or more in nests which they themselves prepare, it is possible that the males provide a social continuum. The aggressive tendencies of males rise when the females are in oestrus and the attraction of males towards females in oestrus may bring male parties and female-with-young parties together in larger aggregations. Fighting is very common and immature galagos are reported to be occasionally killed, presumably by adult males (possibly by suitors of the juveniles' mother).

Ranges may vary with the habitat, a solitary, free ranging pet in a suburb had a total known range of 6 hectares, but spent most of its time in a more restricted area. There are clearly defined routes that are followed, presumably by scent.

This species like *G. senegalensis* may be implicated in yellow fever transmission. The species may be subject occasionally to epidemic diseases, as numbers of galagos have been found dead at one time in South Africa (the disease was not identified). An adult has been found in the gut of a puff-adder, and eagle-owls, snakes and various wild cats are probably not infrequent predators. They are often killed by dogs in towns and villages and are eaten by people in some areas. Bush fires must act as a severe hazard and burnt bushbabies have been reported; the species preference for moist vegetation may be a reflection of this controlling factor.

Births are reported to occur in March in Somalia and between August and November in Tanzania and Zambia, the common factor would seem to be the birth of the young at the end of the dry season. As they start being

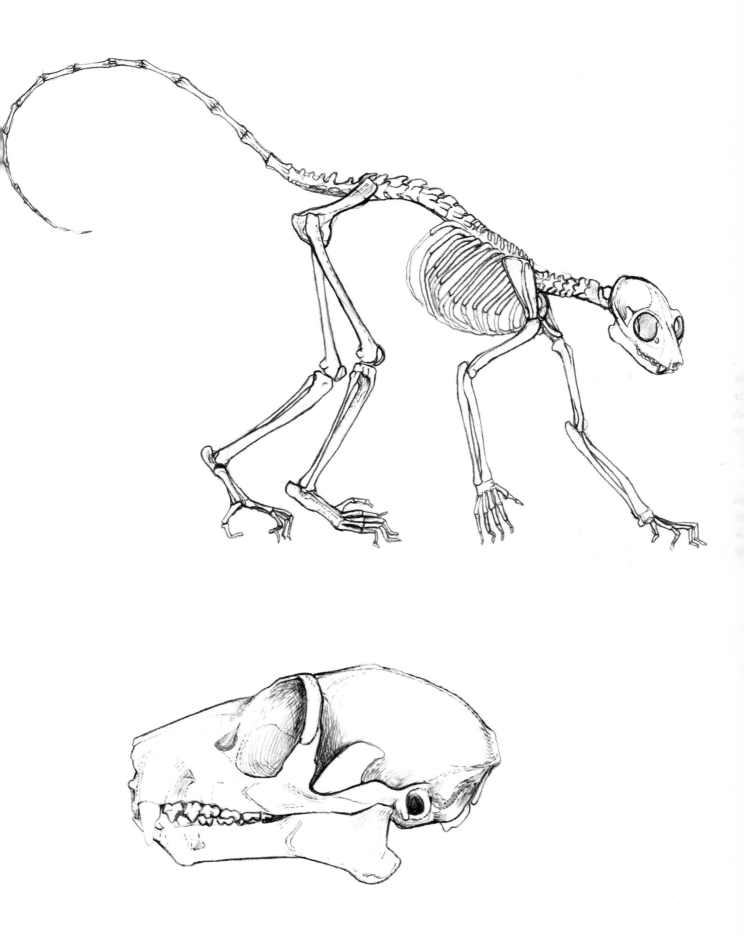

weaned at one month and are not physically fed by the mother, it may be critical that suitable foods are easily found and available at this time, which would be during the early rains. If insects and gastropods are the preferred foods of juveniles (as suggested earlier), there might be a selective pressure for the young to be weaned when these foods are easily obtained, for later in the year insects and snails are not so much in evidence.

Two young are usual but three have been recorded and they weigh approximately 40 g at birth. The female nurses for over three months but the young begin feeding when they are one month old. The young are left in the nest for long periods and the clinging reflex is weaker than in other primates, nonetheless the mother may carry them clinging to her body or by the scruff of the neck.

Sexes are rather similar in size and appearance, although males are somewhat heavier and have broader muzzles. The female clitoris resembles the male penis closely. In captivity they have lived for over 14 years.

Recent observations in Ruanda suggest that a high incidence of melanism in populations of *G. c. crassicaudatus* may not be confined to the Kenya Highlands alone but may also be common in the uplands to the west of Lake Victoria.

**Pigmy Galago,
Demidoff's Galago
(Galago demidovii)**

Family   Prosimii
Order   Primates
Local names

Endinda (Kuamba), Kanduki
(Lukonjo).

**Measurements
head and body**
125—155 mm
tail
150—215 mm
ear
20—22 mm
weight
95—120 g

Pigmy Galago, Demidoff's Galago

This very small galago has a short turned-up nose, large eyes in a round head with rather small ears relative to the other galagos. The spatulate pads on the end of the slender fingers of the hands and feet are very marked.

In *Galago demidovii thomasi* the fur varies between dark greyish or greyish-brown and a bright gingery colour. The differences appear to be due to moulting and the tail in particular may show irregular patches of colour when in moult. The belly fur also varies considerably and may be a bright yellowish colour or pale greyish white. There is usually a dark mask round the eyes and a white stripe down the nose. (There is a superficial convergence with the dormouse *Graphiurus* in colouring and pattern.) There is often some yellow staining over areas of bare skin and it is possible that this influences fur colour.

The species occurs in the main forest belt of Africa. It is found at various altitudes, in lowland and montane forest, bamboo, swamp forest, gallery forest and is particularly noticeable in *Cynometra*. They are amongst the commonest mammals of some Uganda forests. They range throughout the

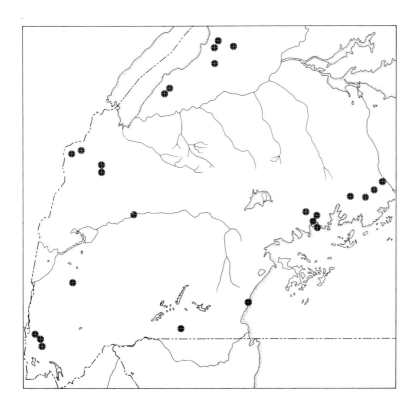

◀ Distribution West Uganda.

301

forest layers and will move from the canopy to the undergrowth and vice versa with the greatest rapidity, generally preferring to run along the branches, but jumping from branch to branch when in the undergrowth or in bamboo. They sometimes bob briefly before jumping, apparently scanning distance.

Their nests are made daily or are augmented with fresh leaves each morning; they may be made at any level, from near ground up to about 18 m and are identifiable as rather untidy collections of leaves, generally built on slender branches where there is a suitably fork. Locally available green leaves or even broad-bladed forest grasses are used and the leaves of a great variety of tree species have been found in the nests.

They are primarily insectivorous. An analysis of 40 stomachs from all times of the year and from five different forests showed that insects made up 78% of the diet. While there were plagues of caterpillars in the *Cynometra* canopy (March—April) the galagos were feeding almost exclusively on these, but when a type of cricket was particularly abundant in the undergrowth (December) these were the principal food. The commonest insects eaten are Coleoptera, Orthoptera and Hymenoptera. In the sample of 40 stomach contents 18% of the diet was gum or resin. This is a common food, having been eaten at all seasons by 15 of the animals examined, the remaining 4% was made up of fruit and seeds. They catch insects by pouncing and grabbing with both hands and in their pursuit the galago displays a speed and efficiency which defies the eye.

In exploratory behaviour, or in normal progression along branches in the forest, this galago scurries along with its back legs doubled up beneath the belly. The small naked heel spots are often rather conspicuously displayed on either side of the tail in this posture and, since these can be displayed or hidden by the surrounding fur, it seems likely that they may have a social function. Whenever two captive galagos meet one another there is a rapid sniffing and scanning of the face, or of the anal region and it appears that the condition of the naked skin may have some special olfactory significance. Naked skin is found on the hands, the feet and the heel spots, on the face and around the genitals; all these areas are prone to a bright yellow stain, which varies in intensity from individual to individual. Of 33 *G. d. thomasi* sampled from five forests throughout the year, all but seven adult males showed this distinct staining; the staining was very intense on ten individuals, of which four were females (either subadult or just pregnant) and six were males, one of which was subadult. The degree of staining shows no correlation with seasons and is only related to sexual condition, inasmuch as no female was found that lacked the colouring altogether.

G. demidovii often lives at fairly high densities and is generally social, parties of eight or twelve animals being not uncommon, but like other galagos they are also very excitable and pugnacious. Some of the animals without yellow colouring were old heavy males with torn ears, suggesting that these were dominant animals. In one party of eight galagos there were four males, the youngest of which had diminutive testes and was very heavily pigmented, two had very large testes and were slightly yellow, while the fourth had scarcely a trace of yellow, also had very large testes, and was the heaviest animal in the party. One of the two second males had a wound on the nose; such wounds are not uncommon in males and are probably due to fighting (perhaps inflicted in this instance by another animal in the party). This evidence suggests that the pigment might be the visual symptom of some biochemical mechanism that might serve to reduce male aggression. The nose to nose greeting and the display of the "heel spot" could therefore allow the pigment to have its appeasing effect on the aggressive male, and nose wounds may be an indication of what ensues when the appeasing action is not sufficiently strong. Dominique has shown that large males play a central role in the social unit or party which disperses and reforms each night. The pigment might therefore protect vulnerable members in a society that assembles and disperses from night to night and which probably has a high rate of turnover from year to year. A marked increase in pigment was particularly noticeable in a captive mother and her female offspring as the young animal approached adult size and the mother ceased to give milk.

The pigmy galago marks with a very rapid transfer of urine from urethra to hand to foot. Captives of both sexes urine mark vigorously in new quarters and a female was also seen to rub her chin on her baby and on branches near the nest.

They have a vocabulary of fruit-bat-like sounds, the most characteristic and most frequently heard call in the forest is a repetitive and shrill rasping chink, which increases in pitch and generally lasts for about four seconds. This call seems to signify excitement but it may be territorial as it is uttered from time to time throughout the night, often without responses. The most vocal period is generally during the first hours of darkness when parties are dispersing for the night and in the morning when they are meeting or reforming. At these times there are numerous phono-responses (Dominique). A similar shrill chinking was made by a subadult captive female when her mother refused to relinquish food.

Pigmy galagos are inquisitive and will come down out of the canopy to investigate and occasionally to scold an observer, making a burring, chittering note which seems to signify both fear and curiosity. The alarm call is very loud and sustained sometimes interrupted by a scolding burr.

They usually emerge from their nests just before dark and are most active during the night, but they are easily disturbed and may come out of their nests during the day. I have seen a party still active at 9 a.m. and another party, that may have been disturbed by a troop of redtail monkeys, at about 11.30 a.m. Three or four animals may share a nest and when first woken they tend to move very slowly and are reminiscent of miniature pottos (*Perodicticus*) in the deliberation of their movements. They threaten crouching or standing upright and opening the gape very widely, meanwhile rocking slightly on their heels.

In a party of seven galagos found in early September near the nest of an eighth female that had just given birth there were four males and two females (the sex of the seventh was not identified). Both females had early foetuses and three of the males had very large active testes, while the fourth was subadult. The party might have been an aggregation of pairs, supplemented by the females' most recent offspring, a social pattern typical of the other East African galagos. They differ from other galagos in being very rarely solitary, and when a lone galago is seen a diligent search will usually reveal another, or others in the immediate neighbourhood. It appears that only the female

separates herself from her fellows when she gives birth to her young. This difference may be due to the poorly defined breeding season. Births or extrapolated births are recorded from all months of the year, so that the females are not all with young at the same time of the year, and I have no records of fully adult females that were neither nursing nor pregnant. Nonetheless there is a definite mating peak in August, with a birth peak in December—February. Males with fully active testes were found in September. Of a large sample of males in December the great majority had only moderately developed, or small testes; females were in various stages of pregnancy or lactating, while five births or extrapolated births are recorded for the period December—February. In February—March there are nine records of sub-adults. It is possible that there is a second birth peak in June.

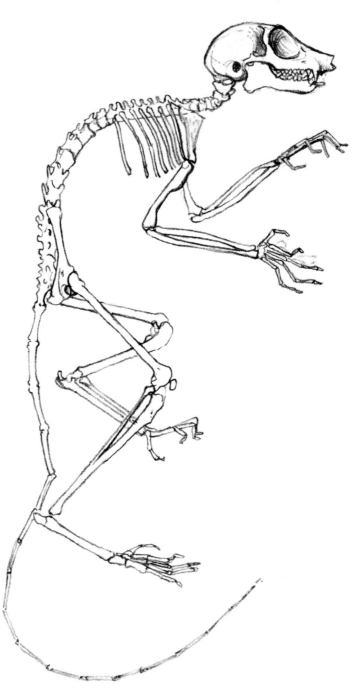

Dominique found localized groups in which females occupied ranges of about 800 sq. m with variable degree of overlap. The males in the group belonged to three classes. The first, "A centrals", were heavy individuals (4 in his study group). They circulated over the range of the females and monopolized all the male–female contacts. "B centrals" were smaller males having contact with these males but not with females. Subadult and other small males formed a "peripheral" class which was often involved in fights. Each localized group or party had little or no contact with neighbouring populations.

G. demidovii has been inoculated with yellow fever virus and, although responses were irregular, this bushbaby alone among the primates of Africa exhibits a considerable natural resistance to infection with yellow fever virus (Simpson, 1965; Burgher, 1951). This animal may be the "oldest" and the commonest primate type living in the canopy; its relationship to yellow fever is therefore of the greatest interest, as the principal vectors of this disease are canopy mosquitoes.

Probable predators of *G. demidovii* are genets, snakes, palm civets, owls and forest hawks.

The courtship may not differ greatly from that of other galagos and the gestation is between three and four months. Blackwell (1969) records 110 days. This species is reported to differ from the other galagos in the implantation of the zygote deep into the mucous lining of the uterus. In other prosimians the zygote is attached to the lining of the uterus, the irregular uterine lining being impressed onto the walls of the blastocyst (Hill, 1953).

Of ten female Uganda galagos five had single foetuses and young, and five had twin foetuses. These results may reflect local genetic tendencies but the species certainly twins with some frequency.

A newly born baby weighs 25 g at birth, with a head and body length of 75 mm and a 90 mm long tail. It was born with a very large head and was rather

uncoordinated and wobbly for the first day. Growth, however, is very rapid in this species and the young are able to clamber about near the nest on the second day. The mother carries the infant in her mouth and can also make long jumps while carrying it. Yellow staining appears on the baby at the age of two weeks; at this age the baby is very active, although still on milk, and will accompany the mother out of the nest. At three weeks it attempts small jumps and may climb about hanging upside down from a branch. It also threatens, bites and can run well. When resting in the nest it clings upside down on the mother's belly. By the time it is a month old the juvenile urine-marks, transferring urine with the hands to the hind feet and will make ambitious jumps, but being still rather clumsy it can occasionally fall and, on the whole, prefers to run. Scraps are stolen or picked up from the mothers' food, but captive mothers were never observed to actively feed their young which tend, however, to be very importunate and scarcely ever leave the feeding mothers' side. By six weeks the young is weaned, but continues to beg and grab food from the mother until it is nine weeks old and starts to find its own food. By this age it is approximately adult in size, is very yellow skinned and is occasionally fiercely repulsed by the mother. Young of this age in the wild would have joined a party, while the female, in the majority of cases would be pregnant again. The infant-to-mother relocation cry develops into an adult call that aids regrouping in the morning (Dominique). The species has lived ten years in captivity (Linn 1970).

Postscript : Since going to press P. Charles-Dominique has published an account of the natural history of this species in *Zeitschrift fur Tierpsycologie* 1970.

Zanzibar Galago
(Galago zanzibaricus)

Family	Prosimii
Order	Primates
Local names	

Komba kidogo (Swahili), Kideri (Kikami).

Measurements
head and body
140—165 mm
tail
200—230 mm
weight
120 g (approx.)

Zanzibar Galago

Synonyms

Galago senegalensis zanzibaricus
Galago demidovii orinus
Galago cocos

This galago was originally described by Matschie (1893) from the Zanzibar Island population and has subsequently been the subject of some confusion. It has been regarded both as a race of *Galago senegalensis* and of *Galago demidovii*.

The type specimen of *Galago demidovii orinus* (Lawrence and Washburn, 1936) was a cinnamon-coloured animal shot in company with two adult female "*Galago senegalensis zanzibaricus*". The collector, Loveridge, correctly regarded the type as a young individual of the same species. These

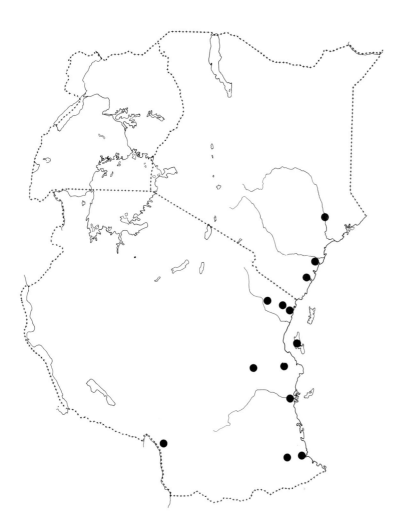

galagos were collected at about 1800 m in the Uluguru Mountains. I am acquainted with the type locality, which Loveridge (1951) described as no more than a "*clump* of giant trees, all that remained of the glorious virgin forest that once covered the hill. The slopes were planted with bananas while between them and the huts was a fringe of small trees smothered in lianas". An unlikely habitat to be shared by two very similar species.

Three principal differences distinguishing the type from the other galagos shot with it led to its later classification as *Galago demidovii*. One was its cinnamon colour, another was an exceptional elongation of the premaxilla and another difference was the lack of swellings at the root of the canines.

Specimens described as *G. d. orinus* in the British Museum collections come from southern Tanzania but they agree closely with the description of the type, particularly in relation to colour, the long premaxilla and the lack of swellings above the canines. However, all are young animals. Their colouring, small size and undeveloped canine roots are features associated with age and are characteristics shared by other young *G. zanzibaricus* collected from several localities down the entire littoral between the Tana and the Rovuma rivers.

I have found the elongated premaxilla to be subject to individual variation. Very marked elongation occurs in several young animals and also in two fully adult *G. zanzibaricus* (both in the upper range of body measurements for the species) from southern Tanzania. The premaxilla is particularly short in all animals from Zanzibar Island and in those from the northern parts of its mainland range. The length of the premaxilla seems to range widely between extremes in the Tanganyika animals. Possibly there is a tendency for young animals to have a longer rostrum to the nose than adults, but it does look as though there might also be a south to north gradient in this feature.

Apart from its small body size, a long turned up nose is perhaps the most striking characteristic of *G. demidovii*. It is therefore not surprising that the first discovery of this feature in a small galago from an eastern forest should lead to its being regarded as an easterly representative of this species. However, the type specimen was also found to have a feature typical of *G. senegalensis*. It had a difference of 0·5 mm between the width of the second and third molars, whereas *G. demidovii* generally shows less difference in the size of these molars. Superficially there are probably equally valid reasons to view this galago either as a forest *G. senegalensis*, or as an isolated eastern population of *G. demidovii*. Against the first view, widely scattered populations of this galago are surrounded by savanna-dwelling *G. senegalensis* with no evidence of hybridization. Hollister (1924) describing this galago said "not intimately related to *Galago braccatus* which occurs less than 15 miles inland from Mazeras" (where *G. zanzibaricus* were collected). Against the second view *G. zanzibaricus* has a distinct resemblance to *G. s. gallarum*.

The taxonomic situation for this galago is similar to that of *Galago inustus* (with features of *Galago elegantulus* and *Galago senegalensis*) in that it is exactly intermediate anatomically between two forms. Some authors have even placed the linked forms in different genera, i.e. *Galagoides demidovii* (Smith, 1833) and *Euoticus elegantulus* (Gray, 1863).

Before discussing the taxonomy further a description will be necessary, which should be read with reference to the two related species. The ears of

Galago zanzibaricus are smaller than those of *Galago senegalensis* and larger than those of *Galago demidovii*. The skull is intermediate in all respects. The profile is concave, but less so than in *Galago demidovii*.

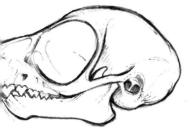

G. senegalensis.

G. zanzibaricus.

G. demidovii.

G. zanzibaricus is intermediate in mean size, although there is in fact a small overlap between typical *G. demidovii* and typical *G. senegalensis*, when the adult size ranges of these galagos are compared.

	G. demidovii (thomasi)	G. zanzibaricus	G. senegalensis
HEAD AND BODY	125—155 mm (mean 145 mm)	140—165 mm (mean 155 mm)	145—200 mm (mean 175 mm)
TAIL	150—214 mm	200—230 mm	205—303 mm
Mean zygomatic breadth	25 mm	27 mm	32 mm

As in *G. demidovii* the fur colour varies considerably; in this the moult probably plays a part. Coastal populations are generally paler and more cinnamon than *G. demidovii thomasi*, but the two species scarcely differ at all in fur texture and in the tendency for some individuals, especially sub-adults, to have a strong yellow suffusion to the fur, and there is generally a greater superficial resemblance with *G. demidovii thomasi* than with *G. senegalensis*. The large eyes are a rich mahogany brown, closer in size and colour to *G. senegalensis* than to the olive coloured eyes of *G. demidovii*.

Its range is very extensive, it occurs in high altitude forest on widely separated massifs in Tanzania, and in lowland forest along the coastal strip from the Rovuma to the Tana and on Zanzibar Island.

The distribution pattern has a certain resemblance with that of other mammals of the "eastern forests", which have affinities with related forms in the main African forest block (*Colobus* spp. *Cercopithecus mitis, Cephalophus* spp. *Anomalurus, Neotragus*). All these species have differentiated and have adapted to decidedly drier conditions than those found in the "western forests".

If for the moment we regard *G. zanzibaricus* as an aberrant, peripheral population of *G. demidovii*, the following remarks by E. Mayr (1954) may have some relevance to the taxonomic situation. He points out that there is a genetic reorganization or "revolution" in peripherally isolated populations, and that every aberrant population is peripheral:

> In fact, peripheral populations are neglected even by the taxonomists of most living faunas because they are comparatively small, isolated, and often in far distant or inaccessible places. . . . It is quite evident that they are not only incipient species but in many instances also incipient genera and higher categories . . . most of these populations will eventually die out without playing a major role in the evolutionary picture. Only a very occasional one will be able to reach a vacant ecological niche. As soon as such a population has completed its genetic reconstruction and ecological transformation it is ready to break out of its isolation and invade new areas. Only then will it become widespread and is likely to be found in the fossil record.

If this galago is an aberrant *G. demidovii* it is also an incipient *G. senegalensis*, but the ecological transformation is incomplete, for while the eastern forests are generally drier than those in which *Galago demidovii* are found they are still forests and the driest habitat in which *Galago zanzibaricus* occurs is the riverine *Acacia* bordering the Tana River. It is only with *Galago senegalensis* that a "genetic reconstruction and ecological transformation" allows the species to break out of its isolation and invade the savanna.

It is therefore of some interest that *Galago senegalensis gallarum*, with the closest resemblance to *Galago zanzibaricus*, is adapted to truly arid conditions and that its range more or less abuts the northern limits of *Galago zanzibaricus*. If *G. s. gallarum* derived from *G. zanzibaricus* the difference in the former's habitat-preference may have been brought about by the isolation of an ancestral population in a very much drier habitat than that occupied by *G. zanzibaricus* today. The climatic fluctuations that might have been responsible for this isolation could have occurred no later than the Pliocene, because *Galago senegalensis* were widespread by the Pleistocene being present as fossils throughout the beds at Olduvai.

After a million or more years it may seem unlikely that *G. s. gallarum* should retain any evidence of its possible primitive status, still less that its contemporary distribution should be of any relevance after the extreme climatic fluctuations of the Pleistocene. However, it is not impossible that a contemporary population should provide some indication of both the nature and, very approximately, the locus of an important evolutionary development.

In eastern Africa, any long-term increase in aridity has tended to proceed from the Somali arid focus, and it is worth noting that the contemporary forests of northeastern Africa exhibit graded successions in which there are ever drier types which eventually culminate in thicket. The driest types tend to be the ones nearest to the horn of Africa.

The opportunities to adapt to drier conditions must have been offered to a large number of isolated populations scattered in numerous forest relics, and the opportunities would have recurred with every major climatic fluctuation.

The present range of *G. s. gallarum* very approximately covers the zone in which such a hypothetical adaptation might have taken place were it in progress today. If this race does in fact represent the primitive type of *G. senegalensis* which adapted to a particular biotope, it would not be extraordinary if the refinement of its adaptations caused a certain conservatism in its morphology and its ecological range; while more advanced populations colonized the various savannas of Africa, this race remained wedded to its original ecological zone.

It has been impossible to discuss this galago without referring in some detail to its related species; its habits also call for comparison. Subjectively, the voice appears to be roughly intermediate, being harsher and more croaky than that of *G. demidovii*, but decidedly shriller than that of the typical *G. senegalensis*. Gucwinska and Gucwinski (1968) describe twitters and a loud mating cry. They also thought this species to be more sluggish during the day than *G. demidovii*. Bouncing from branch to branch is possibly slightly commoner than in *G. demidovii* and it may be more prone to scurry along branches than *G. senegalensis*. Comparative studies might reveal an intermediate position in locomotory patterns. When alarmed one captive male *G. zanzibaricus* showed a penchant for making brief runs down branches back first.

They are reported both to make nests and to sleep in hollow trees, and further documentation of their nesting habits would be most interesting.

I have kept the three species mentioned here over long periods and have attempted to hybridize a male *G. zanzibaricus* with two female *G. d. thomasi*; to date without success, although the three animals have lived together amicably enough for a year.

Foetuses have been found during the long rains in the Uluguru Mountains and near Mombasa during December, a juvenile from the Tana River area was collected in March.

Gucwinska and Gucwinski (1968) have successfully kept and bred this galago in Poland where they have two litters a year in captivity. Copulation takes place soon after giving birth and the female may be mated by more than one male. The males are reported to make a characteristic cry as they chase the females.

Gestation is 120 days; twins or single young may be born. Other members of the captive galago group are reported to take a great interest in the newborn. All animals start crying out loudly when a youngster makes distress calls and even juvenile animals will try to carry an infant in difficulties. Infants are carried in the teeth while rolled up in a ball. They often suckle lying on their back, meanwhile holding the mother's head with their hindlegs.

The young appear to be awkward up to the age of 10 days. At one month they are two-thirds the adult size and eat all types of food. By four months they approximate the adult's size and are sexually active at one year. The young indulge in much play chasing and will slide repeatedly down slippery leaves.

Further details of this species' behaviour and breeding under captive conditions may be found in Gucwinska and Gucwinski's paper (1968) in the International Zoo Yearbook, Vol. 8.

Lesser Galago, Bushbaby
(Galago senegalensis)

Family — Prosimii
Order — Primates
Local names
Komba (Swahili), Idekelwa (Itesot), Adokole (Karamajong), Okodu (Lwo), Kamenamena (Kuamba), Lende (Lugbara), Ngei (Runyoro).

Measurements
head and body
145—200 mm
tail
205—303 mm
weight
108—300 g

Lesser Galago, Bushbaby

Races

Galago senegalensis senegalensis
Grey, large-eared, yellowish-white underparts.

West Uganda

Galago senegalensis braccatus
(*Galago s. albipes* and *Galago s. sotikae*)
Grey, smaller-eared in the east, larger-eared in the west, with bright yellow thighs.

East Uganda, South Kenya, North and Coastal Tanzania

Galago senegalensis gallarum
Brownish, smaller-eared, yellowish limbs.

Northeast Kenya

Galago senegalensis moholi
Grey, with brownish rump, large-eared with yellow thighs.

Tanzania

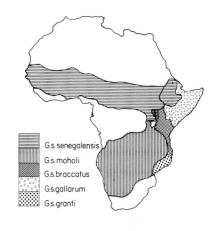

G.s.senegalensis
G.s.moholi
G.s.braccatus
G.s.gallarum
G.s.granti

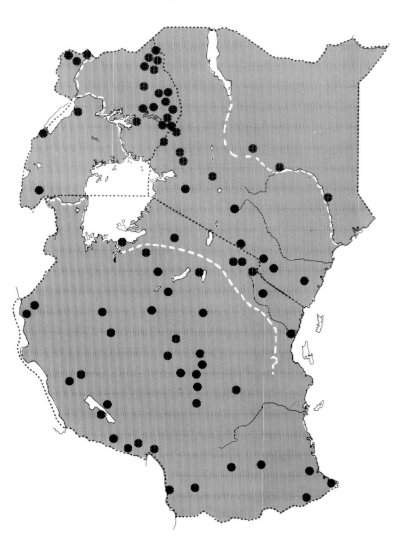

317

Several races of lesser galago have been described from East Africa. They include two very widely distributed subspecies; *Galago senegalensis senegalensis*, which ranges throughout the northern savannas and *Galago senegalensis moholi* from the southern savannas. The northern Kenya race, *Galago senegalensis gallarum* was described by Schwartz (1931) as having a similar range in East Africa to that of Grevy's zebra. Rode (1937) includes all other East African varieties in *Galago senegalensis braccatus*, a race which tends to merge with *G. s. senegalensis* in Uganda. The boundaries of *G. s. braccatus* and *G. s. moholi* await definition as do the northern and eastern margins of the latter race. Boundaries suggested on the distribution map are tentative and do not take account of the mixed or graded types which undoubtedly occur.

The lesser galago ranges throughout the savannas, bushland and woodlands of Africa, but is not generally found above about 2,000 m. It is most abundant in *Acacia* and *Combretum* bush and in *Isoberlinia* woodland. The common shelter in these habitats is inside the hollow branches or trunks of trees. Some species like *Terminalia, Adansonia, Isoberlinia* and *Combretum* are very prone to being hollow and the availability of such shelters may be a very important determinant of the density of population. Nests are sometimes made in tangles or in forks of branches but are less common than natural shelters (where nesting material may or may not be used). The preference for hollow trees and for localities where grass is thin and burns off without too violent a conflagration reflects the frequent hazards faced by this species from bush fires. Lesser galagos are said to be almost impossible to smoke out of their hollows and this behaviour may well be an adaptation to annual or biannual burning. In the *Brachystegia* woodlands they are frequent tenants of the local beehives, which are hollowed logs or bark drums hung from branches.

Their habitat suffers great seasonal changes which are reflected in their diet which includes a wide range of foods. There are also regional differences according to the area and the vegetation in which they live. Insects are probably the single most important item of the diet, but they are not commonly available throughout the year; for instance, caterpillars are only common while there are fresh, green leaves and galagos are usually only able to eat them in large numbers during the wet season. Similarly, scarab beetles are numerous and active in the rains and at times may constitute the bulk of the galagos food in areas where these beetles are common. While *Piliostigma* and *Adansonia* are in flower, galagos visit the blossoms for nectar and the petals of *Acacia, Hybiscus* and cotton flowers are also eaten. A dominant tree in semi-arid areas is *Balanites aegyptiaca* (heglig, desert date) and while this tree is fruiting its fruit may become a principal food. Other fruits eaten are *Tamarindus* and *Sclerocarya*, also the seeds in immature cotton bolls. During the dry season, galagos have been seen picking termites off bark and also turning over stones; scorpions, spiders and other cryptic insects are probably found in this way. Many species of *Acacia* and *Albizzia* exude resin (gum-arabic) and while the resin is flowing it is a major food for galagos, indeed they are capable of subsisting on it alone for long periods. The association of *G. senegalensis* with *Acacia* is a close one and the species' reliance on gum at times when and in places where other foods are scarce undoubtedly contributes to this species' success in what are sometimes unpromising environments. Other foods recorded for galagos are seeds, nestlings, lizards, mice and growing shoots. The young are reported to be initially very frightened of emperor moths and hawk moths; aposomatic insects are avoided or spat out with every sign of distaste.

Both the eyes and ears are important in locating prey and a shrilling cicada secured by a string is a device used to trap these galagos. When pouncing on an insect, both hands are used; they can make enormous jumps which allow them to grab prey from a distance with the greatest speed and precision. A measured jump of about seven metres is recorded. The lesser galago therefore has an ability to locate prey and a striking range comparable to that of birds and this has undoubtedly contributed to the species' successful adaptation to open savanna. When on open ground the galago may make kangaroo hops but it tires easily and the jump is primarily an adaptation to secure prey, escape enemies and circumvent obstacles and is not its sole mode of locomotion. In suitable vegetation and undisturbed conditions the animal employs an unhurried quadrupedal walk or run.

The galago is very vocal, the most commonly heard call being a stereotyped croaking which appears to be territorial, this is made by both sexes but is usually maintained longest by the male. Intruding galagos or other animals may be clucked at (low intensity), chattered at (medium intensity) or a shrill whistling note may distinguish a high intensity alarm. A variety of quiet contact notes between individuals can also be heard.

The urine-marking characteristic of all galagos is present in this species to a marked degree. Captives always urine mark new objects or cleaned surfaces that appear within their cage or home range and like other species they leave by identical routes from their hole or nest in the evening, using exactly the same leaping perches and branches. Each animal urine marks prior to setting out in the evening, and sometimes may urine mark again before making a leap, the routes are therefore constantly rescented and such branches acquire a distinct "patina". Sauer and Sauer (1963) observed rivalry between males which ostentatiously urinated on the borders of their territory. Male urine marking is associated with meeting a strange animal, with the approach of a female, when the animal enters its nest and when it smells the infants.

After a very active period immediately after sunset, this galago has numerous short bursts of activity throughout the night (Haddow and Ellice, 1964). These are interspersed with periods during which animals are inactive. It is rarely seen on cold, windy or wet nights and seems to be more difficult to find on bright moonlit nights. Activity patterns also vary according to the reproductive cycle, and sexually excited males are particularly vocal and restless. They may rest in their holes or on a favourite roost nearby, their croaking calls are generally made during otherwise inactive spells. This species is nocturnal, but they have been seen just outside their holes during the day and are capable of accurate leaps after insects in daylight.

When resting, the large, delicate ears fold back flat along the side of the head and this is also observable when the animal is in thorn bushes or thick tangles. Torn ears are probably a mark of fighting.

Aggression seems to depend to a very large extent on the galagos' reproductive condition. Males often fight unfamiliar animals but their tendency

to fight increases when the females are in oestrus, the females on the other hand can become very aggressive just before or after the birth of their young, in captivity this may even lead to her injuring the male (Doyle and Bekker). Where aggregations of galagos are brought about by captivity or occur naturally, the capacity for aggression is often overcome, at least during the rest periods, by a love of contact, so that a number may share a hollow tree trunk, box or nest. Such sleeping parties usually split up into smaller units at night only to come together in the early morning to sleep. I have seen over twenty *G. senegalensis* in a large baobab, *Adansonia*, just before dawn, chasing one another about the branches before finally disappearing into the big tree, which presumably provided the only suitable shelter in a wide area of very thinly bushed country. The chasing was perhaps symptomatic of an ebbing

320

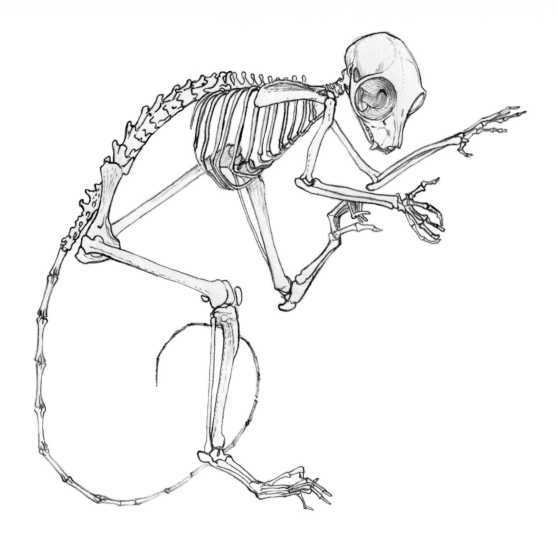

aggression that became overcome by the need for shelter as the light became stronger. The association of such large numbers of galagos is probably a relatively rare phenomenon, partly encouraged by an absence of large trees, but other observers have seen the galago in equally large numbers (Durrell, 1954).

The social organization can be characterized as an extension of the parent-offspring relationship in that the female with her juvenile offspring comes into oestrus, whereupon she attracts and quite possibly seeks the adult males.

There are regional differences in the breeding pattern of *G. senegalensis*. Galagos from Zambia and southern Tanzania tend to bear their young towards the end of the dry season between August and December, with a birth peak in September/October—the gestation period being four months. This means that the mating season coincides broadly with the end of the rains, when the animals are in best condition. At present there is no evidence of breeding at other times of the year. There are advantages for the young to be born at the beginning of the rains, and the breeding season may also coincide with the onset of oestrus in the females that were born eight months before. Twins are usual in the south. In northern Uganda there are two rather ragged breeding seasons and a single infant is commoner than twins. The first birth season is in December—February and the second is in June. The seasonal correlation that was suggested for the southern areas of *G. senegalensis*' range are rather different in the north where there are two wet seasons. The

December—February births are at the height of the dry season and the June births also occur in a second dry spell. During their second or third month of life the young born in both seasons have a period when food is relatively abundant, while the mating peak about September follows five or six months of adequate rain and nutritionally ideal conditions.

Butler (1967) in the Sudan found anoestrus periods in female *G. senegalensis* of up to three months and reports on periodicity of oestrus in this species' range down to less than twenty days. Oestrus therefore appears to be highly variable and its timing may be affected by environmental factors. Males remain interested in females throughout the year but reproductive behaviour in males is stimulated by the appearance of a white vaginal discharge which may last 5—10 days and may attract more than one male to follow the female. The male or males meanwhile "clucking" and occasionally urine marking.

Copulation is accompanied by a loud call by the male and is sometimes followed by grooming. Females in captivity may form attachments to one male and pairs are common in the wild. When giving birth the female is generally on her own, in captivity the young may be killed by the male when he is not separated before birth.

The new-born galago weighs about 12 g. At birth the mother grooms it, picks it up in the hands and the infant immediately seeks contact with the mother. She scarcely moves during the first three days but thereafter when she goes out for food she will either leave the infant or infants in the nest, or carry them in her mouth or clinging to her back or belly. The young suckle for six weeks and are able to feed themselves at two months, their growth is

322

very rapid and when two large young are attached to the mother she is quite unable to jump and can only walk slowly and awkwardly. When she has twins, she may sometimes alternate the young she takes with her. The young are playful and indulge in much hopping, chasing, wrestling and imaginary catching of food.

Galagos are popular pets and in spite of their small size are caught for food in some areas. They have lived 14 years in captivity.

The female's isolation at parturition may cause temporary disintegration of small units into solitary individuals or, in larger groups, may cause all-male or predominantly male units. Such units are known, and where there is a marked seasonal breeding peak this may amount to a typical seasonal decline of social activity in the birth season and a peak of sexual and territorial behaviour when the majority of females are in oestrus. Males may tolerate the presence in the nest of subadult animals accompanying their now pregnant mothers and the tolerance often extends to include other females and their young and more rarely other adult males. The period of greatest social tolerance would appear to be between the mating and birth peaks, but more evidence is necessary before this can be shown to be the case.

Eagle owls, *Bubo*, and genets, *Genetta*, are recorded as eating galagos and large snakes may also take them.

The importance of this species for man lies in its possible involvement in yellow fever epidemiology (Haddow and Ellice, 1964). The species is capable of circulating yellow fever virus after experimental infection, and yellow fever antibodies have been discovered in wild populations. This galago is therefore capable of infecting mosquitoes if it is bitten while the virus is circulating, and the animal came under particular suspicion in relation to a great yellow fever epidemic in the Sudanese Nuba Mountains in 1940, an area where *G. senegalensis* is extremely numerous. Curiously enough, an inverse relationship between immunity in galagos and monkeys has been observed, this might reflect a local pattern in the habits of these primates or of the mosquitoes.

323

Needle-clawed Galago
(Galago inustus)

Family   Prosimii
Order   Primates
Local names

Katinda mugogo (Runyankole).

Measurements
head and body
160—195 mm
tail
195—260 mm
weight
170—250 g

Needle-clawed Galago

This galago is anatomically intermediate between *Galago senegalensis* and the West African *Galago elegantulus*; its fur colour and texture resembles that of *Galago demidovii thomasi* with a yellowish tinge to the fur on the shoulders and a very dark mask round the eyes. *G. elegantulus* has the largest eyes among galagos, it has a specialized alimentary tract (Hill, 1953) and teeth and has developed pointed nails. The significance of these specializations is not known. *Galago inustus* shows some features which anticipate the specialized condition of *G. elegantulus*. The nails are keeled and pointed and the eyes have a clearly defined ridge along the upper margin of the orbit. In the lower jaw the molars have a more linear alignment and are less broad than those of *G. senegalensis*. The mandibular ramus is shallow and splayed so that it intrudes less into the orbital cavity, a development that is clearly an accommodation for the large forward-looking eyes of this species. No trace of yellow skin pigment has been observed. The eyes are larger and yellower than those of *G. senegalensis*.

The species is restricted to forests in central Africa and its anatomical resemblances suggest that the larger *G. elegantulus* evolved in isolation in the West African forests, having derived from a stock resembling *G. inustus*, which in turn might have been an aberrant population of *Galago senegalensis* returned to a forest habitat. In spite of its great resemblance with *G. senegalensis*, *G. inustus* may reveal something of the specialized nature of its West African relative when more is known of its habits.

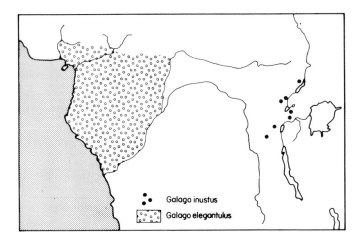

The animal is found in medium altitude forest where *Parinari excelsa* is the dominant tree, it also occurs along the forest margins. It moves freely between the canopy and thick tangles at lower levels. It is sympatric with *Galago demidovii*, but the latter species appears to be relatively rare or inconspicuous in the forests where *G. inustus* is common. Recent observations suggest that this species may be widespread in southern and western Uganda; its eastern range possibly reaching the Victoria Nile.

There is some evidence of seasonal changes in diet. Specimens collected in the November rains had only fed on insects, while others in February had fed principally on fruit and resin. The February sample was collected on a single night together with a specimen of *G. demidovii* from the same locality, this galago had only eaten insects so that a seasonal shortage of insects alone is unlikely to explain the difference. A third specimen collected in July had fed equally on fruit, resin, caterpillars and very small beetles. A specimen from the Congo (Rahm, 1966) had eaten insects in September.

G. inustus has a very shrill, loud alarm call similar to that of *G. senegalensis*, it also makes a hoarse insect-like churr.

A male and three females collected in western Uganda during November were in breeding condition, another female was lactating at this time. In February, a female from the same locality was pregnant with a single 50 mm foetus. A large male collected in July had small testes. There may therefore be a breeding peak in November—December.

G. senegalensis.   *G. inustus.*

 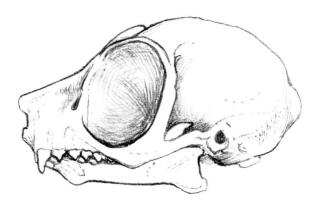

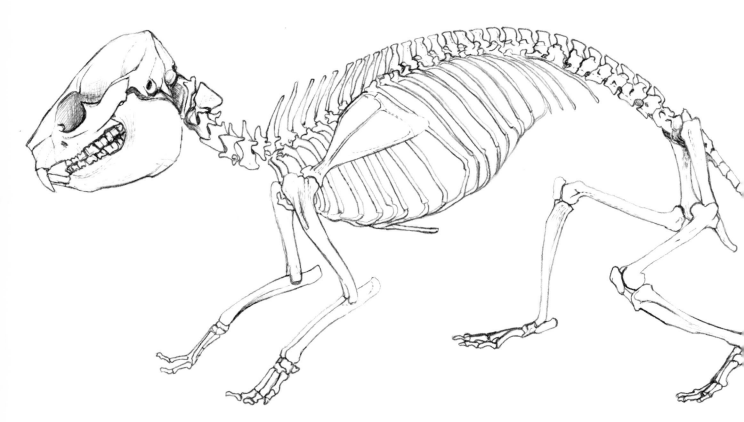

Procavia.

Hyraxes

Procaviidae

The hyraxes have had a particular interest for zoology ever since it was discovered that they have similarities with the primitive Eocene ungulates known as *Paenungulata*, the stock from which the elephants and the sea cows are thought to have developed. Numerous fossil finds (largely from the Oligocene of Egypt and the Lower Miocene of East Africa) show that they have been a local but very successful group in the past. The early forms diverged in their adaptive structure and in size, for instance a large bear-sized hyrax, *Megalohyrax*, is known from the Lower Miocene of Kenya.

Both fossil and living forms are only known from Africa and the eastern Mediterranean. Simpson (1945) remarked on this odd situation, "since the group is old, has been locally abundant and associated with mammals much more widespread, this implies some peculiar ecological limitations the nature of which is not clearly apparent".

Anatomical peculiarities of the group are the absence of sweat glands and of a gall bladder (Grasse, 1955). The olfactory system possesses nasal glands and several sensory organs, "Stenson's Channel" and the "Organ of Jacobson"; the brain is of primitive ungulate type. The lower incisors in all modern species are modified to act as a fur comb, while two of the upper incisors serve as canine-like tusks. Vestigial teeth found in the milk dentition of some individual hyrax are thought to represent canines originating as involuted premolars (A. Brauer, 1913). Probably correlated with their powerful voice is a sort of guttural pouch in the Eustachian channel which would seem to act as a resonator. The hyrax cry probably has both a territorial and a sexual function. They have a bicornuate uterus and a similar placenta to that of Lemuroidea. They have rete mirabile in their limbs. The pads on the feet are pliant and rubbery but their efficacy as climbing members seems to depend primarily on a strong grip and supple articulation of the limbs. The hyraxes, particularly the arboreal species, are marvellously agile and adept climbers, but the structure of the feet cannot be described as a primary adaptation for climbing and appears to be a phylogenetically late makeshift.

Hyraxes have thick furry coats with scattered long sensory vibrissae. All genera have dorsal sebaceous glands which are probably used for marking of the home range; it is possible that the gland may also advertise sexual condition. In addition to the auditory and olfactory "markers" mentioned above, the hyrax habit of always depositing their dung and urine in "lavatories" is probably also territorial in function.

A similarity between the Pliocene *Pliohyrax* of Samos and *Procavia* has been remarked on by Withworth (1954), but this large form is not regarded as being directly related. A closer resemblance to living hyrax is discernible in two Pliocene hyracoids excavated at Ngorora near Baringo (these have not yet been described but are dated at about 10 million years (W. Bishop, personal communication).

Sale (1965) has suggested that the very long gestation of living species and the heavy weight and complete development of the young at birth are evidence that modern hyrax have become smaller.

The decline of the hyracoids since the Miocene coincides with the rise of the bovids which have come to fill almost every ungulate niche (see p. 55). The feeding of all hyrax species is clumsy as virtually all cropping is done with the molars, by turning the head sideways and opening the very wide gape a large mouthful of vegetation can be seized, but the mechanism is clearly not as effective as that employed by ungulates. This functional inadequacy, together with poor temperature regulation and short plantigrade feet, has probably placed limitations on the hyracoidea and might explain the decline of this once numerous group in Africa.

The living forms seem to represent a relatively late radiation which initially escaped competition by adapting to the rocky thickets which are found scattered over so large an area of Africa. Such outcrops have a tendency to conserve water and vegetation and would have provided these primitive

Dissection of newborn *D. arboreus*.

animals with the constant environmental conditions they need. Their physiological vulnerability is borne out by frequent observations that captives die of heat stroke if exposed for too long in the sun. They are completely dependant on natural shelter for their existence, as they are unable to dig. They are only able to run fast over a short distance and have no more than their tusklike incisors and well-organized "social threat" for defence.

The generally smaller *Heterohyrax brucei* is limited to eastern and southern African but it is the dominant hyrax in less arid areas, it is certainly the commonest species in East Africa. In the Sahara, a species of *Heterohyrax* appears to have been confused with *Procavia ruficeps bounhioli* (Allen, 1939). The species in question, *Heterohyrax antinae*, may be of doubtful validity (Hatt, 1936) but it serves to illustrate that the distinctions between these genera can become very slender.

Broken rocks raise difficulties for all modern ungulates except for the klip-springer, while climbing trees is almost beyond the bovids adaptive powers; these two habitats have therefore been free from the competition of more developed ungulates, and in spite of inadequate feet (relatively to truly arboreal forms like monkeys and squirrels) hyraxes exploit these niches very successfully.

The body is much longer than it appears and when running the gait is a bouncy gallop with the back flexed considerably for each leap. The normal walk is creeping and plantigrade. While sitting or feeding the back is arched and the animal consequently looks fat and round.

The three genera have been erected on ill-defined differences in skull structure, and in a very variable group these criteria are easily challenged. Notwithstanding this, some recognition of the distinctness of the three groups will be necessary even if the present generic structure is abolished, for the genera very roughly coincide with three phases of development.

Dissection of a hyrax (*Dendrohyrax dorsalis*).

Procavia is the more robust with larger tusks, the least marked diastema, a shorter muzzle and a post-orbital bar which is partly cartilage. The dental peculiarities of *Procavia* may have to do with its predominantly grazing habits, nonetheless, this genus probably retains the most primitive features; it is the most widely distributed species and is especially successful in the drier areas of northern and southern Africa.

Heterohyrax brucei is largely a browser and tree climber, and ancestral *Heterohyrax* may have first gained advantage over a larger parent stock (probably closely resembling modern *Procavia*) by climbing out further on branches to feed on buds and leaves. Although still tied to rocky areas for shelter, this type appears to have started an orthoselective trend towards more arboreal habits and ultimately to the invasion of true forest.

The habits and the skull structure of *Heterohyrax* suggests an intermediate condition between *Procavia* and *Dendrohyrax*, and one race from Malawi, *Heterohyrax brucei manningi*, appears to have similarities with *Dendrohyrax validus*. An animal that might be a hybrid between these species has been collected at Morogoro (see p. 341). It is very likely that *Dendrohyrax* evolved from *Heterohyrax* and considering the range of these genera and the climatic changes that would have encouraged this development, it probably occurred in eastern Africa. The *Dendrohyrax* species have achieved complete emancipation from rock shelters, although in many mountain localities the colonization of alpine areas by *Dendrohyrax* has led to a return to rock dwelling.

In spite of their ecological and geographic preferences all the three hyrax genera are adaptable and will survive on almost any vegetable food. They are very successful colonizers where their basic shelter requirements can be met. The opportunity to colonize a new habitat has occurred on several mountain ranges in East Africa, where glaciers have retreated or broken ground and lava flows have been invaded by plant growth.

There are highland areas of East Africa where a different genus, species or subspecies of hyrax has in each case succeeded in colonizing this habitat when it became available. Where the mountains were surrounded by forest as on Ruwenzori, the Bufumbira volcanoes and the Usambaras, a *Dendrohyrax* species has colonized this habitat. On Mt Kenya, an arid belt on the north of the mountain opened the alpine habitat to *Procavia johnstoni*. In the Livingstone Mountains and Mt Rungwe, *Heterohyrax brucei lademanni*, and on Mt Elgon *Heterohyrax brucei kempi* have been successful, suggesting that this was the species at hand when the habitat first became available.

Each hyrax type adjusts its habits to the environment, and as physical conditions are similar, the mountain races often have a closer superficial resemblance to one another than to their conspecific lowland populations. The successful colonization of the mountains might have chronological implications, for instance the alpine habitat on Mt Elgon might have become available to *Heterohyrax* before *Dendrohyrax* had evolved, or it may be a direct product of competition. Coe (1962) has suggested the latter course for the Mt Kenya hyrax. It is particularly interesting that *Heterohyrax* flourish between 2,700 and 3,000 m on Elgon, *Dendrohyrax arboreus* in the montane forest, while *Procavia habessinica* occurs at 2,000 m and *Heterohyrax* is found again in the foothills.

Interspecific competition between hyrax in East Africa is a refined and widespread process. *Dendrohyrax arboreus* has replaced *Dendrohyrax validus* along the eastern seaboard and *D. validus* is now confined to Zanzibar and Pemba. However, this species has resisted competition in the montane forests of Uluguru, Usambara, Pare, Kilimanjaro and Meru. *Dendrohyrax arboreus* has in turn been ousted in Uganda by *Dendrohyrax dorsalis* but this species has been limited to lowland rain forests and has failed to gain ground in montane areas or where forest is less developed. The figure below is a graphic representation of this aspect of hyrax distribution in East Africa.

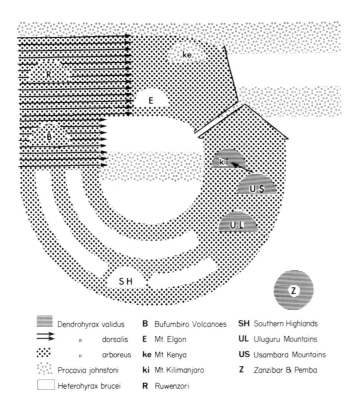

▤	Dendrohyrax validus	**B**	Bufumbiro Volcanoes	**SH**	Southern Highlands
→	" dorsalis	**E**	Mt. Elgon	**UL**	Uluguru Mountains
∴	" arboreus	**ke**	Mt. Kenya	**US**	Usambara Mountains
⬚	Procavia johnstoni	**ki**	Mt. Kilimanjaro	**Z**	Zanzibar & Pemba
☐	Heterohyrax brucei	**R**	Ruwenzori		

In spite of their biblical reputation as "feeble folk" hyrax are successful animals. Their capacity to eat almost any type of vegetation has assisted their success, and in some habitats the only competition they face is that of the other hyrax species. On farmland in the Karoo in South Africa reduced competition and the extermination of many predators has allowed *Procavia capensis* to increase enormously. It invades cropland and pastures and has become an agricultural pest as much as the rabbit was in Australia.

Procavia johnstoni mackinderi.

Rock Hyrax,	**Family**	Procaviidae	**Measurements**
Rock Rabbit	**Order**	Hyracoidea	Vary with species
(Procavia)	**Local names**		and races
	Pimbe (Swahili), Okille (Lwo).		**total length**
			450—550 mm
			weight
			2,500—3,700 g

Rock Hyrax, Rock Rabbit

Procavia. This hyrax is recognizable by its rounder head and short muzzle. The molar teeth are heavy and hypsodont. The premolar tooth row 1—4 being much shorter in length than the molars 1—3.

Species:

Procavia johnstoni, Procavia habessinica.

Procavia johnstoni : dorsal spot black or yellow. Black patch behind ears.

Procavia johnstoni matschiei	North Tanzania
Procavia johnstoni lopesi	West Uganda
Procavia johnstoni mackinderi	Mt Kenya

Procavia habessinica : dorsal spot black or yellow or with mottled hairs.

Procavia habessinica daemon	West Kenya
Procavia habessinica jacksoni	Central Kenya

This genus is more widely distributed than the other hyraxes and is found in northern and southern Africa as well as Sinai, Israel, Syria and Arabia.

The two species are found in drier areas of Kenya, Uganda and northern Tanzania, and one race, *Procavia johnstoni mackinderi*, has colonized alpine Mt Kenya between 10,500—15,500 ft (3,200—4,650 m).

Procavia live in both very hot and very cold country, and members of the genus live within a total temperature range of 41·6° C—minus 5° C. However, Sale (1966) found that, inside the holes, there is a smaller temperature range (between 3 and 10° C). He also found that *Procavia* avoid isolated holes, those

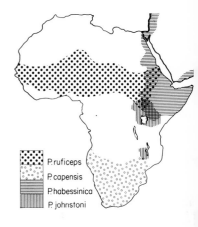

P. ruficeps
P. capensis
P. habessinica
P. johnstoni

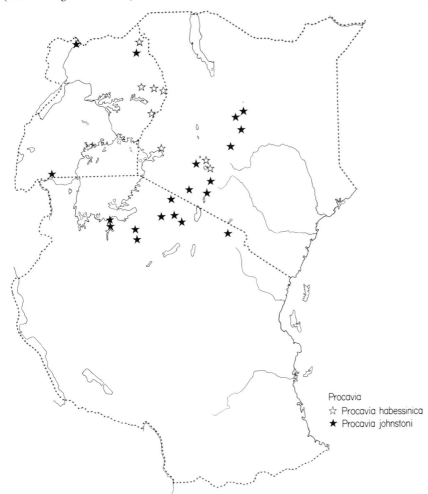

Procavia
☆ Procavia habessinica
★ Procavia johnstoni

facing the prevailing wind and those large enough for predators to enter. These observations probably hold true for other genera of hyrax living in rocky habitats.

The food is variable, sedges, herbs and tree leaves are eaten, but grasses are predominant. The most common foods of the Mt Kenya hyrax are moss, *Philonotis*, tussock grass, *Festuca*, and a composite, *Haplocarpha rueppellii*. The dry country species eat a variety of short grasses, also *Boscia* spp., *Grewia* spp., *Croton*, *Commiphora* and fallen *Acacia* pods. Sale has observed one colony feeding almost exclusively on the highly poisonous leaves and shoots of *Phylotacca dodecandra*. Tests have shown the plant contains a high proportion of haemolytic saponin. A physiological explanation for this immunity has not been suggested. Lang (in Hatt, 1936) reports *Procavia* wading into puddles to feed on the sprouts of a flowering water plant.

They feed very rapidly, sometimes for little more than half an hour if they are undisturbed; the principal feeding periods are during the early morning and again shortly before sunset. They are usually within reach of shelter and are led out, often in single file, by large old animals. Like other hyrax they follow well-defined tracks; once settled they may form "fans" feeding in a group with their backs to one another.

The species is very alert, particularly while away from their holes, I have found hyrax colonies round Lake Victoria almost impossible to approach due to a very noisy cliffchat, *Thamnolea cinnamomeiventris*, which shares the hyrax habitat and betrays one's presence. Another species of chat is reported to act as an alarmist in South Africa. In Serengeti, the tree-climbing *Heterohyrax* performs a similar function for the *Procavia* which are enabled thereby to graze further from their rock shelters than if they were without such well-placed sentries.

The commonest sound made by this species is a twittering contact call; there is a low intensity whistle which inhibits feeding and alerts the group and a loud roar of alarm which precipitates headlong flight. After being chased an old animal may turn about at the entrance of the hole, to threaten, growling

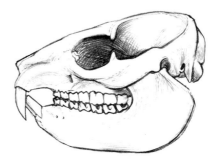

Procavia johnstoni.

and grinding its teeth. The upper incisors or tusks have a defensive function and can inflict deep gashes. The lower incisors are used as a comb to clean the fur. *Procavia*, like other hyrax, may croak loudly, this very carrying noise is nearly always produced near the entrance to a hole with the animal facing out. Its head is raised and the lips slightly parted. The hyrax is less alert while

Procavia habessinica.

making this cry and can often be approached closely. It can be heard in the morning and evening, not infrequently on moonlit nights when it may be answered by the calls of distant animals.

They enjoy sun-bathing and also rolling in dust. A colony usually smells strongly, dung and urine are deposited in "latrines" and the urine forms a caked deposit on rocks that is sought after for folk medicine to cure kidney and bladder complaints. This material was once called "Hyraceum" and was thought to soothe hysteria, it is still sold as "Dassiepiss" in South Africa.

In a hyrax colony there are between 25—60 animals, made up from family groups with a dominant male in each. There is usually a single old male which is the leader and the most watchful animal in the colony. They are very territorial and will threaten intruders of the same species or will attack enemies as a group. In South Africa, Hanse (1962) watched fourteen adult hyraxes converge on a young jackal and succeed in intimidating it with sudden

Procavia johnstoni mackinderi.

movements and barking threats. He also saw a strange hyrax killed by the dominant animals of the colony. I have seen numbers of hyrax come to the entrance of their holes and threaten dogs with barks and chattering teeth.

Hawks, eagles, snakes, various cats, leopards, genets, wild dogs, jackals, baboons and owls prey upon rock hyraxes. Occasional epidemics are reported to kill large numbers of these hyraxes and *Procavia* have been reported to die of bubonic plague. The species is eaten in some areas but is not normally hunted except by small boys.

The gestation is about 214—225 days and usually one or two young are born. Up to five young born at one time have been reported. They are active from birth and start to nibble green food within a day or two of birth. This hyrax has been found to breed seasonally in South Africa. Colonies of *Procavia*

338

habessinica in the Kenya rift valley have been shown by Sale (1969) to have distinctly different breeding patterns depending on whether they live out on the rift floor or on the precipitous rift wall. The former have a slight birth peak between August and November, while the latter have a very marked peak in June and July. Sale suggests that the differences are due to a local climatic pattern, in which rain is relatively continuous over the rift floor, but falls mainly from March to May (and again in November) along the rift wall. The mountain race, *P. j. mackinderi*, gives birth between August and January.

Sale observed that pregnant females approaching parturition form temporary nursing groups. The young are born easily and quickly and are on their feet within minutes; there is a very marked urge to climb in Hyrax, and the young try to climb onto the mother's back as soon as they are born. By this means the young come into contact with the dorsal gland and become marked with its scent. Other significant functions suggested by Sale for this behaviour are: cementing the mother-young bond, insulation from the cold rock and preparation for adult watchfulness. The young make a bird-like chirrup distress call, and twitter when suckling and on regaining contact with the mother, this later becomes the adult contact note. The weight of *P. j. mackinderi* at birth is 240—405 g; the animal has quadrupled its birthweight at three months by which time the young is weaned. Thereafter growth is steady, about 150 g per month. At fifteen months the animal is mature.

Yellow spotted Rock Rabbit, Bruce's Hyrax
(Heterohyrax brucei)

Family   Procaviidae
Order   Hyracoidea
Local names

Pimbe (Swahili), Orsolee (Galla),
Mutiliondet (Sebei).

Measurements
total length
305—465 mm
weight
1,020—2,210 g

Yellow spotted Rock Rabbit, Bruce's Hyrax

Heterohyrax: Yellowish dorsal spot and relatively small size. The face has a "peaky" look, as the head is narrow and the short muzzle carries weaker incisor teeth than the other hyrax genera. In the adult toothrow the length of P 1—4 is equal to or just less than the length of M 1—3.

Species:

Heterohyrax brucei	
Heterohyrax brucei bakeri	West Nile
Heterohyrax brucei kempi	Elgon
Heterohyrax brucei hindei (*albipes*)	Northeast and Central Kenya
Heterohyrax brucei diesneri (*Vic. nyansae*)	Northwest and North Tanzania
Heterohyrax brucei munzneri	Central-West Tanzania
Heterohyrax brucei prittwitzi	Central-East Tanzania
Heterohyrax brucei songeae (*frommi*)	South and Southeast Tanzania
Heterohyrax brucei lademanni	Southern Highlands

Note:

A large white-bellied hyrax collected near Morogoro and named *Dendrohyrax validus schusteri* might be a hybrid between *Dendrohyrax validus* and *Heterohyrax brucei*.

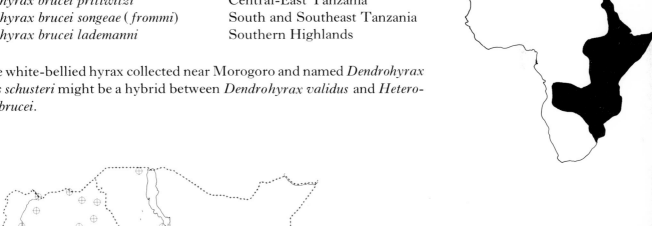

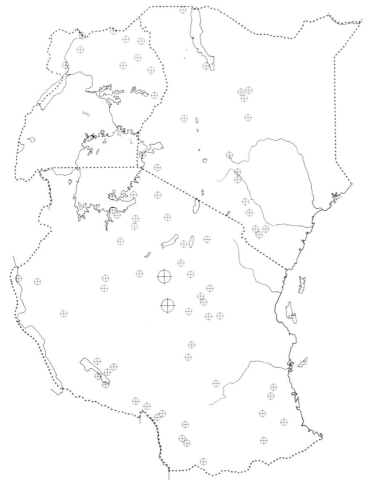

This species is found in rocky outcrops throughout East Africa, like all hyrax it is adaptable and covers a considerable range of altitudes from sea-level to 3,800 m. It fills the same alpine niche in the Southern Highlands and on Mt Elgon that is occupied by *Procavia* on Mt Kenya and by *Dendrohyrax* on other mountain blocks. Elsewhere it lives in a variety of vegetation zones but usually near woody growth, in rocky outcrops with extensive natural cracks, holes and cavities that provide safe shelter.

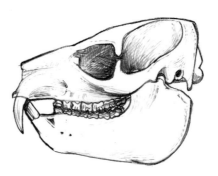

In some areas this hyrax is a very catholic feeder and browses on a wide variety of plants, eating leaves, fruit, stems, twigs and bark, it is also reputed to eat locusts. The animal feeds from 7.30 a.m.—11 a.m. and again from 4 p.m.—6 p.m.; it may also be active at night, particularly in moonlight. On the Serengeti outcrops Turner (1965) found that 90% of the *Heterohyrax* diet was *Acacia tortilis* and *Allophylus rubifolia* and that grass was never eaten. Where there were no *Acacia tortilis* trees in what were otherwise apparently suitable situations, *Heterohyrax* were absent.

It is a capable climber and will jump out of tall trees without injury. Group composition and social behaviour would seem to resemble *Procavia*, living in large colonies they suffer from similar limitations imposed by the habitat and by predators. They sometimes feed at some distance from their rock shelters and are taken by leopards, other carnivores, large snakes and also eagles. This species co-exists with *Procavia johnstoni* in many of the outcrops around the southern shores of Lake Victoria and on the Serengeti plains. Coetzee (1966) found an anatomical difference in South African species which may be sufficient to maintain sexual isolation: the penis of *Heterohyrax* is situated along the belly (rather as in a bull) whereas that of *Procavia* emerges near the anus.

Turner (1965) has investigated the nature of these mixed colonies and shown that different diets allow minimal overlap and competition for food resources and also that the *Heterohyrax*, feeding high up in the branches of the trees are effective alarmists for *Procavia*, which consequently have a greater safety margin and may forage further from the rock shelters. The dispersal of the *Procavia* relieves pressure on those food plants that are shared by *Heterohyrax* in the immediate vicinity of the colony and competition for the same food is only apparent in the dry season.

Activities such as sun-bathing and grooming, usually in the early morning, may occur in rather closely packed groups in which the species mix, and mixed groups also rest in the shade between 11 a.m. and 4 p.m.

In some areas there does not seem to be any clearly defined breeding season, as young of all sizes are to be seen at a time, but *Heterohyrax* near Nairobi have a marked breeding peak, with most young born in February and March, very shortly before the rains.

In addition to noting the obvious advantage of abundant food during the weaning-growing period, Sale (1968) points out that thermo-regulation in juvenile hyraxes is extremely poor and that there might be some advantage in the young being born at the hottest time of the year.

Dendrohyrax arborcus ruwenzori.

Tree Hyrax
(Dendrohyrax)

Family Procaviidae
Order Hyracoidea
Local names
Perere (Swahili), M'ha (Chagga),
Echiyama (Lukonjo), Maparaka
(Runyoro), Njoga (Luganda),
Uruwango (Runnyankole).

Measurements
vary with species,
races and sex
total length
405—600 mm
weight
1,500—3,350 g

Tree Hyrax

Dendrohyrax, generally nocturnal and arboreal. In the adult toothrow P 1—4 exceeds the length of M 1—3.

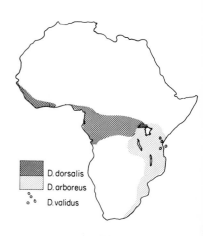

D.dorsalis
D.arboreus
D.validus

Species

Dendrohyrax validus, Dendrohyrax arboreus, Dendrohyrax dorsalis.

Dendrohyrax validus, cinnamon dorsal spot (glandular patch 20—40 mm), furred nose, without chin spot, belly orange, long soft hairs with dark bases, milk canines absent, upper diastema 15·2—16·1 mm.

Dendrohyrax validus validus	Mts Meru and Kilimanjaro
Dendrohyrax validus terricola	Pare, Usambaras, Ulugurus
Dendrohyrax validus neumanni	Zanzibar and Pemba

Note :
A large white-bellied hyrax collected near Morogoro and named *Dendrohyrax validus schusteri* might be a hybrid between *Dendrohyrax validus* and *Heterohyrax brucei*.

Dendrohyrax arboreus, cream-coloured dorsal spot (glandular patch 23—30 mm), furred nose, without chin spot, long soft fur, milk canines present, upper diastema 15·2—15·8 mm.

Dendrohyrax arboreus ruwenzorii	Ruwenzori
Dendrohyrax arboreus stuhlmanni	Tanzania and South Kenya
Dendrohyrax arboreus adolfi-friederici	Kigezi
Dendrohyrax arboreus crawshayi	Uganda and Kenya

Dendrohyrax dorsalis, long pale dorsal stripe (glandular patch 42—72 mm long), naked nose, white chin spot, milk canines absent, upper diastema 16·9—18·8 mm, disproportionately large head, may be anything between cream colour and near black.

Dendrohyrax dorsalis marmota	Uganda

Known to hybridize with *Dendrohyrax arboreus crawshayi* in Buganda.

The tree hyraxes are adaptable and variable. The species and races listed above are, however, quite distinct and closer examination reveals a pattern to their distribution which confirms the usefulness of this nomenclature.

The mountain forests of Pare, Usambara, Uluguru, Kilimanjaro and Meru, and the islands of Zanzibar and Pemba are occupied by a distinct species, *Dendrohyrax validus*. This distribution suggests a relic status and the species may represent the earliest hyrax type adapted to live in true forest. It probably derived from a population of *Heterohyrax*.

Dendrohyrax arboreus is found over a very extensive area of eastern Africa and probably represents a type that developed later than *D. validus*. *Dendrohyrax arboreus stuhlmanni* survives in gallery forest, riverine strips and montane relics, and occurs in a number of rather dry localities in Tanzania, a northern extension of this population is adapted to even drier conditions and lives in *Acacia* groves growing on the lava fields at Kilwezi.

A Late Pleistocene wet period probably established the forest race, *Dendrohyrax arboreus crawshayi*, across Uganda and into Kenya. This population meets the southern *Dendrohyrax arboreus stuhlmanni* in the vicinity of Nairobi, here the two populations dominate highland and lowland habitats

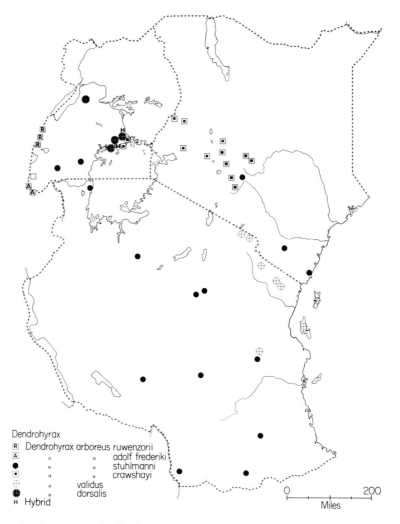

Dendrohyrax
R Dendrohyrax arboreus ruwenzorii
A   "   "   adolf frederiki
●   "   "   stuhlmanni
■   "   "   crawshayi
⊕   "   validus
         "   dorsalis
H Hybrid

0   200
    Miles

respectively, an ecological division that is not true for other parts of their range, for *Dendrohyrax arboreus crawshayi* is found in low-lying forest in Uganda and *D. a. stuhlmanni* inhabits montane areas in other parts of its range.

It would seem that the last great expansion of forest allowed a West African species, *Dendrohyrax dorsalis*, to reach East Africa. It is possible that the limited distribution of *D. dorsalis*, which apparently only reaches the Nile, might reflect a time lag between the expansion of forest over Kenya and the arrival of this western species. Its distribution overlaps that of *D. a. crawshayi*, but isolated records of *D. a. crawshayi* against numerous records of *D. dorsalis* suggests that it is in the process of replacing it. The species are also interbreeding, as hyraxes with intermediate characters have recently come to light. The race, *D. d. marmota*, is principally distinguished

Dendrohyrax dorsalis.

by softer fur than that of the more westerly populations and this difference in fur texture could be the result of long term hybridization with *D. a. crawshayi*.

Although *D. dorsalis* is the dominant type in lowland forest, *D. arboreus* seems to have held its own in the mountainous areas of the eastern Congo and West Uganda; *Dendrohyrax arboreus ruwenzorii* is a numerous animal throughout the alpine, subalpine and forest zones of Ruwenzori, it lives among rocks, is partly diurnal and has colonial habits on the higher reaches of the mountain, but is arboreal and nocturnal in the forest. It is a larger, woollier animal than the other races but otherwise resembles *D. a. crawshayi* closely. On the Bufumbira volcanoes is a pale, soft-furred race, *Dendrohyrax arboreus adolfi-friederici*, which has colonized the alpine lava fields and upper montane forest. In north Kigezi there is a form resembling *D. a. ruwenzorii*.

Tree hyraxes can be found from sea level to 4,500 m in all types of forest, in riverine vegetation of various types and among rocks, particularly at high altitude. The food range is equally great, for it can exploit the leaves, fruit and twigs of the forest canopy as effectively as the grasses and sedges of treeless mountain screes. Among many other food plants the leaves of *Ficus*, *Chlorophora*, *Ricinodendron*, ferns, *Asplenium*, *Senecio*, various *Graminaceae*, *Ranunculus*, *Alchemilla*, *Acacia* and the exotic *Eucalyptus* have been noted, also insects have been recorded as being eaten. They are rapid feeders, and eat a third of their own body-weight in food a day. Bees were found in one stomach, but they were almost certainly accidentally swallowed, having been combed out of the hair by the lower incisors.

347

The habitat clearly influences their behaviour considerably. The rock-dwelling populations of *D. arboreus* and *D. validus* may live at a high density with many family groups living close together. Populations living in forest, particularly *D. dorsalis* are more dispersed: pairs, small families or occasionally solitary animals occupy a small area of forest, usually centred on a single home tree. Forest-dwelling tree hyraxes are very dependent on lianas and the tree hyrax may be expected to suffer a decline in managed forests since these plants are a major weed to foresters and are being poisoned out. Tree hyraxes follow definite paths both in open country and in the trees. Even in the forest they descend to the ground with some frequency. In southern Tanzania and in Ankole, *D. arboreus* have occasionally been found to shelter in termitaries.

They can be very noisy and their calls often appear to follow a frequency pattern. My own experience suggests that the most vocal spells might follow intensive feeding periods, i.e. between 8 p.m. and 11 p.m. and between 3 a.m. and 5 a.m. Richard (1964) studying *D. dorsalis* in Gaboon recorded changes in the timing of calls. In December the maximum vocal output was between 7 and 8 p.m., in January it was about 2 a.m. In February between 10 and 11 p.m. and in March about midnight. The cry is made by both sexes but is very much more powerful in the male, this may be due to the larger larynx and guttural pouches of the male. It probably has both a territorial and sexual function and Mollaret (in Richard, 1964) reports a tree hyrax screaming while mating. The call is most noticeable on dry nights and during the dry season. Each animal builds up to a strained crescendo of screams and a calling animal appears to initiate responses from its neighbours so that on a suitable night there may be concert periods of croaking screams ringing out through the forest. The call usually appears to be made from the same vantage point which is a feature characteristic of territorialism. An adult male *D. dorsalis* in Uganda never called while it was in captivity but shortly after escaping could be regularly heard calling from some riverine vegetation where it took up residence (A. Walker, personal communication). Richard found his captive *D. dorsalis* going on rut without displaying any territorial behaviour, he never heard a captive utter its cry. I have heard *D. arboreus ruwenzorii* make a piercing alarm note to which all animals within earshot responded immediately. Richard records a low neighing sound in *D. dorsalis*, which seems to originate as an infantile distress call when the newborn is separated from the mother. This sound is uttered by adults with raised head and open mouth and is associated with great activity, trying to escape, and in a male as a preliminary to copulation. The mother responds to the infant call with a sound similar to that of a piglet and makes a "flapping" noise when disturbed with its young.

Richard reports that the males become very aggressive with the rutting season but may be tame at other times. A male *D. dorsalis* in Uganda exhibited very fierce behaviour when caught (see drawing) and fanned its dorsal spot on which secretion could be seen and smelled. It made threatening wheezing growls, ground its teeth meanwhile exposing the nictitating membrane of the eyes and made short fast rushes with snapping teeth. Richard noticed that a female produced the sebaceous secretion (smelling like cinnamon) on its dorsal gland for a few days preceding copulation. These two observations suggest that the gland has an important role in sexual behaviour as well as in

348

territorial marking.

A nervous animal will open its mouth wide and the display of its long teeth may have something in common with the primate "yawns" that are typical of male patas monkeys and baboons. *D. dorsalis* also shares sexual allometry of the skull with these primate species.

The excreta are a useful clue to the resting place of the tree hyrax, as the small hare-like droppings accumulate at the base of a well-used tree. Most species are nocturnal, particularly *D. dorsalis*, but the mountain races may feed in the early morning or evening and are fond of sun-bathing near the entrance of their holes in the middle of the day.

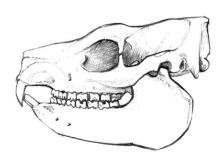
D. validus.

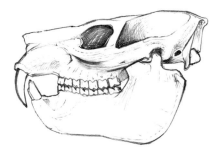
D. arboreus.

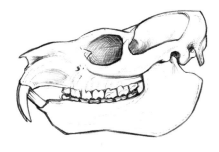
D. dorsalis.

An amoeba of large dimensions is found in the gut of lowland hyraxes, but the alpine races have been reported to be largely free of both internal and ecto-parasites.

Patterns of interspecific competition among *Procaviidae* have been discussed earlier; it seems that it is only where *Heterohyrax* or *Procavia* have been absent that the tree hyrax has successfully colonized rocky habitats. In this competition the later evolution of *Dendrohyrax* may be an important factor. However, whatever the species, once established, they are highly successful animals. In lowland and middle altitude forest *D. dorsalis* would seem to be the dominant species and represents the most wholly arboreal animal of the genus. The limbs of *D. dorsalis* are not noticeably more specialized than in the other species and genera of hyrax but Richard notes that it can climb very easily and quickly up the edge of an open door, or up a perfectly smooth trunk with a 50 cm diameter. The grip is very strong and the feet are capable of revolving on their axis and assume a position equivalent to supination. These hyraxes can also perform a half turn while proceeding upside down along a wire and can exactly counterbalance movement when the wire is jogged. While this species normally moves cautiously and with three of its limbs clinging firmly to a support, it may on occasion leap across spaces and rely on hooking a paw around some branch. In regaining control its long supple body effectively comes into play.

There are many predators of the tree hyrax, chief of which are hawk-eagles, leopards, golden cats, genets, servals and pythons. *Dendrohyrax* are hunted for their fur and flesh, and the large liver is a great delicacy. They are very hard to see at night as they usually keep quite still when disturbed and their eyes reflect very little light, nonetheless, the Bakonjo, the Abaluhya

and the Chagga are adept hunters, the latter find the hollow tree in which they hide and extract them by means of a cleft stick which tangles their fur. Skins are sewn together to make large karosses.

They have a long gestation, between seven and eight months. The Ruwenzori race appears to have young all the year round, with a peak birth period in June—July, so that both mating and birth peak occur during the driest time of the year. Births are recorded in May and August for the lower Congo and March and April for Gaboon. An increased noisiness in the dry season has been widely noted, suggesting that there might be an increase in territorial and sexual behaviour at this time. They may have between one and three young which are born hairy and well-developed, the average number of young is lower than in the other genera.

The permanent dentition in these hyrax develops very late and they may be already breeding before the third molar is in place.

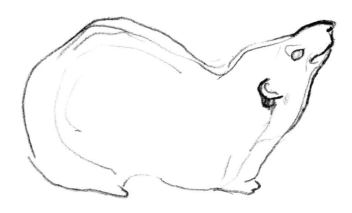

Anteaters

Pangolins (Manidae) and *Orycteropus* are not related, but were once classified together with the American edentates (Edentata) on the basis of common features related to their special diet.

An exclusive diet of ants and termites calls for a very high degree of specialization and presumably a very long period of time is needed to evolve anatomical and physiological adaptations suited to this diet, but a pre-eminence in this particular niche, once evolved is probably virtually incontestable. This would tend to stack the odds in favour of phylogenetically older groups in which ant-eating specialization had proceeded furthest. This would seem to have ensured the survival of very ancient groups all over the tropical world. Orycteropus and pangolins (Manidae) in Africa, pangolins in Asia, armadillos (Dasypodoidea) and anteaters (Myrmecophagoidea) in America, and the spiny echidna (*Tachyglossus*) in Australasia are all primitive groups without living relatives.

Top left : Manis temmincki. Top right : Tamandua.
Left above : Orycteropus. Right : Dasypus.
Bottom right : Tachyglossus aculeatus.

Skulls of anteaters.

352

Pangolins

PHOLIDOTA

Manidae

Pangolins are a widely distributed group of specialized, strictly insectivorous mammals found in Eurasia (three species) and Africa (four species). Their resemblance to other ant-eating animals is purely convergent and they probably derive from an ancient Mesozoic proto-insectivore. Fossil fragments referable to the Manidae are known from Oligocene and Miocene Europe.

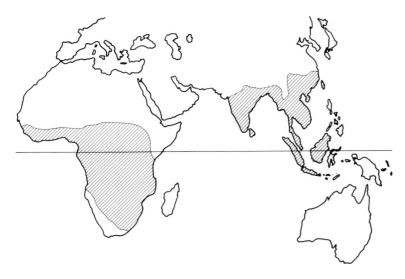

World distribution of Manidae.

A fossil resembling the African *Manis gigantea* is known from the Indian Pleistocene, suggesting that the differentiation observable in African pangolin species today may have occurred before the connections between the continents were broken. Fossil species that were larger than any of the forms living today are known.

The skull is massive and lacks several cranial bones including the zygoma. The brain is primitive. There are no teeth but a gizzard-like pyloric region in the stomach grinds the ants and termites they eat with small stones and sand. They depend upon their protective scales for defence, they can also emit unpleasant odours and, at least one species, hisses in a remarkably snake-like manner. All species have large anal glands and their social life appears to be dominated by the sense of smell.

The tongue is very long and thin (see overleaf) and, when not extended, a section of the tongue folds back through a "window" into a pocket in the throat, consequently when pangolins are feeding the throat can be seen to bulge whenever the fold of the tongue fills the pocket. The last pair of cartilaginous ribs forms an extension to the posterior end of the sternum;

these ribs have become separated from their vertebral attachments and have joined up at their posterior end to form a free-floating spatulate attachment for long muscles that are continuous with those of the tongue. This xiphisternum is of a different structure in each of the three species. This allows an enormous extension of the tongue mechanism, so that the tongue and its accessory structure stretches from near the kidneys, in a wide curve round the lower abdomen along the base of the sternum to the throat, where the tongue develops a kink when at rest. The length of the tongue and its accessory musculature therefore exceeds the length of the animal's head and body

Dissection of tongue and associated structures.

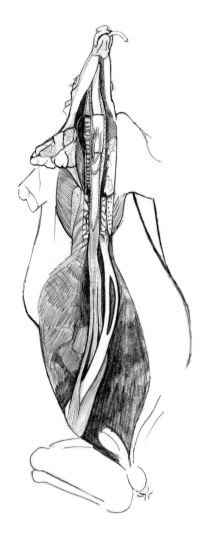

M. tricuspis.

M. gigantea.

(see dissections). *Retia mirabilia* have been found in the limbs of *Manis tricuspis* and are presumably an adaptation to maintain muscle tension and circulation while the animal is curled up for long periods. There is also a peculiar arrangement of parallel blood vessels running down the length of the vertebral column, the significance of which is unknown.

The three East African species show interesting differences in the emphasis, size and role of the feet. In juvenile *Manis tricuspis* (see drawing) the

355

◀ *Manis tricuspis.*

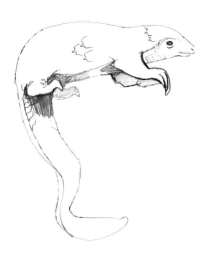

Juvenile *M. tricuspis*.

forelegs are larger and are utilized more than the hindlegs, as these members are the principal clasping limbs while the tail is the most important posterior "limb". In adults the balance between fore and hindlegs is more even, but the light weight of the animal does not necessitate graviportal hindlegs like the giant pangolin. *Manis gigantea* is a capable digger and has powerful claws and shoulders, the gait is generally quadrupedal so that forelegs and hindlegs, although differentiated in function, are approximately equal in size and strength. The ground pangolin, *Manis temmincki*, does very little digging and has relatively insignificant forelegs, this species is capable of bipedal walking and running and carries most weight on the hindlegs. An opportunity to appreciate the relative unimportance of digging for this species was offered when I was sent an old female *M. temmincki* that had lost a front leg at the elbow. The stump had healed perfectly and was clearly several years old, notwithstanding this she was fat, healthy, pregnant and also with a young one at heel.

The locomotory differences between these species show up in a comparison of their pelvises. In *Manis temmincki* the structure is most vertical, the tuber coxae being particularly prominent, while the lower edge of the pubic symphysis is further forward. In *Manis tricuspis* the tuber coxae is more horizontal and the pelvis is very lightly built. In *Manis gigantea* the entire pelvis is massive, compact and more horizontal than in *M. temmincki*.

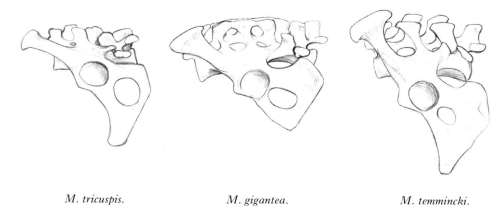

M. tricuspis.   *M. gigantea.*   *M. temmincki.*

Drawn to approximately the same scale.

The protective scales in the tree pangolin are light and thin and this can be correlated with the diminished danger from predators in a largely arboreal existence and the need for flexibility and light weight in trees. The skin and scales in relation to the body weight are nonetheless heavy compared with that of other mammals; an animal weighing 2,210 g having a skin and scales weight of 450 g, i.e. over one fifth of the body weight. The ground pangolin, *M. temmincki*, has an even higher proportion; an animal of about 16 kg bearing over 5 kg of skin and scales. When an animal of this species is in poor condition half its weight may be scales and skin. The dermal muscles of the flanks are exceptionally large in all species.

356

All three species show no regional variation throughout their extensive ranges and are very constant in specific size. Being an animal of very exacting habits, whose survival depends on an armour of scales, the precision engineering of this feature seems to leave little room for variety and presumably selection would tend to stabilize the genetic make-up of pangolins.

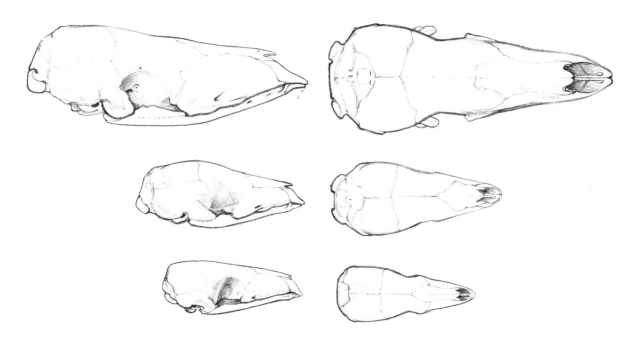

Pangolin skulls: *M. gigantea* (*above*), *M. temmincki* (*middle*), *M. tricuspis* (*below*). All are half life size.

Giant Pangolin
(Manis gigantea)

Family   Manidae
Order   Pholidota
Local names

Olugave (Luganda), Wakonga
(Runyoro), Okong (Lwo), Amikek
(Karamajong), Nkaboso (Kuamba),
Ekisirisiri (Lukonjo), Wakawaka
(Madi).

**Measurements
total length**

1,185—1,650 mm
1,370—1,530 mm (male)
1,185—1,365 mm (female)

tail

550—700 mm

weight

33 kg (one specimen)

Giant Pangolin

The giant pangolin is the largest and heaviest pangolin with very thick scales and a longish snout. It is hairless, except on the eyelids and round the auditory meatus. Its large anal glands are filled with a white waxy substance with a powerful odour like stale chicken dung.

The species has some resemblance with the ground pangolin but giant pangolins are more powerful diggers and do not tolerate any degree of aridity; indeed their range, which is from West Africa to Karamoja, appears to depend upon a high rainfall without severe dry seasons and a high density of termites. Its fondness for water is evidenced by tracks not infrequently leading into rivers and swamps, also by the daily drinking of captives which find water from some distance (presumably guided by smell).

The animal is found over most of Uganda and has recently been found to occur in western Kenya and western Tanzania. It inhabits both forest and savanna at low and medium altitudes and is widely distributed in the forest-savanna-cultivation mosaic typical of western and southern Uganda, where

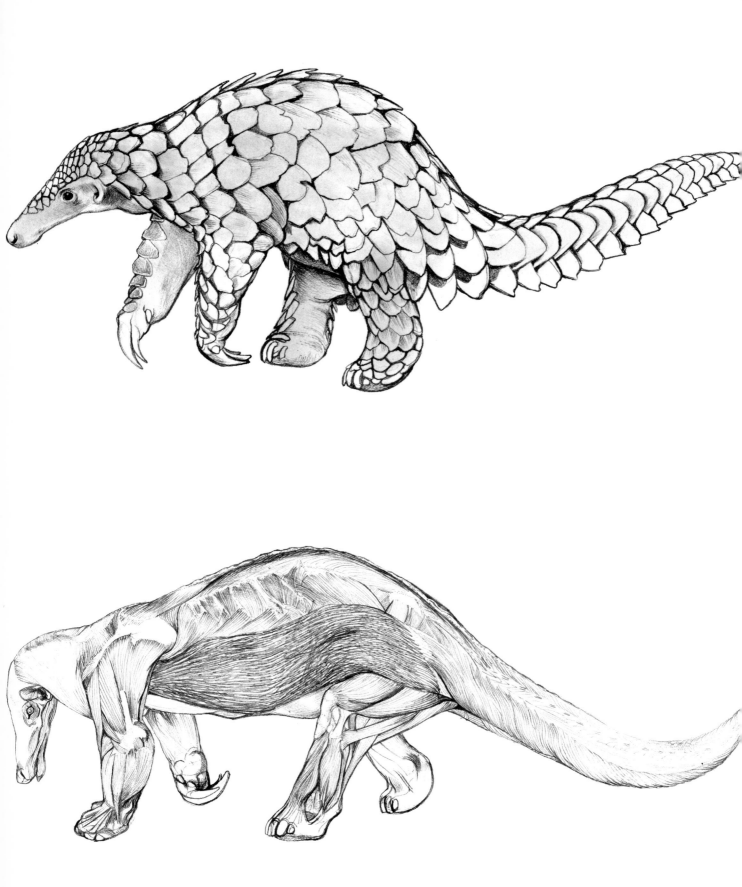

food is sought in both wooded and open country. Here termites of various species flourish, harvester termites consume the grass trampled by cattle on the grassy hilltops and other species occur in the thickets and forests. Termitaries are numerous everywhere even amongst the cultivation which tends to follow the valley slopes. Along the valley floors there are papyrus swamps fringed by forest and this vegetation is also visited by the giant pangolin. There may be some seasonal preference for habitats, but in Ankole and other localities the animals have no choice as they live in open grassy savanna where the only woody vegetation is clumps of thicket.

Like other pangolins, much food is gathered by superficial scratching and the picking up of foraging ants and surface feeding termites. However, termite nests, both subterranean and in mounds, provide a larger and more concentrated source of food for this large animal, and numerous nests and termitaries show signs of extensive damage by giant pangolins. When opening a large mound the animal often takes its entire weight on the tail and tearing the earth apart with its claws kicks out the broken clods with its hindlegs. When walking most weight is taken on the columnar hindlegs, and the animal can walk bipedally freeing its forelegs to dig. When the body is extended the forelegs take the weight with the claws curled inwards, so that the outer edge of the wrist is placed on the ground, not the palm. The spoor frequently shows the mark of the tail dragging on the ground.

The only noises I have heard have been the usual sniffing and a loud menacing hiss, which has the same intimidating effect as a snake and is made when the animal is disturbed. If the pangolin is approached at night, it may after locating the source of disturbance take two quick steps towards the intruder, and throwing its head between the forelegs present its armoured shoulders meanwhile hissing loudly. If it is approached closely it curls up but with the tail loose and continues to hiss, if bullied it curls up completely

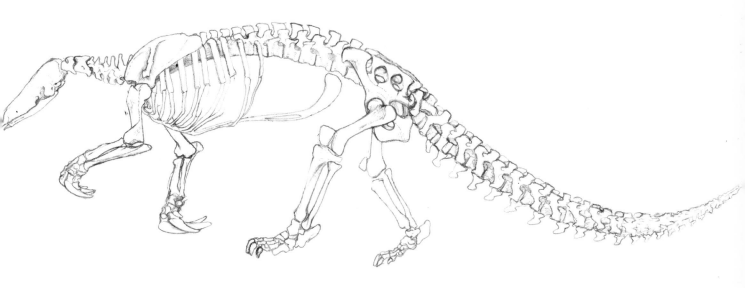

using its tail to seal up the ball and it is then silent. If the dorsal surface of the tail is tickled it unrolls that part of the tail and may try to nip with the scales. Any touching of the ventral surface causes a very fast rolling up reaction. Attempts to restrain a pangolin may be met by a club-like swing of the tail. The sharp scales at the base of the tail can inflict real damage, if this organ closes round the limb of any man or beast interfering with a giant pangolin. Booth recounts (1960) how a Ghanaian hunter found a giant pangolin dragging the freshly dead body of a leopard, its neck held in the pangolin's tail.

The range of a pangolin is probably little greater than a night's ramble from its hole or holes. In Buganda a male was known to have occupied a burrow for two years, a chamber had been excavated in a termitary which was situated near a hut between cultivation and the river. The animal was well-known to local people and appeared to live a solitary existence, it was reported to be conservative in its habits and paths and was seen at the same time and place over long periods. A pair has been found together with a young one in a burrow. Social life in the giant pangolin is unlikely to embrace anything more than these "family" associations. Their burrows vary from long tunnels of up to 40 m in length and 5 m below the surface to a hole excavated for only a metre or so. The resting place is usually an enlarged chamber frequently placed in the middle of a termitary, and short galleries may lead off the main tunnel. Disused orycteropus burrows are also used and, like this animal, it sometimes seals the entrance. A curled-up animal places its nose against its own anal gland.

They may stay at home in a burrow for weeks at a time and the capacity to do without food for long periods is remarkable. Judging from captives, the animal is most active between midnight and 5 a.m., it is therefore not surprising that little is known of the giant pangolin even in areas where it is reasonably common.

These pangolins may be burnt by fires (the individual figured here had been badly burnt) and they are occasionally killed on the roads or for food, but the greatest threat to their existence is the purveyor of charms, who can sell their scales at prices that make a whole pangolin skin worth a small fortune. In Buganda, these scales are believed to have two functions and their magic is associated with women. First mixed with the bark of a tree they make a "mayembe" to neutralize evil spirits. Secondly a scale buried with the assistance of a diviner under a husband's or lover's doorstep is supposed to make him incapable of denying his woman's wishes and he will, under the influence of this scale, buy her new dresses or indulge any other of her desires; naturally these are popular charms.

Two birth records from Buganda are 25th September and 25th October. The young measure 450 mm at birth and weigh just over 500 g. The baby is born with soft scales but its eyes are open and it is very active although quite unable to take its weight on the legs, instead it scrambles spreadeagled. A one day old baby put near the mother worked its way onto the base of the mother's tail immediately, it had a very strong clinging reflex. Its tail was very prehensile, the tip hooked over a finger allowed it to dangle in much the same way as a tree pangolin.

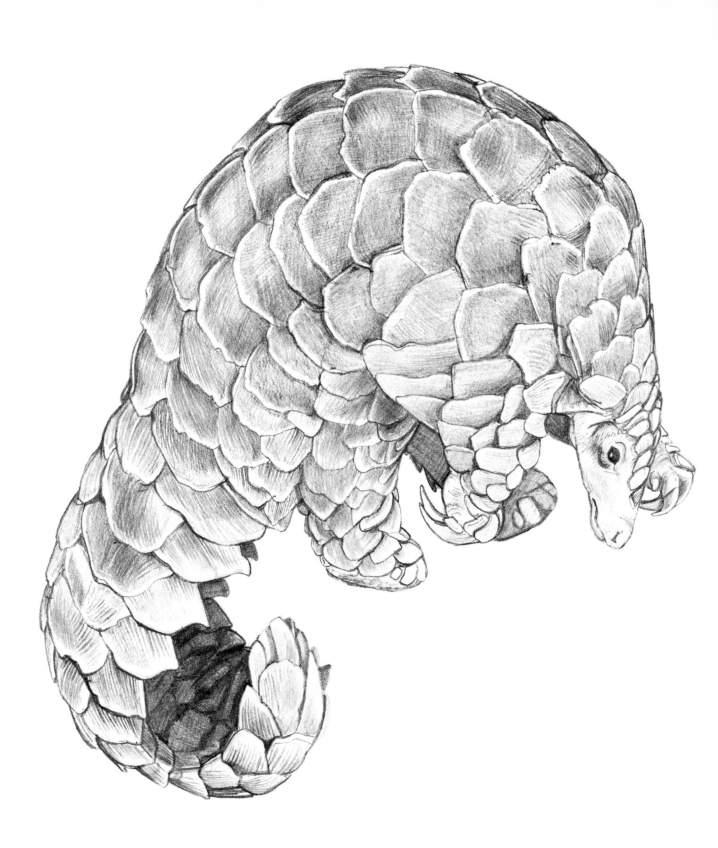

Ground Pangolin

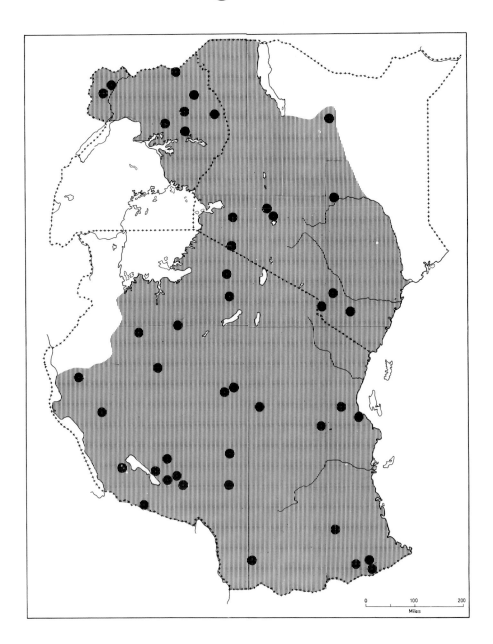

Ground Pangolin
(Manis temmincki)

Family   Manidae
Order   Pholidota
Local names

Kakakuona (Swahili), Okong
(Lwo), Amikimek (Ateso), Amekek
(Karamajong), Wakawaka (Madi).

Measurements
total length
690—1,070 mm
head and body
340—610 mm
tail
350—475 mm
weight
7—18 kg (approx.)

365

The ground pangolin ranges from the Cape to the southwestern Sudan and Oubangui Chari, below about 1,700 m. The animal is apparently absent west of Lake Victoria from the Ugalla-Moyovosi Rivers to the Victoria Nile. It is found in woodland and savanna within reach of water and shows some preference for sandy soils. The animal is very widespread in East Africa but is relatively numerous in some localities and very rare in others; this may be largely determined by the species of ants and termites that it feeds on and the soil conditions with which it has to cope.

This pangolin does not normally expend great energy in excavating but prefers to dislodge detritus and scrape the soil superficially for its food. It is highly selective, feeding mainly on the juvenile stages of specific ants and termites, the most important of which are *Crematogaster*, *Odontotermes*, *Microcerotermes*, and also *Microtermes*, *Amitermes* and *Ancistotermes*. Sweeney (1956) who patiently accompanied a tame Sudanese pangolin on its nightly walk noticed that the animal would frequently stop and sniff, and would very rarely dig fruitlessly or for the wrong species. It avoided the very common *Trinervitermes*, which have a strong smell, and it generally ignored *Macrotermes bellicosus*, this latter species makes hard mounds some way from the surface and is protected by formidable soldiers. This pangolin has frequently been observed to grub in animal dung for termites and often seizes cow-pats

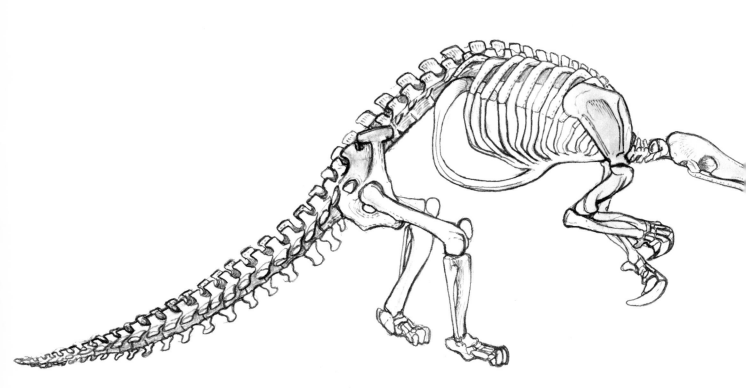

or pieces of termite-ridden wood and, rolling onto its back, breaks them up meanwhile catching the termites on its chest with rapidly darting tongue. Sweeney found his pangolin investigated every cow-pat it could find but only touched those that contained termites; subterranean ants living in cracked clay soils were licked out of their hiding places. While searching for food it sniffs continually and the sniffing intensifies in frequency and volume while it is digging or feeding. While foraging a pangolin will not infrequently rear up on its hindlegs and peer round, listening and sniffing before proceeding, but if at all suspicious it will "freeze" and may subsequently roll up in alarm. It swallows a lot of soil and debris with the ants and termites and depends on eating grit to grind food in its gizzard-like stomach.

367

Like the giant pangolin it is fond of water and is a capable swimmer. In wet soil it may scoop out a wallow for itself by gyrating its body, the scales helping to scrape out the soil. It may also drink water from its wallow.

It can climb well, if slowly, using its tail in a clumsy imitation of the tree pangolin. Its hind feet leave a rounded impression, while the nails of the fore-legs make a small sharp scuff; the dung is dark and earthy with a strong smell. A pangolin walks several kilometres each night but its range is probably fairly limited, as the same hole may be used for many months.

A pangolin may dig itself a hole, usually in a termitary but it frequently uses abandoned *Orycteropus* burrows. It is not adapted for rapid and efficient digging as the orycteropus and giant pangolin are, also it lives in country where for a large part of the year most soils are drier and harder than those found in the giant pangolin's moister habitat, so that its survival depends upon its being able to find an adequate supply of ants and termites close to the soil surface.

Disease and natural predation are unlikely to take a great toll of this species, although a specimen from South Africa was reported to have measle-like spots on the lungs and liver, and another was found emasculated, apparently by a ratel. They have been found killed or badly burnt by bush fires and like the other pangolins they are killed frequently for the scales (as charms and dance ornaments) and for meat. They are sacrificed for rain-making ceremonies in some areas. I found burnt scales of this animal in an old camp fire in Acholi and learnt that the smoke of burning scales is said to keep lions away.

In Botswana, the health of cattle is thought to be improved by burning pangolins. In Tanzania, the animal is sometimes known as the doctor, "bwana mganga", as every part of the body is thought to have some healing property, for instance: nose bleed is treated with pulverized scales (Wright, 1954).

The gestation is about 140 days. In the West Nile district, two newly caught pangolins mated immediately when they were put together in a large box, the male coiled his tail round that of the female as he mounted her with his hindquarters under her and to one side.

Claws of foetus.

The young are 150 mm long and weigh 340 g at birth, their pale soft scales harden by the second day. The baby is kept folded up in the coiled body or in the mother's lap. The mother may leave it while feeding and starts to carry it when it is a month old; at this time the baby begins to eat termites.

It rides on the base of the tail, but slips under into the coiling body if the mother becomes alarmed. A lactating female from south-west Tanzania was accompanied by a half-grown young while at the same time carrying a well-developed foetus (see drawing).

Captive animals distinguish between people they know and strangers; they tame rapidly.

Tree Pangolin
(Manis triscuspis)

Family   Manidae
Order   Pholidota
Local names

Nkaboso (Kuamba), Olugave
(Luganda)

Measurements
total length
775 mm (630—1,027 mm)
head and body
380 mm (300—430 mm)
tail
463 mm (350—607 mm)
weight
1·60—3 kg

Tree Pangolin

The tree pangolin resembles the larger ground species in its general appearance and in features such as the moist nose, peculiar retractable eyes with heavy lids and the valvular nostrils, but it has many distinctive features. The forefeet are placed palm downwards instead of the weight being taken on the outer edge of the wrist as in the heavier ground pangolin, and the hindlegs are relatively long-toed and well provided with claws. The slender tail is longer than the body and has a finger-like pad at the tip; this organ provides the tree pangolin with a sensitive "fifth limb". The scales are very numerous, thin and leaf-like possessing marked striations, they have tricuspid tips while the animal is young, but become worn so that this character is lost with age. The weight of the scales and skin in an adult constitute about one fifth of the total body weight. The smooth underside of the tail (with beautiful reticulate scales) and the belly are usually pale white, the back is dark. The belly is covered with sparse white hair, while the legs and face have a thicker covering of black hair. The number of scales and details of colouring vary from individual to individual, but East African animals are indistinguishable from those in West Africa.

The animal ranges across the continent throughout the lowland forest but has not been found in montane areas. In East Africa it is limited to Uganda forests and Kaimosi in western Kenya. It is commonest in secondary growth where both ants and termites abound.

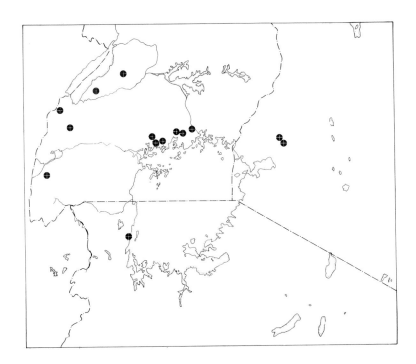

Tree pangolins feed on forest species of ants and termites; army ants, *Dorylus*, and *Mirmicaria* are a principal food, and the termites, *Nasutitermes*, *Microcerotermes*, and their nymphs are also important. Although one species of ant, *Megapomera foetens*, has been seen to attack a tree pangolin and put it to flight, swarming ants crawling over the animal when it is feeding are generally thrown off by the continuous movement and shivering of the scales. The animal eats very rapidly, darting the tongue feverishly and sweeping with the edge of the tail to concentrate the ants. It is likely that much of its food is taken at ground level where bulk requirements are more easily met with. This pangolin is generally active from dusk to dawn, and the ant species eaten are presumably also active at night.

The pangolin is very sensitive to vibrations and sound, and will start or roll up at any sudden noise. They sniff incessantly but seem to see only at very close quarters.

In the trees the prehensile tail with its sensitive finger-like tip plays an important role, clinging to branches, probing for a support, balancing or twining round lianas. On big trunks the long, flat lower surface of the tail and the points of the scales along the lower edges press against the bark and take the weight of the animal. When climbing the fore and hind limbs are moved as pairs, both clawed forelegs probing the surface for a purchase, the tail then takes the weight below, while the hindlegs move up together. Descending a large tree, the animal spirals down the trunk at an angle. Walking on the ground, the back legs take most of its weight, the forelegs act as momentary supports. It can run in a clumsy gallop.

The tree pangolin preens itself by scratching with the hindlegs, lifting its scales to allow the claws to reach the skin. It also removes insects from under the scales with its tongue.

Wild animals which have coiled up on being approached respond consistently and apparently automatically to the following treatment: a pangolin lifted gently by the tip of the tail will curl the tip over a human finger and thus support the weight of its body, it will then uncoil and seek for a purchase with hooking forefeet and may subsequently climb up its own tail. If the tip of the tail is taken when the animal is rolled up on the ground and tilted from side to side, the animal uncoils and starts to walk; the stereotyped nature of these responses is remarkable. The only noise known is a wheezing snort often made when it is touched or surprised.

Two adult male pangolins I watched smelled one another carefully on meeting (particularly in the anal region) and then hastily departed in opposite directions. On being put into the same small cage one of these pangolins curled up, the other, sniffing vigorously, dug into the fold between the curled tail and the back legs with its claws and succeeded in cutting the skin. Later this male made several synchronized slashes with the foreclaws, which were raised behind the head and brought down very suddenly, hitting the other pangolin on the head and shoulders; the less aggressive animal tried to flee after this display. Two fights between captives in Gabon (Pages, 1965) were fatal, the blows were aimed at the opponent's belly and eyes. These incidents may have been induced by captivity.

The animals were found in some numbers when secondary growth in Mabira had been cleared for plantations, suggesting that there can be a fairly

high density of the species. In Gabon, Pages (1965) often found couples together, generally in the hollows left by torn-off branches at 15 m height or
more. On one occasion a pair with a weaned juvenile were found together.
Males are commonly solitary, but Pages found that five or six captives would
sleep together, and sometimes walk in single file, occasionally climbing on
one another's tail. An individual was found in Uganda returning to the same
diurnal resting place (in a tree fork) for over a fortnight. A young captive which
had formed an attachment to its keeper was observed to follow a very irregular
path when searching for her after being parted, but it eventually succeeded in
finding her; when separated again, the animal repeated every detail of its

former path but with greater speed and assurance. Scratch marks on tree-trunks made by pangolins repeatedly following the same route are reported (Pages, 1965). These observations are indicative that this species like the other pangolins tends to have regular habits and follow well-worn tracks by scent; in the wild, marking might employ glandular scent, urine or excreta.

The defensive behaviour of this pangolin includes the use of perineal glands (which exude a revolting secretion), squirting urine and defecating, meanwhile wheezing vigorously. The urine is apparently an irritant to mucous membranes and this battery of noxious scents has been seen to be an effective deterrent to dogs. I have however found the remains of a tree pangolin killed by a carnivore which had carefully stripped away the skin and viscera. This pangolin was subadult and the majority of those found and offered for sale in Kampala are young; possibly young animals are less cautious and have a less developed arsenal of scents to defend themselves.

Tree pangolins are not infrequently hawked in Kampala and their scales are sold as charms in the markets, but the pangolin clan in Buganda has on one occasion acted in the defence of pangolins, when the elders of the clan remonstrated over a scientist placing an advertisement for these animals in a local paper.

An adult pangolin released in suburbia near Kampala regularly visited its former captors over a period of about six months. It fended for itself and was not attracted to the verandah by food or drink, it apparently appreciated being scratched and tickled, although it always urinated if picked up. Its visits were during daylight and its behaviour is difficult to interpret.

The gestation period for the species is not known but is probably in the region of four months. The species breeds continuously but from seven records of embryos and very young animals there might be a birth peak in Uganda and East Congo between November and February. Pages in Gabon found that females were seldom not pregnant, suggesting that the period of reproductive inactivity between pregnancies is very short.

At birth the young are pink and hairless, except for a ring of hair round the eyelids, hair growth is slow and first appears on the body after three weeks. They are about 290 mm long at birth and weigh about a hundred grams; they are born well-developed and active.

In Nigeria, a newly caught pangolin gave birth two days after capture, it left the baby in a corner when it took food in the evening. Most of the day the young suckled while contained in the curled body of its sleeping mother.

The mother continues to care for the young after weaning and young animals will seek contact eagerly from any other pangolin, a female will curl round its juvenile even after it has been dead some time. The young also travel occasionally on the tail of the mother where they may face in either direction.

Protoungulates

Superorder: PROTOUNGULATA
Order: TUBULIDENTATA

Orycteropus is the only living representative of what is considered to be the oldest stock of ungulates which Simpson (1945) has called the proto-ungulates.

Like other primitive relics, it has survived by becoming a specialist.

The skeleton shows a remarkable resemblance to a condylarth, *Phenaco-*

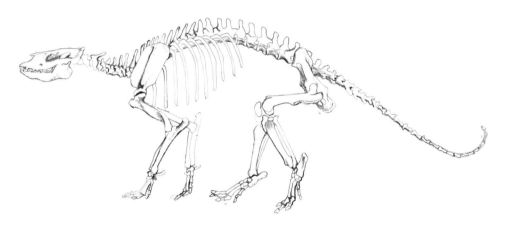

Phenacodus.

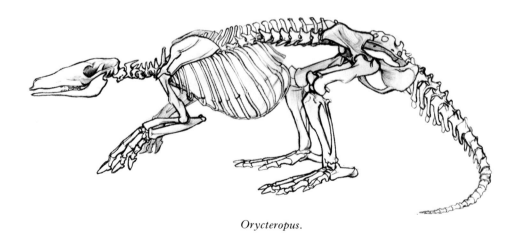

Orycteropus.

dus, from the Lower Eocene. The cranium also retains the general outline and form of that of *Phenacodus*, and differences principally concern the teeth and the olfactory apparatus. The skull and teeth of *Phenacodus* prohibit its

376

consideration as an ancestor, but an immediate precursor might have been related to both forms. Fossils belonging to the *Orycteropus* group have been found in Pliocene deposits in Eurasia, Madagascar and Africa and in Miocene deposits in East Africa.

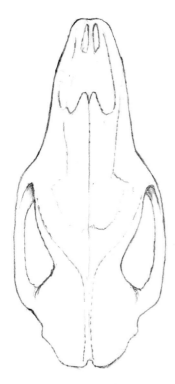

Phenacodus

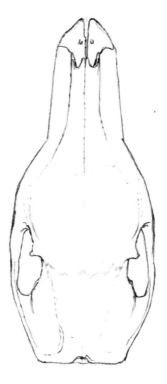

Orycteropus.

Ant Bear,
Aardvark,
Orycteropus
(Orycteropus afer)

Family   Orycteropodidae
Order   Tubulidentata
Local names

Muhanga, Kukukifuku (Swahili),
Ngari ya Kithaka (Kikuyu),
Nyamulimi (Luganda and
Runyoro), Mwok (Lwo), Etukuton
(Karamajong), Awandidu (Galla),
Ngaya (Kinyanturu), Kinyastit
(Sebei).

Measurements
total length
1,610—2,000 mm
head and body
1,000—1,580 mm
tail
445—610 mm
weight
50—82 kg

Ant Bear, Aardvark, Orycteropus

Orycteropus does not really resemble any other living mammal but it is sometimes likened to a pig, hence its Afrikaner name *Aard-vark* or "earth-pig". Only the imperative need for a familiar precedent when trying to describe the unprecedented could have led to the inappropriate English name "Ant-Bear". The Continental vulgarization of the scientific name is followed here.

The most superficial peculiarities of this animal are the long muzzle, soft mobile snout, spade-like nails or "hooves" and the very thick, muscular limbs and tail.

Unlike the pangolins which swallow their food whole, orycteropus uses its teeth to masticate. Young orycteropus have a full complement of rudimentary milk teeth (Broom, 1909) and even adults have vestigial tooth cavities for canines and premolars. The peg-like molar teeth are at the back of the long palate, they vary in number and are unlike those of any other mammal, in being made up of large numbers of very fine columns of dentine in a matrix of pulp; enamel has been eliminated.

Having no teeth in the muzzle, the slender snout requires no buttressing and is merely a lightly built tube containing conventional turbinals. The size and length of the muzzle are subject to great individual regional variation, suggesting that exact proportions in this feature are not of vital importance to the species.

The orycteropus skull is greatly swollen behind the muzzle, and the area of greatest breadth and depth is immediately in front of the eyes. This swelling is due to a layer of minutely convoluted bone which completely surrounds the olfactory bulbs. Delicate scroll-like bones radiate out from the olfactory bulbs to a thickness of 10—15 mm in adult animals. The surfaces are apparently covered with olfactory epithelium and served by olfactory receptors, which are linked to the lobes through the highly perforated cribriform plates which

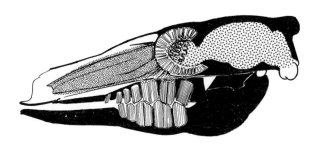

surround them. The total volume taken up by the olfactory bulbs and this accessory structure is large and gives the orycteropus' head its peculiar profile. The total surface area served with olfactory receptors must be immense and the structure clearly enhances the very highly developed sense of smell.

Another unique feature is the presence of thick fleshy tentacles on the nasal septum, which probably have a sensory function. Very different is the dense hair surrounding the nostrils, which seals them while digging and

Nostrils of *Orycteropus*.

perhaps assists in filtering dust out of the air while breathing. It has glands on the elbows and on the hips, and the male has a glandular sac on the prepuce which secretes a powerfully scented yellow musk. The ears resemble those of ungulates; they can fold back and close, probably to exclude dirt while digging.

Today, orycteropus are confined to Africa where they range from the Cape to West Africa. A forest race has been described and they may be found in a very wide range of vegetation types. They are very local in distribution and this is undoubtedly determined by the abundance of the termites and ants on which they feed. Many of the termite species most favoured are not mound builders, but it is generally true to say that an abundance of termitaries characterizes all areas where the animal is common, and here most mounds will show signs of orycteropus claws.

In the areas where I have watched this animal, mounds are excavated anywhere there has been fresh activity; the same mound may be visited frequently and repeatedly, the orycteropus maintaining a sort of "living sore" in the termitary.

Where there are large concentrations of wild ungulates or domestic stock the biomass of termites becomes enormous. The quantities of trampled grass and dry dung are attacked by species of *Odontotermes*, *Microtermes* and *Pseudacanthotermes* which live just beneath the surface. In dry country there are harvester termites, *Hodotermes*, and these insects come to the surface at night to eat the trampled grass and dry dung. Under these conditions,

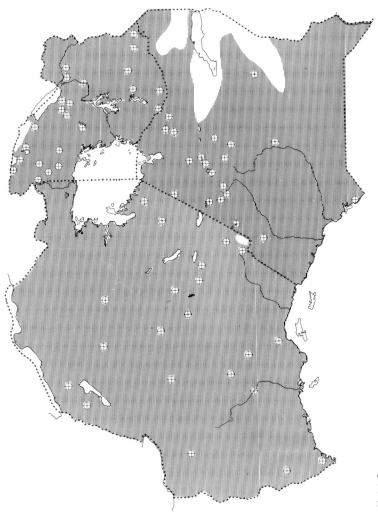

Orycteropus range in E. Africa. Distribution is more local than this map suggests.

orycteropus have no difficulty in finding adequate food. It is perhaps worth mentioning that on the slopes of Mt Elgon orycteropus is reputed to be a new arrival, although it is thought always to have been common in the Kitale corridor below. The build-up of large herds of cattle on the mountain has opened up the country, and has no doubt altered termite populations, thereby creating conditions suited to the orycteropus. Outside highland areas *Macrotermes* species are an important food. In parts of South Africa, pasture and cereal crops have suffered enormous depredations from termites of the genera *Hodotermes* and *Trinervitermes*, which in a dry year may remove over 60% of the standing ley. It is perhaps significant that in these areas orycteropus and many other insectivorous mammals and birds have been exterminated. Orycteropus also eat large quantities of larvae, the well-known scarab beetle buries its eggs in pellets of dung, which are stored in caches some 30 or 40 cm below the surface. One orycteropus stomach contained over 40 scarabaeid pupae eaten in a single night, giving an indication of the numbers of larvae available and the powers of detection possessed by the orycteropus.

381

Seasonal changes affect the abundance of all insects, termites no less than others, and in the dry season many termites remain quiescent. There is also less evidence at this time of orycteropus activity. Animals collected at the beginning of the rains in eastern Congo were noted by Verschuren as being lean, while the Uganda female figured here, was very fat in November after an exceptionally long and wet rainy season. From the period of the beginning of the rains until the grass is burnt, traces of orycteropus are very much in evidence. Their weaving tracks are easy to follow and may cover, within a limited area, a distance of up to 10 km. In some places the animal may stop every few yards to scratch the soil, in most cases a mere mouthful of termites are obtained but on the termite mounds the digging may be more concerted and the rewards greater. During the rains, the fungus gardens are rejected by the orycteropus and piles of broken fungus may lie in the excavations, but in the dry season these would seem to be eaten.

There are usually pockets of insects remaining as a result of the animal's digging, these are often discovered and attacked by ants, and I have seen baboons picking up these termites in the early morning. It is possible that the aardwolf, *Proteles*, may follow the orycteropus and benefit from its wasteful feeding. Locusts are also occasionally eaten. There is a single report of a mouse being taken. Deviations from a purely animal diet have been reported from Kenya (MacKay, personal communication), from the Congo (Verschuren, 1958) and from Zambia (Mitchell, 1965). The seedlings of wild cucumber, *Cucumis humifructus*, have been found growing from the dung of orycteropus and its seeds in orycteropus stomachs. These plants often grow round the entrance to orycteropus holes and are superficially a normal type of cucumber. The extraordinary habit of this plant is that the newly fertilized seeds are forced as much as 30 mm into the ground by a downward growth of

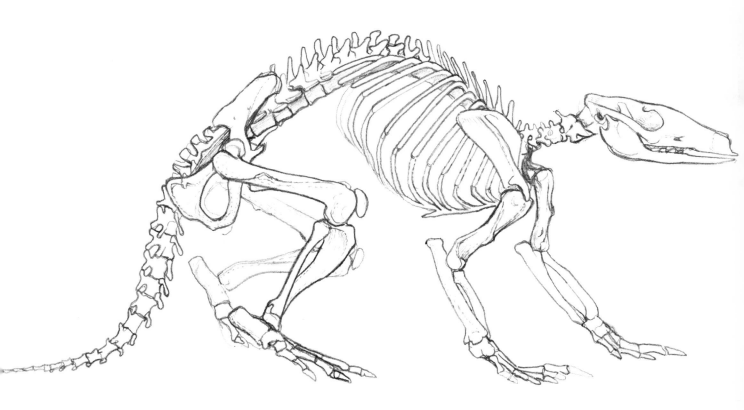

the stem; there it develops rather like the groundnut. The dispersal of this plant would seem to be dependent on the aardvark, which probably eats them for moisture and does not damage the seeds with its teeth. As the orycteropus generally covers its dung with earth this may also assist their propagation.

It drinks where water is available and probably cannot survive in totally dry areas. An exceptional drought in Kenya brought out unusually large numbers of orycteropus in daylight. One that was released in the town at Naivasha at this time seemed blind in the sun and walked into objects in its path.

When searching for termites, the nose is trailed along close to the ground and is frequently pressed flat onto the ground after several vigorous sniffs. The curious fleshy tentacles mentioned earlier are probably highly sensitive to chemical stimulus, or perhaps to minute vibrations set up by the insects' reaction to the animal's sniffing. The pauses following the sniffing would seem to be moments of sensory reception and are followed by rapid digging or by the animal walking on. These fleshy tentacles probably perform a similar function to the extraordinary nose appendages of the North American star-nosed mole, *Condylura cristata.*

When feeding, the animal pauses frequently to listen and may even sit up kangaroo-like and sniff the wind (Fossati, 1937). The sticky tongue is rapidly darted in and out of swarming insects and can extend 30 cm. In opening mounds, orycteropus may dig quite narrow holes with vertical sides which look as though they were dug with a post holer. The spade-like form of the claws seems to affect their gait when running, for they make clumsy bounds, striking the ground hard and noisily. Their normal gait is a slow walk on the toes. A night's walk usually follows a zig-zag course, which may bring the animal back to the same burrow or to a hole within a short distance of its starting point. During the rains it may dig itself a fresh hole almost nightly and several sleeping holes may be started and then abandoned before a night is out.

They make a muffled grunt while walking and have been noticed to grunt before entering a burrow. A bleating calf-like bellow accompanies any sudden fright or pain.

Their excreta are large earth-coloured cakes which are generally deposited near the entrance of a hole or in the earth accumulated by digging; the animal usually covers its excreta with earth. Their spoor is unmistakable and the blade edges of the claws make a strong impression even on quite hard soil.

Orycteropus usually leaves its hole at about 8 p.m., emerging and standing in the entrance listening and sniffing before leaving, however, Verschuren (1958) describes one coming out tail-first with a great flurry of dust. In bright moonlight they are very active, and two animals have been seen at full moon gambolling round the burrow's entrance (M. Adams, personal communication), when disturbed one animal watched the observer from his hole at quite close quarters. Possibly the play was a preliminary to mating as this behaviour is a contrast to their normal caution.

They return to their holes or dig a fresh one before dawn, but there are numerous observations of their coming out to sun themselves. A free-ranging tame orycteropus regularly sunned itself for about an hour each day, at about 10 a.m. (H. Bowker, personal communication).

They sleep curled in a tight circle, with the snout covered by their tail and hind feet. Before sleeping the orycteropus blocks the entrance to the hole, leaving a very small opening at the top. Hediger (1951) wondered how the animal could survive on the little oxygen that must be available in deep, sealed burrows and suggests that there may be a partial arrest of circulation as has been noted in the armadillo (Scolander, 1943; Hediger, 1951). Urbain (in Verschuren, 1958) also remarks on the exceptional resistence of orycteropus to being smoked out.

When undisturbed orycteropus may dig itself a sleeping chamber little larger than the size of its body, less than about a metre from the entrance; similarly it may sometimes sleep near the entrance of an extensive burrow. The narrowest part of the burrow is the entrance beyond which there are passages of varying length. Along the passages there are chambers, sometimes found at three or four metre intervals and these probably allow the animal to turn round easily. Sometimes eight or ten entrances may be scattered over a small area, these may interconnect and some burrows certainly have several exits. Three men entered and were lost in a very extensive and deep burrow in Zambia. After ten days digging only one body was recovered, indicating the extraordinary extent of these burrows (Foran, 1961). I have excavated a warren where the animal had tunnelled about 6 m deep below the ground level. If an attempt is made to excavate the animal, it will dig rapidly, blocking the passage behind it at two metre intervals. It uses its tail and hind feet to throw out the earth.

The animal is normally solitary and the sexes spend most of the year in separate burrows, but where their density is high, several animals may meet in the course of a night, and may feed in the same area.

The animal is preyed upon by the large carnivores, lion, leopard, cheetah, hunting dogs and python are all known to kill it, but man and hyaenas are probably the most important enemies. There is evidence that hyaena frequently take the young.

When pursued in the open, the animal makes for a burrow or hole where it bolts or starts to dig furiously, throwing earth out with vigorous kicks of its hindlegs. When attacked by dogs, it has been seen to turn a somersault. This effectively throws off the attackers, after which it proceeds throwing somersaults whenever touched. This behaviour is probably a nervous reflex, as captive and free-ranging tame orycteropus have been seen to throw somersaults whenever suddenly frightened or alarmed (A. Archer, personal communication). A calf-like bellow usually accompanies the somersault which seems to be assisted by the downward thrust of the tail (H. Bowker, personal communication).

Inhabited burrows are invariably given away by flies around the entrance, these may be the flies that lay eggs on the skin of the orycteropus (probably when it sunbathes), and give rise to the bots that often cover their backs.

Apart from the flesh which is greatly relished by some tribes, bits and pieces of orycteropus are used as charms; the teeth are put on wristlets to prevent illness and bad luck, while the claws, put in a basket while termite collecting, help the catch. The claws are also a hoeing charm connected with the harvest, the hairs are chopped up and used as a food poison, while the tip

of the tail has some symbolic function I have been unable to find out. The animals have been kept up to ten years in captivity and probably live longer in the wild.

There is evidence to indicate that the breeding season is initiated by the rains. In the area where most of my observations have been made, the rainy season causes flooding over large areas of grassland where the orycteropus generally roam, this causes a marked wet-season concentration of these animals on higher ground which may assist the process of finding a mate. It is also interesting to note that the observation of pairs gambolling (mentioned earlier), falling into a well or entering burrows together, were all noted as being in the rain season, while females collected in the eastern Congo during April and May were neither pregnant nor lactating.

I have found the tracks of three or perhaps more orycteropus together in the rains, possibly males pursuing a female.

The male has dark hair on the head, while the female has light cheeks, the female also has a bright white tip to the tail, a feature that is less marked or absent in the male, perhaps this "flash" may assist the young to follow the mother.

The gestation is about seven months. One authentic birth record from Singo, Uganda, is from early November.

Sampsell (1969) reported that an orycteropus weighed 1·87 kg at birth and that the young animal, which was hand-reared followed its keeper very closely. It was apparently comfortable and content when the temperature was between 29·4° and 30·2° C, but the animal became restless when it fell below 23·9° C. At two weeks of age it would dig a hole to bury its excreta and always scratched with its hind-legs before eliminating. It first ate ants at the age of three months.

The young starts to follow its mother out of the burrow at the age of two weeks, when it is very vulnerable. A young orycteropus thought to be 4—6 months old and weighing about 14 kg when caught, was allowed to free range on a farm in western Kenya, where it dug a burrow in the lawn. It would follow its captor and came for a bottle of milk every evening, but otherwise fed itself on termites (H. Bowker, personal communication).

The ancient Egyptian god Set (Greek, Typhon) was represented as an animal with a tubular snout, big upright ears and a stiff tail. He was associated with desert animals and he devoured the moon each month in the guise of a black pig. He was rough and wild and his skin was compared to the pelt of an ass. He was supposed to have hunted in the swamps of the Nile delta. Orycteropus is the most likely identification for this god and unmistakable representations of this animal painted on a pre-dynastic Egyptian vase confirm that it ranged to the Mediterranean within historical time.

Set.

Sea Cows

SIRENIANS

Judging from the numerous fossil forms that have been found, the sea cows appear to have been a widespread and successful group in the past. The fossils span a great period of time and an interesting series of forms have been found that are more or less ancestral to the dugong. These are *Eotheroides* from the Eocene, *Halitherium* from the Oligocene, *Hesperosiren* from the Miocene, and the Pliocene *Felsinotherium* (see illustration).

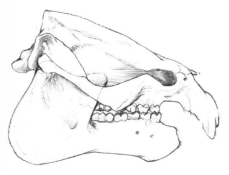

Eocene <u>Moeritherium</u>

Eocene <u>Eothcroides</u>

Oligocene <u>Halitherium</u>

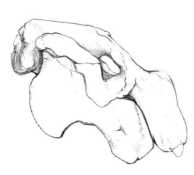

Pliocene <u>Felsinotherium</u>

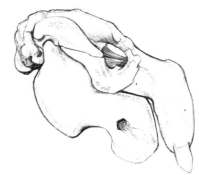

Recent <u>Dugong</u>

The Eocene *Moeritherium* which, for many years was thought to be an ancestral proboscid, may in fact be a primitive sirenian.

The *Proboscidea*, *Sirenia* and *Hyracoidea* are all thought to have derived from a common stock of Palaeocene terrestrial mammals named *Paenungulata* (Simpson, 1945), and although the three orders had diverged by the Eocene, the fossils of that period still retain common features.

388

Unlike other aquatic mammals, which are carnivorous, the sirenians subsist almost entirely upon flowering plants growing in the sea, in estuaries or in rivers.

Living sirenians are the manatee, *Manatus*, from tropical riverine waters around the Atlantic basin, and the dugong, *Dugong*, from the Indian Ocean and the western Pacific coasts. An extinct Arctic form, *Rhytina*, was exterminated in the 18th century.

The sirenians have many special features correlated with their aquatic habitat but they also retain many primitive features indicative of their early departure from the *Paenungulata*. Thus the brain is small and foetal-like, but like the cetaceans it has developed the cerebellum which is associated with balance and orientation. There are numerous ribs as in elephants, but these, in common with all sirenian bones, are short, thick and heavy, lack marrow and resemble the foetal bones of other mammals. It has been suggested (Harrison and King, 1965) that these characteristics may be connected with an apparently rather inactive thyroid, which in turn may affect the metabolism and behaviour of sirenians.

Dugong
(Dugong dugon)

Family
Order
Local names

Nguva (Swahili).

Dugongidae
Sirenia

Measurements
total length
2,500—4,006 mm
weight
360—1,016 kg

Dugong

The entirely aquatic dugong is a large, grey torpedo-shaped animal with horizontal tail flukes similar in shape to those of whales (with which they have no relationship). Externally there is no suggestion of neck or upper arm, both being contained within the regular cylindrical form of the body. The forearm and hand form simple paddles which have a stabilizing function and may occasionally be used to free food plants of sand. The main propulsive force is provided by movements of the tail and body. Correlated with this mode of locomotion the back muscles are highly developed and run from the head to the tail, the latter is made up of skin and tissue. The diaphragm is almost horizontal instead of vertical, allowing the lungs to expand back beneath the vertebral column.

The functional teeth in the dugong are reduced to several peg-like molars which are successively shed until only two remain, the last molar gradually acquires a figure of eight section as it wears away.

The first pair of upper incisors is also shed when the animal is young but the second incisor develops a tusk-like form and is contained in a very extended down-turned premaxilla. This tusk is present in both sexes but only projects from the bone in fully adult males, they become worn on their outer surface, presumably through the action of sandy foods gathered in by the lips, and the wear creates a sharp edge along the back of the tusk. This is the only potential weapon possessed by dugongs and it is restricted to old males.

The lower jaw has similar molars to the upper jaw, rudimentary incisors and canines are occasionally found in young animals and these occur in the down-turned and extended front of the jaw which is vertically flattened. The bone surrounding the rudimentary tooth cavities forms a platform for a flat horny plate which is firmly anchored to the jaw by numerous ligaments which are attached in the cavities themselves.

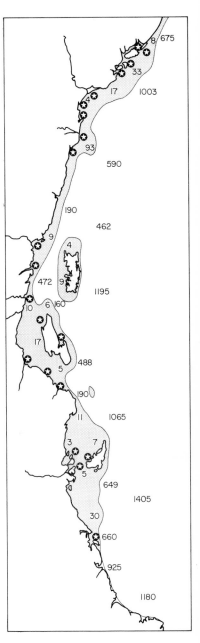

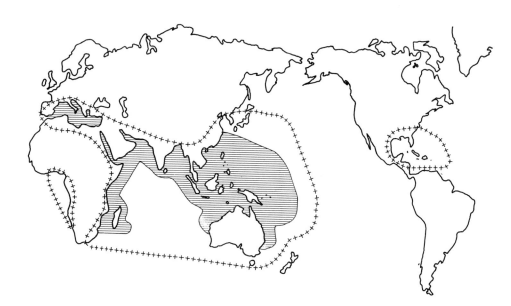

World distribution of dugong (shaded) and *Potamogetonaceae* + + +. 391

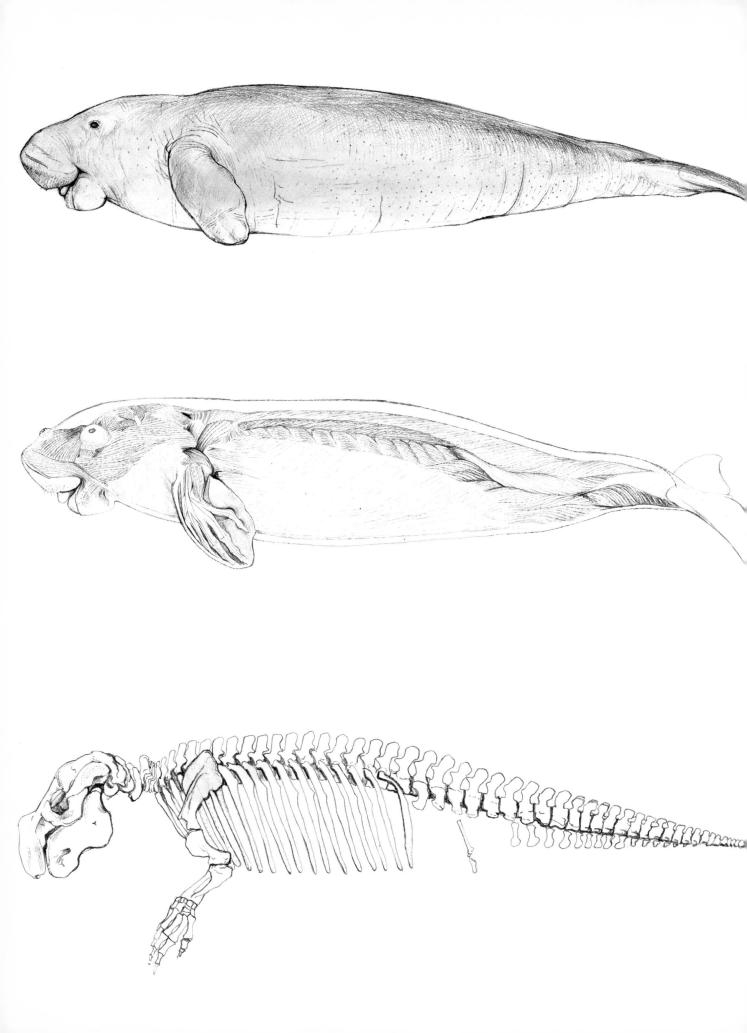

The phylogenetic extension of this peculiar muzzle and the lower jaw is well illustrated in the fossil series (p. 388). Osman Hill (1945) remarked that the peculiar muzzle of the dugong struck him as a truncated proboscis rather than a modified upper lip, and other observers have remarked on the great mobility of the upper lip. However not all the muscles in the dugong's muzzle and the elephant's proboscis are homologous, nonetheless the resemblances between skulls are still sufficient to betray their distant relationship. The bulk of the muzzle is made up by a greatly developed levator labii muscle terminating in a bulky orbicularis oris. Embedded in the great circular upper lip are thick bristles which act as food "rakes". The longest and thickest bristles surround the rostrum of the premaxilla, which has a tough gristly knob on its tip. On the lower lip, thick short bristles occupy an analogous position to teeth. These are able to interlace with those in the upper lip (and around the rostral plug) when the lips close in over a mouthful of food. There is a hard palate and a rough, short tongue which is fixed in the mouth, has little flexibility and cannot be protruded (see drawing).

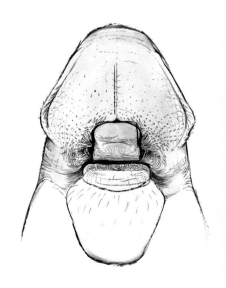

The eyes are small and are surrounded by sphinctered eyelids. The depth of blubber and skin surrounding the skull necessitates the eyeball being projected well out of the bony socket beneath it in order to reach the surface. There are nictitating membranes and an oily gland beneath the eyelid produces a secretion which flows copiously when the animal is out of the water or excited.

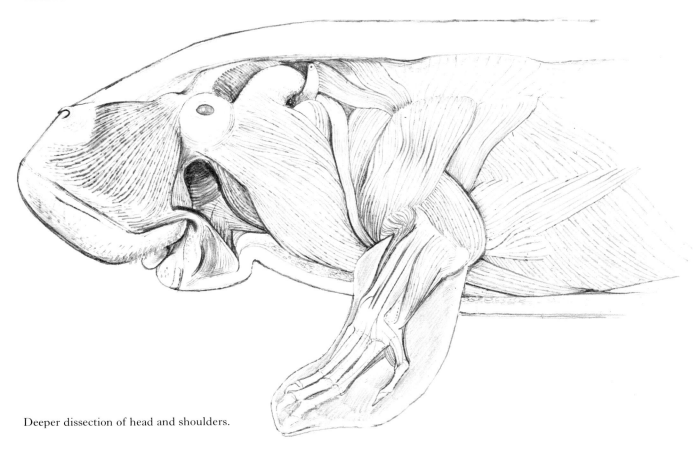

Deeper dissection of head and shoulders.

393

The dugong can only breathe through the valvular nostrils which are situated above the rostrum of the premaxilla and form small horseshoe-shaped depressions. The female has nipples beneath the flipper. Osman Hill (1945) found a small fish inside the prepuce of a male dugong and suggested this might be a case of commensalism.

The species was formerly distributed through all the tropical waters of the Indian and West Pacific Oceans and once occurred in the Mediterranean. Their past range was more or less co-incidental with the tropical and sub-tropical distribution of their food plants, the *Potamogetonaceae*, and *Hydrocharitaceae*, Except for the Atlantic coasts where the manatee, *Trichechus*, is found (see map). On the East African coast they are found near the mouths of larger rivers and in sheltered bays near islands. There are three principal centres for dugong populations, these are the Lamu-Malindi area, the coasts of the Pemba-Zanzibar channel and the Rufigi-Mafia area. Although they may sometimes be seen at any point along the coast, the narrow width of the continental shelf (3—6 km) and the absence of suitably sheltered bays and sandbanks are a deterrent to their continuous distribution. The principal natural factors determining their distribution are the presence of sandbanks where their food plants can grow and tolerable safety from rough conditions. Although they may feed and are seen in deep waters beyond the reefs they probably need shallow sheltered waters at certain times of the year.

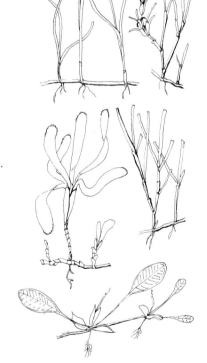

Syringodium.

Zostera capensis.

Halodule uninervis.

Cymodocea ciliata.

Halophila.

Marine *Angiosperms* often grow best in the alluvium-fed sandbanks near the mouth of rivers. The animal is only capable of feeding off the substrate and cannot eat floating vegetation, the whole plant is raked in by the muscular,

394

bristly snout, shaken free of sand and eaten; the flippers, with which the animal drags itself along the bottom, may be used to brush sand off a plant held in the mouth. When dugongs have been feeding in the shallows, the generally irregular course of their grazing can be followed at low tide, as they leave a trail or a series of patches of rootled mud or sand. Crabs have been found in dugong's stomachs and captives will eat clams, so that fishermen's claims that dugongs eat oysters may be true. Reports of their browsing on mangrove leaves are very unlikely and the marine Angiosperms undoubtedly make up the bulk of their diet. Adults eat 25—30 kg (wet weight) daily.

Fishermen report that herds of dugongs move through the feeding grounds in ranks and files. Groups will rise simultaneously during feeding and will often peer around briefly, the surfacing is less frequent than when on the move and they remain grazing for five to ten minutes without breathing. Their principal food plants are species of *Syringodium* and *Cymodocea*, which grow mostly in deeper water and are found beyond the reef wherever there is sand to put down their extensive and (for the dugong) edible roots. These species may well be the dugong's staple diet being an abundant plant and available in bulk. There are vast undersea meadows of these plants off-shore from the Lamu archipelago and in other coastal localities, but due to their inaccessibility it is difficult to know exactly how much they are utilized by the dugong.

Through the kind offices of the Kenya Fisheries Department I have conducted a survey on dugong sightings, numerous fishermen have been interviewed and have answered questionnaires. The fishermen's answers suggest a pattern which appears to be related to seasonal changes in marine flora, in water temperatures and in the roughness of the seas. These factors are inter-related, so it is difficult to be sure that any one factor alone is decisive. Dugongs are reported to be numerous in deeper water off Kiwaiyu during the Kasikazi, North-East Monsoon (November—March) and so may be feeding principally on *Syringodium* and *Cymodocea ciliata* beyond the reef. At this time the winds are gentle and the surface waters are warm and not rough. Later with the Kusi, the South-East Monsoon (April—October), the seas become rough and the dugongs seek the more sheltered bays behind the islands. Most records of catches and sightings in the Lamu area are from between May and August and intermittently on to December, when they are caught in shark nets or seen in the shallows within the Lamu inland sea. Coincidental with the dugong's need for shelter in the worst period of the South-East Monsoon there is a rich growth of sea grasses, *Halophila* spp., *Zostera* spp. and *Halodule uninervis* (the latter is reported to be the exclusive food of dugongs in the Red Sea). These plants which are greatly favoured by dugongs grow in extensive meadows at this time but mostly disappear during the North-East Monsoon, so that the rarity of dugongs in the inland sea during this period is almost certainly due to a shortage of food. Sightings at Mafia suggest that a similar pattern might occur there also, for dugongs are reported to be common during the latter part of the South-East Monsoon.

The dugong's sensitivity to weather was apparently made use of at one time by Arab fishermen in the Persian Gulf who took them in their boats for their "uncanny foreknowledge of approaching storms" (J. H. Williams, 1960). Fishermen have said that the dugong "dislikes rough weather" and

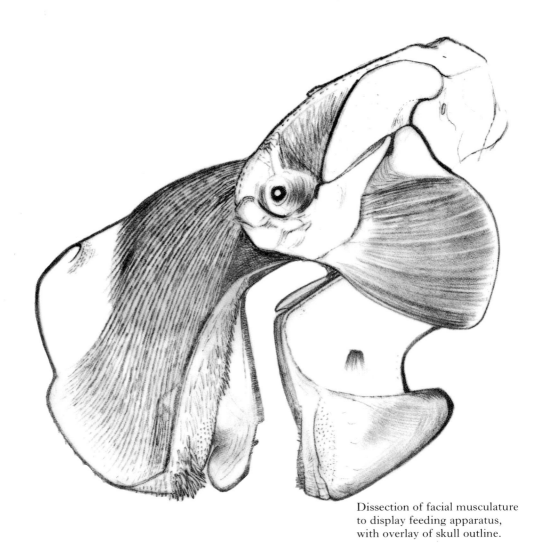

Dissection of facial musculature
to display feeding apparatus,
with overlay of skull outline.

also that it avoids the coral reefs. Perhaps there are real dangers for dugongs in rough weather, since a large number of dead dugongs were reported washed ashore on the South Indian coast after a cyclone in 1954 (S. Jones, 1968). Movement into sheltered bays or deeper water may both be equally effective means of avoiding danger in rough weather and there may therefore be some seasonal movements between the shallows and the deeper sea. There is no direct evidence so far that this involves any long distance migrations, but the appearance of dugongs in large numbers in localities where they are not commonly seen (on the Australian coast in the past and at Mombasa in recent years) may indicate that individuals or families that are normally scattered periodically aggregate; this may be caused by seasonal and localized sources of food.

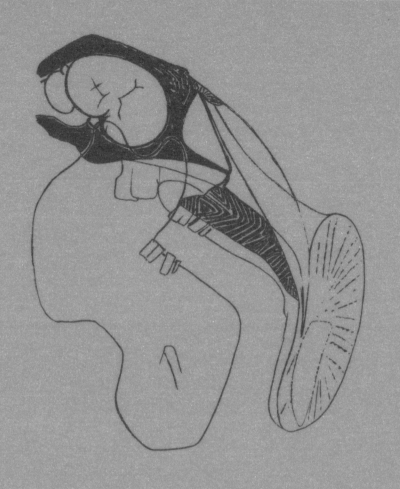

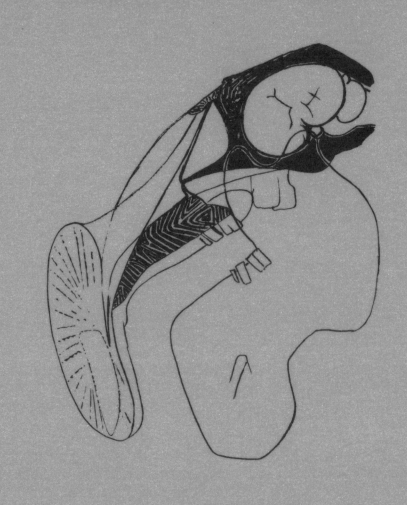

The dugong's day to day movements are determined by tides, winds and human disturbance or interference. They prefer to feed in the shallows on a rising tide and will move in with the tide and follow it out with the ebb, unless disturbed they will feed in one area for some days. Reports on day and night movements are conflicting, but they are most frequently reported to be seen during the late morning or midday; it seems unlikely, however, that daylight is as important a factor as tides. Where there is more human activity and fishing in the day dugongs might prefer to feed inshore at night.

Dugongs are seen singly and in pairs or threes with about equal frequency, the latter are said by the fishermen to be mated pairs with young. Jones (1968) reports that a captive pair in South Ceylon was very attached and that the male was active to avoid separation and would struggle to lie side by side with the female when the tank was emptied. There are several eyewitness accounts by fishermen of large male dugongs attempting to interfere when a young one or a female was being pulled in after being netted. The behaviour is reminiscent of that of the extinct giant sea cow (Steller, 1751, *trans. in* Wendt, 1956). "If one of them was harpooned all the others tried to save him. Some formed a ring round their wounded comrade and endeavoured in this manner to keep him from the shore. We also observed with astonishment that a male came on two successive days to his dead mate lying on the beach as though to enquire after her wellbeing." This behaviour might be thought to be indicative of some intelligence in sirenians, but a captive dugong after six and a half years appeared incapable of recognizing her keeper.

The frequent sightings of solitary animals or pairs and their occasional aggregations do nonetheless suggest an extraordinary capacity for assembly, dispersal and independence of behaviour in so primitive an animal. Their ability to find mates and to congregate at certain times probably depends upon inherited behaviour patterns which are closely co-ordinated with a variety of hidden stimuli deriving from the cycles and changes of their watery environment. Their congregations must also depend upon a certain minimum of numbers so that a drastic reduction in population would probably lessen the individual's chances of meeting up with other dugongs, for in spite of keen hearing and a good sense of smell their capacity for communication must be very short ranged. They make squeaky calls and have been positively tested for ultrasound; extrapolating from detailed studies on manatees (which live in very muddy waters) the sound is unlikely to be used as navigational sonar but only for communication. The fishermen say that they make a whistling sound when distressed and this sound may be a stimulus for the male's defensive action.

The principal natural enemy is the shark, one fisherman saw a dugong being pursued by a shark near Shimoni in June, and another claimed that the hammerhead shark was a predator. Another fisherman found a dugong almost torn in two apparently by a shark. Presumably "buffeting" in the only defence available to the dugong. A dugong in Ceylon was caught dying from the spine of a sting-ray lodged in the peritoneum.

The dugong has been eliminated from many areas where it was formerly common and has been ruthlessly exploited for its hide, meat and oil. A dugong of 200—300 kg may yield 25—56 litres of oil and the meat (which is not forbidden to Moslems) is said to be very palatable. In recent years one fisherman at Kilifi was said to have killed an entire herd which had become rather tame and was often seen by passengers from the Kilifi ferry. When caught the dugong is either speared or drowned by means of a heavy stone tied to the head.

There are many legends attached to the dugong. A Lamu story holds that dugongs are the descendants of some women who were lost at sea in ancient days. Classical authors described the siren and gave it the attributes of a seductive girl. This is interesting as to this day fishermen in Zanzibar that have caught a female dugong have to swear that they have not interfered with it. Throughout the ages the exposed nature of the genitals and the two breasts under the flippers seem to have been a sufficient stimulus for sailors' erotic imagination, and with time the stories became elaborated into classical legends. Three thousand years ago in the eastern Mediterranean fishing licences were a legal obligation connected with the cult of Atargatis a Syrian deity represented with a fishes tail and it is probable that this mermaid goddess derived from the dugong. The dugong formerly occurred in the Mediterranean and was known to the Egyptian, Greek and Phoenician seamen, but it was certainly eliminated at an early date.

In Indonesia, where the name dugong originates, the meat is thought to give strength, the fat is used in obstetrics and cooking, the siren's "tears" are a love potion and the incisors are thought to render their owner invulnerable. With sophistication the incisors have retained their prestigeous value but are made into ivory cigarette holders; the dugong is consequently thought to be very rare in Indonesia now. Likewise in all Arabian and Indian waters the animal is becoming rare; the increasing difficulties for scattered individuals to congregate and breed must hasten their extinction. In eastern and northern Australia they are now protected and are said to be recovering somewhat after heavy exploitation in the past (Bertram and Bertram, 1966). In spite of being fully protected by law, in East Africa the increasing use of nylon nets and the development of fisheries will pose a real threat to the species in future for numbers of animals are caught accidentally each year, and judging from bones in fishing villages the number caught each year is quite significant. Their continued survival may depend on a proper survey of their habits, range, day to day and seasonal requirements. Herds of hundreds of dugongs have been seen in the past in Australia and over a hundred have been seen together in the Lamu area. Scientifically and possibly economically a long-term survey of dugongs in the Lamu area would be very well worth while (see Bertram and Bertram, 1968).

Both copulation and the birth of young are reported to take place in shallow sheltered bays. Courtship has been seen in captivity (Jones, 1968). The male became excited and would nudge and muzzle at the female's neck and turn belly up for a few minutes; the female also turned belly up on occasion and both animals would roll together from side to side. This play was seen on several consecutive days. The muzzling of the male was finally followed by his turning across the female and lying upside down. This was repeated and copulation was presumably effected. The female appeared to be exhausted; the animals acquired abrasions on the skin at this time.

Gestation is 11—12 months. The embryo has abundant hair which falls out during development; in the adult, hair roots are discernible as pimples.

One or very rarely two young are reported. Births have not been observed but probably resemble the manatee where the young is born under water and is immediately helped to the surface by the mother, where it may ride on the mother's back for three quarters of an hour. Initially immersions are short but they gradually become more prolonged and soon the infant manatee takes breath every few minutes. At first only the flippers are used in movement, while the foetal position with the tail curled under the body may be retained for some hours.

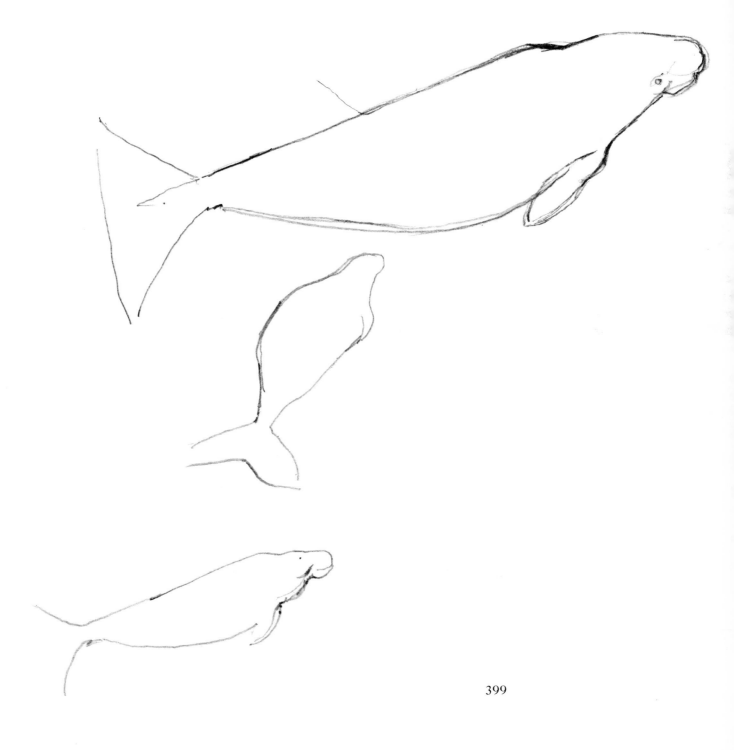

Appendix III
Preliminary Data Sheet

Nomenclature (order, family, genus, species)

English Name

Vernacular Names

Description : Measurements (Total length, total weight, head and body, tail, height).

Anatomical notes and Adaptations.

Paleontology : Notes on any related fossils.

Distribution :
a) Geographical.
b) Ecological (includes notes on physical, climatic, floral habitats and on the microhabitat for small mammals).

Food : a) Composition. b) Mechanics and feeding behaviour.

Behaviour : (includes senses, gait, age, communication, excreta and spoor).

Activity

Social : (includes population structure, sexual composition, density, territory, home range, movements and migration, limitations on a species, notably through the habitat, predation, competition, disease and relations with man).

Breeding : (includes gestation, seasons, shelter, sexual characteristics, courtship and the characteristics and development of the young).

References and Bibliography.

Families and approximate number of species of mammals found in East Africa

INSECTIVORA
- *Potamogalidae* 2
- *Chrysochloridae* 1
- *Erinacidae* 1
- *Macroscelididae* 6
- *Soricidae* 24

CHIROPTERA
- *Pteropodidae* 16
- *Rhinopomatidae* 1
- *Emballonuridae* 6
- *Nycteridae* 7
- *Megadermatidae* 2
- *Rhinolophidae* 8
- *Hipposideridae* 8
- *Vespertilionidae* 32
- *Molossidae* 18

PRIMATES
- *Lorisidae* 1
- *Galagidae* 5
- *Cercopithecidae* 13
- *Pongidae* 2

PHOLIDOTA
- *Manidae* 3

LAGOMORPHA
- *Leporidae* 4

RODENTIA
- *Sciuridae* 17
- *Anomaluridae* 2
- *Pedetidae* 1
- *Muridae* 69
- *Rhizomyidae* 1
- *Muscardinidae* 1
- *Hystricidae* 3
- *Thryonomyidae* 2
- *Bathyergidae* 4

CARNIVORA
- *Canidae* 5
- *Mustelidae* 6
- *Viverridae* 18
- *Hyaenidae* 2
- *Protelidae* 1
- *Felidae* 7

TUBULIDENTATA
- *Orycteropodidae* 1

PROBOSCIDEA
- *Elephantidae* 1

HYRACOIDEA
- *Procaviidae* 6

SIRENIA
- *Dugongidae* 1

PERISSODACTYLA
- *Equidae* 2
- *Rhinocerotidae* 2

ARTIODACTYLA
- *Suidae* 3
- *Hippopotamidae* 1
- *Tragulidae* 1
- *Giraffidae* 1
- *Bovidae* 43

West
Nile

UGANDA

Imatong
Mts.
Kidepo
Madi
Lomunga
Acholi
Turkana
L. Rudolf
Aswa Lolim
Murchison
falls N.P.
Bunyoro
Karamoja.
Moroto
Marsabit
Budongo
Forest
Napak
Debasien
Northern
Bugoma
L. Kyoga
Sebei
Mt.
Elgon
Cherengani
Mts.
Frontier
Semliki River
Toro GR
Bwamba
Bukedi
Mt. Urguess.
Ruwenzori Mts.
Kibale
Singo
Buganda
Busoga
Busia
L. Baringo
Samburu
Q. Elizabeth NP
Kampala
Mabira
Jinja
Kakamega
Kaimosi
Isiolo
Kalinzu
Entebbe
Busoga
Kisii
Sotik
Nyanza
Fort
Ternan
Mau
Nakuru
Aberdare
Mts
Mt.
Kenya
Maramagambo
Ishasha
Ankole.
Sesse Is.
Lolui.Is.
Kayonza
Kigezi
Nkosi Is
Rusinga Is.
Fort
Ternan
Forest.
Suswa
KENYA
Kisoro
Kagera River
Sango Bay
Lake
Kikuyu
Nairobi
Bufumbira
Bukoba
Mara Masai
Athi
plains
Victoria
Olorgesailie
Tana
River
Lamu
Biharamulo
Serengeti N.P.
L Natron
Ol Doinyo Lengai
Amboseli
Tsaro N.P.
Olduwai
Meru
Kilimanjaro
Galana River
Malindi
Ngorongoro crater
Arusha
Voi
Sokoke
forest
Gedi
Kilifi
L. Eyassi
L Manyara
Shinyanga
Tarangire
Pare
Mts
Shimba
hills
Mombasa
Wembere
Usambara Mts
Shimoni
Gombe stream
Amboni
Tanga
Pemba Is
Kigoma
Uvinza
Malagarasi
Tabora
Pangani
Kungwe Mt.
Mahari Mts.
TANZANIA
Itigi
Nguru Mts,
Zanzibar Is
Dodoma.
Mafwomera
Bagamoyo
Katavi
Rungwa
Rubeho
Mts
Morogoro
Uluguru Mts.
Dar es Salaam
Mikumi N.P.
Ufipa
L. Rukwa
Southern
Ruaha N.P.
Iringa
Luhombero Mts.
Rufigi River
Mafia Is
Uzungwa Mts.
Kalambo
Chunya
Usangu
Highlands
Mbeya
Poroto
Rungwe Mt.
Livingstone
Mts.
Southern
Liwale
Kilwa
Songea
Region
Rondo
Lindi
Rovuma River

0   100   200
Miles

Gazetteer

Abercorn Zambia 8° 50′ S 31° 25′ E
Aberdare Mts C. Kenya 1° 20′—1° 40′ N 36° 40′ E
Acholi N. Uganda
Albert, Lake W. Uganda
Amboni N.E. Tanzania 5° 3′ S 39° 4′ E
Amboseli S. Kenya 2° 40′ S 37° 10′ E
Angola S.W. Africa
Ankasa Forest, Ghana
Ankole S. Uganda
Arusha N. Tanzania 3° 23′ S 36° 43′ E
Aswa-Lolim G.R. 2° 35′ N 31° 45′ E
Athi Plains C. Kenya 1° 30′ S 36° 50′ E
Avakubi Zaire 1° 30′ N 27° 25′ E
Bagamoyo Tanzania Coast 6° 25′ S 38° 54′ E
Baringo Lake 0° 40′ N 36° 5′ E
Biafra W. Africa
Biharamulo N.W. Tanzania 2° 38′ S 31° 19′ E
Botswana Southern Africa
Budongo W. Uganda 1° 45′ N 31° 40′ E
Bufumbira Mts E. Congo W. Uganda
Buganda S. Uganda
Bugoma W. Uganda 1° 20′ N 31° E
Bukedi E. Uganda 0° 40′ N 33° 50′ E
Bukoba N.W. Tanzania 1° 21′ S 31° 48′ E
Bunyoro W. Uganda
Burundi C. Africa
Busia E. Uganda 0° 30′ N 34° 8′ E
Busingiro (see Budongo) W. Uganda
Busoga S.E. Uganda
Bwamba W. Uganda 0° 40′ N 30° E
Cameroons W. Africa
Central African Republic or Congo Brazzaville

403

Congo River C. Africa
Dabaga C. Tanzania 8° 7′ S 35° 55′ E
Dar-es-Salaam Tanzania Coast 6° 48′ S 39° 18′ E
Debasien 1° 45′ N 34° 50′ E
Dja River W. Africa
Dodoma C. Tanzania 6° 11′ S 35° 45′ E
Edward, Lake W. Uganda
Elgon, Mt E. Africa 1° 10′ N 34° 35′ E
Entebbe S. Uganda 0° 3′ N 32° 28′ E
Ethiopia N.E. Africa
Ethiopian Faunal Region African Continent + Arabia
Eyassi, Lake C. Tanzania 3° 40′ S 35° E
Fayum Egypt
Fernando Po Is W. Africa
Fort Ternan W. Kenya 0° 20′ S 35° 25′ E
Gabon Gaboon W. Africa
Gedi Kenya Coast 3° 18′ S 40° E
Ghats "African Ghats" E. Tanzania
Gombe Stream Reserve W. Tanzania 4° 59′ S 30° 57′ E
Greek River S.E. Uganda 1° 28′ N 34° 35′ E
Gregory Rift Eastern Rift Valley, Kenya
Horn of Africa N.E. Africa
Imatong Mts S. Sudan 4° N 32° 50′ E
Iringa S. Tanzania 7° 47′ S 35° 42′ E
Ishasha W. Uganda 0° 48′ S 29° 35′ E
Isimila S.W. Tanzania 7° 48′ S 35° 41′ E
Isiolo G.R. 0° 22′ N 37° 35′ E
Itigi Thicket C. Tanzania 5° 40′ S 34° 30′ E
Ituri E. Congo
Ituri Maniema E. Congo
Ivory Coast W. Africa
Jinja S. Uganda 0° 25′ N 33° 15′ E
Jozani Forest S. Zanzibar Island
Kabale W. Uganda 1° 18′ S 30° E
Kagera River Uganda Tanzania border
Kaimosi W. Kenya 0° 18′ N 34° 55′ E
Kaiso W. Uganda 1° 35′ N 30° 58′ E
Kakamega W. Kenya 0° 20′ N 34° 45′ E
Kalahari S.W. Africa
Kalambo Falls S.W. Tanzania 8° 35′ S 31° 13′ E
Kalinzu W. Uganda 0° 30′ S 30° E
Kampala S. Uganda 0° 20′ N 32° 35′ E
Karamoja E. Uganda
Kasai River Congo basin
Katavi Plain W. Tanzania 6° 30′—7′ S 31° E
Kayonza W. Uganda 1° S 29° 35′ E
Kazinga W. Uganda 0° 10′ S 30° E
Kenya Highlands C. Kenya
Kenya, Mt E. Africa 0° 10′ S 37° 25′ E

404

Kibale Forest W. Uganda 0° 30′ N 30° 35′ E
Kidepo Valley N. Uganda 3° 50′ N 33° 55′ E
Kigezi W. Uganda 1° S 30° E
Kigoma W. Tanzania 4° 52′ S 29° 38′ E
Kihura (see Budongo) W. Uganda
Kikuyu C. Kenya 1° 20′ S 36° 40′ E
Kilifi Kenya Coast 3° 40′ S 39° 50′ E
Kilimanjaro N. Tanzania 3° S 38° E
Kilwa Tanzania Coast 8° 45′ S 39° 25′ E
Kinkizi Kigezi W. Uganda 0° 50′ S 29° 50′ E
Kinshasha Congo Kinshasha
Kisoro W. Uganda 1° 18′ S 29° 40′ E
Kisii W. Kenya 0° 40′ S 34° 45′ E
Kivu, Lake E. Congo
Kivu Province E. Congo
Koru W. Kenya 0° 15′ S 35° 20′ E
Kungwe, Mt W. Tanzania 6° 8′ S 29° 48′ E
Kyoga, Lake C. Uganda 1—2° N 33° E
Laetolil near Olduvai N. Tanzania
Lamia River, Bwamba Uganda
Lamu Kenya Coast 2° 20′ S 40° 58′ E
Liberia W. Africa
Lindi Tanzania Coast 10° S 39° 45′ E
Livingstone Mts S.W. Tanzania
Liwale S.E. Tanzania 9° 47′ S 37° 58′ E
Lolui Is. Lake Victoria 0° 10′ S 33° 45′ E
Lomunga G.R. 3° 35′ N 31° 30′ E
Luhombero Mts C. Tanzania
Mabira S. Uganda 0° 30′ N 32° 50′ E
Madi N.W. Uganda
Mafia Is. Tanzania Coast
Mafwamera E. Tanzania 6° 50′ S 36° 35′ E
Mahari Mts W. Tanzania 6° S 30° E
Makunduchi S. Zanzibar Is.
Malagarasi W. Tanzania 4° 5′ S 29° 30′ E
Malawi C. Africa (also Lake Malawi)
Malindi Kenya Coast 3° 8′ S 40° 10′ E
Manyara, Lake N. Tanzania 3° 30′ S 35° 50′ E
Maramagambo W. Uganda 0° 30′ S 29° 50′ E
Mara Masai G.R. S.E. Kenya
Marsabit N. Kenya 2° 25′ N 38° E
Mau Forest C. Kenya 1° S 0° 35° 25′—36° 20′ E
Mbeya S.W. Tanzania 8° 54′ S 33° 26′ E
Meru, Mt N. Tanzania 3° 15′ S 36° 45′ E
Mfwanganu Is. Kenya 0° 30′ S 34° E
Mikumi Nat. Park E. Tanzania
Mombasa Kenya Coast 4° S 39° 35′ E
Momella, Lake N. Tanzania 3° 13′ S 36° 52′ E
Mongiro Bwamba Uganda

Morogoro E. Tanzania 6° 48′ S 37° 40′ E
Moroto E. Uganda 2° 30′ N 34° 45′ E
Mozambique E. Africa
Moyovosi River N.W. Tanzania
Muongoni Zanzibar Is.
Murchison Falls Nat. Park W. Uganda 2° 20′ N 31° 40′ E
Nairobi C. Kenya 1° 20′ S 36° 50′ E
Nairobi Nat. Park
Naivasha, Lake C. Kenya 0° 45′ S 36° 20′ E
Nakuru C. Kenya 0° 20′ S 36° 5′ E
Namuli, Mt Mozambique
Napak E. Uganda 2° 8′ N 20′ 34° E
Natal S. Africa
Ngorongoro Crater N. Tanzania 3° 10′ S 35° 35′ E
Ngurdoto Crater 3° 13′ S 36° 52′ E
Nguru Mts E. Tanzania
Nigeria W. Africa
Nkata Bay Malawi
Nkosi Is. Lake Victoria 0° 45′ S 32° 22′ E
Northern Frontier District (N.F.D.) Kenya
Nyanza S.W. Kenya
Nyasa, Lake (Lake Malawi) Central S.E. Africa
Nuba Mts Sudan
Oldonyo Lengai, Mt 2° 46′ S 35° 55′ E
Olduvai Gorge N. Tanzania 3° S 35° 20′ E
Olorgesailie S. Kenya 1° 40′ S 36° 28′ E
Omo S. Ethiopia 5° N 36° E
Pangani River Tanzania Coast 5° 26′ S 38° 58′ E
Pare Mts N. Tanzania
Pemba Is. 4° 50′—5° 30′ S 39° 45′ E
Popokanyiki S. Zanzibar Is.
Poroto Mts S.W. Tanzania 9° S 34° E
Queen Elizabeth Nat. Park W. Uganda 0° S 30° E
Rondo Plateau S. Tanzania
Rovuma River S. Tanzania 10° 45′ S 33° 40′ E 10° S 48° E
Ruaha Nat. Park C. Tanzania
Rubeho Mts E. Tanzania
Rudolf, Lake N. Kenya
Rufigi River E. Tanzania 8° 31′ S 37° 22′ E 8° S 39° 25′ E
Rukwa, Lake S.W. Tanzania 7° 35′— 8° 32′ S 31° 48′— 35° 52′ E
Rungwe, Mt S. Tanzania 9° 8′ S 33° 40′ E
Rusinga Is. Kenya 0° 25′ S 34° 18′ E
Ruwenzori Mts W. Uganda
Sango Bay S. Uganda 0° 55′ S 31° 45′ E
Sankuru River Congo basin
"Sclater Line" dividing line between the N. and S. Savannas from Congo
    basin to Tana watershed
Sebei E. Uganda 1° 20′ N 34° 35′ E
Semliki Forest Reserve W. Uganda 0° 4′ N 30° E

Semliki River W. Uganda and E. Congo 1° N 30° 20′ E
Sempaya in Bwamba W. Uganda
Senegal W. Africa
Serengeti Nat. Park N. Tanzania
Serengeti Plains N. Tanzania 2° 25′ — 3° 5′ S 34° 40′ — 35° 20′ E
Sesse Is. N. Lake Victoria
Shimba Hills 4° 20′ S 39° 35′ E
Shimoni S. Kenya Coast 4° 45′ S 39° 25′ E
Shinyanga C. Tanzania 3° 33′ S 33° 25′ E
Sierra Leone W. Africa
Singo County C. Uganda
Sokoke Forest 3° 15′ S 39° 55′ E
Somalia N.E. Africa
Songea S. Tanzania 10° 42′ S 35° 38′ E
Sotik W. Kenya 0° 40′ S 35° 15′ E
Southern Highlands S.W. Tanzania
South West Africa
Suam River, Mt Elgon into Turkwell Gorge E. Uganda
Sudan N.E. Africa
Suswa C. Kenya 1° 15′ S 36° 24′ E
Tabora C. Tanzania 5° 2′ S 32° 48′ E
Tana River E. Kenya 0°   2° 50′ S 38° — 40° 30′ E
Tanga Tanzania Coast 5° 4′ S 39° 6′ E
Tanganyika, Lake E. Africa
Tarangire Nat. Park Tanzania
Taveta S.E. Kenya 3° 20′ S 37° 40′ E
Teso E.C. Uganda
Tokwe Bwamba, Uganda
Toro W. Uganda
Tsavo Nat. Park E. Kenya 3° S 38° 30′ E
Turkana N.W. Kenya
Uelle N.E. Congo
Ufipa S.W. Tanzania 7° — 8° S 31° 30′ E
Ugalla River N.W. Tanzania
Uluguru Mts E. Tanzania 6° 51′ S 37° 44′ E
Upper Guinea W. Africa
Urguess, Mt or Garguess or Mt Warges N. Kenya 0° 58′ N 37° 28′ E
Usambara Mts N.E. Tanzania 5° 6′ S 38° 38′ E
Usangu or Buhoro Flats S.W. Tanzania 8° 30′ S 34° 5′ E
Usungwa Mts C. Tanzania 7° 45′ — 8° 35′ S 45° —
Uvinza W. Tanzania 5° 7′ S 30° 22′ E
Victoria, Lake E. Africa
Virunga Volcanoes W. Uganda and E. Congo 1° 30′ S 29° 35′ E
Voi E. Kenya 3° 28′ S 38° 35′ E
Wembere Depression C. Tanzania 3° 45′ — 5° 10′ S 33° 40′ 34° 45′ E
Zambia C. Africa
Zanzibar Is. 5° 43′ — 6° 28′ S 39° 11′ — 39° 41′ E
Zululand S. Africa

Bibliography

Chapter 1: An Introduction to Method

Allen, G. A. (1909). Mammals from British East Africa collected by the Tjader expedition of 1906. *Bull. Amer. Mus. Nat. Hist.* **26**.

Allen, G. M. (1939). A checklist of African mammals. *Bull. Mus. Comp. Zool. Harvard* **83**

Ansell, W. F. H. (1960). "Mammals of Northern Rhodesia." Government Printer, Lusaka.

Bere, R. (1962). "The wild mammals of Uganda." Longmans, London.

Bourlière, F. (1955). "Mammals of the world: their life and habits." Harrap and Co., London.

Bourlière, F. (1955). "The natural history of mammals." Harrap and Co., London.

Dekeyser, P. L. (1955). "Les mammifères de l'Afrique noire française." Inst. Fr. Afr. Noire.

Ellerman, J. R., Morrison-Scott and Hayman, R. W. (1953). Southern African mammals 1758–1951: a reclassification. British Museum (Nat. Hist.), London.

Flower, W. H. and Lydekker, R. (1891). "An introduction to the study of mammals, living and extinct." London.

Grassé, P. P. (ed.) (1955). "Traité de zoologie," Vol. 17 "Mammifères systématique." Masson et Cie, Paris.

Haagner, A. (1920). "South African mammals." Witherby, London.

Hill, J. E. and Carter, T. D. (1941). The mammals of Angola, Africa. *Bull. Amer. Mus. Nat. Hist.* **78**.

Hollister, N. (1924). East African mammals in the U.S. Natural History Museum. *Bull. U.S. Nat. Hist. Mus.* **99**.

Huxley, J. (1943). "Evolution, the modern synthesis." London and New York.

Jeannin, A. (1951). "La faune africaine." Payot, Paris.

Malbrant, R. and Maclatchy, A. (1949). "Faune de l'équateur africain français." "Mammifères," Vol. **2**. Lechevalier, Paris.

Mayr, E. (1963). "Animal species and evolution." Harvard University Press, U.S.A.

Meester, J. (ed.) (in press). "Preliminary identification manual for African mammals." Smithsonian Institution U.S. Nat. Mus. Washington.

Percival, A. B. (1928). "A game ranger on safari." London.

Portman, A. (1952). "Animal forms and patterns." Faber and Faber, London.

Rensch, B. (1959). "Evolution above the species level." Methuen, London.

Roberts, A. (1951). "The mammals of South Africa." Trustees of the "mammals" of South Africa.

Roosevelt, T. and Heller, E. (1915). "Life histories of African game animals." London.

Rosevear, D. R. (1953). "Checklist and atlas of Nigerian mammals." Lagos.

Roure, G. (1962). "Animaux sauvages de la côte d'Ivoire." Abidjen.

Schouteden, H. (1948). Faune du Congo Belge et du Ruanda-Urundi. I. Mammifères. *Ann. Mus. Roy. Congo Belge Sér. 8 Sci. Zool.* **1**.

Setzer, H. W. (1956). Mammals of the Anglo-Egyptian Sudan. *Proc. U.S. Nat. Mus.* **106**.

Shortridge, G. C. (1934). "The mammals of South-west Africa." Heinemann, London.

Simon, N. M. (1962). "Between the sunlight and the thunder: the wildlife of Kenya." Collins, London.

Simpson, G. G. (1945). The principles of classification and a classification of the mammals. *Bull. Amer. Mus. Nat. Hist.* **85**.

Stevenson-Hamilton, J. (1947). "Wildlife in South Africa." Cassell, London.

Swynnerton, G. H. and Hayman, R. W. (1950). A checklist of the land mammals of the Tanganyika territory and Zanzibar protectorate. *J. E. Afr. Nat. Hist. Soc.* **20**.

Taylor, H. (1955). "Art and the intellect." Museum of Modern Art, New York.

Thompson D'Arcy, W. (1942). "On growth and form" (2nd ed.). Cambridge University Press, London.

Walker, E. P. (1964). "Mammals of the world." John Hopkins, Baltimore.

Williams, J. C. (1967). "A field guide to the national parks of East Africa." Collins, London.

Young, J. Z. (1962). "The life of vertebrates." Oxford University Press, London.

Chapter II: East African Environment

Asdell, S. A. (1964). "Patterns of mammalian reproduction." Cornell University Press, New York.

Aubreville, A. (1949). Climats et désertification de L'Afrique. *Soc. Edit Geogr. Marit. Colon Paris.*

Baker, S. J. K. (1958). East African environment *In* "History of East Africa," Vol. 1 (Oliver, R. and Matthew, G. eds.). Oxford University Press, London.

Brooks, A. C. (1961). "A study of Thompson's gazelle (*G. thompsoni*) in Tanganyika." H.M.S.O., London.

Butler, H. (1960). Some notes on the breeding cycle of the Senegal galago (*Galago senegalensis*) in the Sudan. *Proc. Zool. Soc., London* **135**.

Butler, H. (1967). Seasonal breeding of the Senegal galago (*G. senegalensis*) in the Nuba Mountains, Sudan. *Folia Primat.* **5**.

Darlington, P. J. (1959). Area, climate and evolution. *Evolution, Lancaster Pa,* **13**.

Delaney, M. J. (1964). Forest rodents in Uganda. *Rev. Zool. Bot., Brussels.*

Estes, R. D. (1966). Behaviour and life history of the wildebeeste (*Connochaetes*). *Nature* **212**.

Flux, J. E. (1966). Timing in the breeding season in the hare (*Lepus europaeus*). *In* "Biology of reproduction in mammals." Blackwell, Oxford.

Gregory, J. W. (1921). "The rift valleys and geology of East Africa." London.

Harrison, H. (1936). The Shinyanga Game experiment: a few of the early observations. *J. Anim. Ecol.* **5**.

Henderson, J. P. (1949). Some aspects of climate in Uganda. *Uganda J.*

Herter, K. (1965). "Hedgehogs." J. M. Dent, London.

Holthum, R. E. (1953). Evolutionary trends in an equatorial climate. *Symp. Soc. Exp. Biol., Cambridge.*

Ionides, C. J. P. (1965). "A hunter's story." W. H. Allen, London.

Jackson, S. P. (1961). "Climatological atlas of Africa." CCTA/CSA, Nairobi.

Kellas, L. M. (1954). Observations on the reproductive activity, measurement and growth rate of the dik-dik (*Rhynchotragus kirkii*). *Proc. Zool. Soc., London* **124**.

Kendrew, W. G. (1961). "The climates of the continents." Clarendon Press, Oxford.

Laws, R. M. and Clough, G. (1966). Hippo biology in Uganda. *Symp. Zool. Soc., London* **15**.

Laws, R. M. and Parker, I. S. C. (1968). Recent studies on elephant populations in East Africa. *In* "Comparative nutrition of wild animals." Zoological Society Symposium No. 21. Academic Press, London and New York.

Laws, R. M. (1969). Reproduction in the elephant. *In* "Biology of reproduction in mammals." Blackwell, Oxford.

Marshall, I. and Corbet, J. (1959). The breeding biology of equatorial vertebrates. Reproduction of the bat *Chaerophon hindei* (*Tadarida pumila*) at lat. 0° 26′ N. *Proc. Zool. Soc., London* **132**

Moreau, R. E. and Pakenham, R. H. W. (1941). The land vertebrates of Pemba, Zanzibar and Mafia. *Proc. Zool. Soc., London* **110**.

Mutere, F. A. (1965). The biology of the fruit bat *Eidolon helvum*. Ph.D. Thesis, University of East Africa.

Mutere, F. A. (1966). Breeding pattern of some bats in Uganda. *E. Afr. Inst. Virus Research, Entebbe. Annual Report.*

Mutere, F. A. (1968). The breeding biology of the fruit bat *Roussettus aegyptiacus* living at 0° 22′ S. *Acta Tropica* **252**

Mutere, F. A. (1969). Reproduction in two species of free-tailed bats (*Molossidae*). *Proc. E. Afr. Acad.*

Parker, A. H. (1962). Elephants breeding in Uganda. *Acta Tropica* **19**

Percival, A. B. (1928). "A game ranger on safari." London.

Saggerson, E. P. (1962). The physiography of East Africa. *In* "The natural resources of East Africa." (Russell, E. W., ed.) E. Afr. Lit. Bureau.

Sale, M. (1969). Reproduction in Hyrax. *In* "Biology of reproduction in mammals." Blackwell, Oxford.

Sansom, H. W. (1954). The climate of East Africa. *E. Afr. Met. Dept. Memoirs* **3** (2).

Steel, R. (ed.) (1964). "Geographers and the Tropics." Liverpool Essays. Longmans, London.

Talbot, L. M. and Talbot, M. H. (1963). The wildebeeste in Western Masailand, E.A. *Wildl. Monogr.* **12**.

Trewartha, G. T. (1961). "The earth's problem climates." University Wisconsin Press, Wisconsin, U.S.A.

Vesey-Fitzgerald, D. F. (1965). The utilization of natural pastures by wild animals in the Rukwa Valley, Tanganyika. *E. Afr. Wildl. J.*

Chapter III: Vegetation

Anderson, G. D. and Talbot, L. M. (1965). Soil factors affecting the distribution of the grassland types and their utilization by wild animals on the Serengeti Plains, Tanganyika. *J. Ecol.* **53**.

Aubreville, A. (1949). Climats, forêts et désertification de l'Afrique. *Soc. Edit. Geogr. Marit. Colon., Paris.*

Beaton, W. G., Pereira, H. C., Swift, L. H., Talbot, L. M. and Van Den Berghe, L. (1962). "Wildlife development in the savanna lands of East and Central Africa." New York United Nations Special Fund.

Bogdan, A. V. and Pratt (1961). "Common acacias of Kenya." Government Printer, Nairobi.

Bourlière, F. and Verschuren (1960). "Introduction à l'écologie des ongules du parc national Albert." Inst. des Parcs Nat. Congo Belge, Brussels.

Bourlière, F. (1964). Densities and biomass of some ungulate populations in eastern Congo and Rwanda. *Zool. Afr.*

Brosset, A. (1968). Permutation du cycle chez *Hipposideros caffer* au voisinage de l'équateur. *Biologia Gabonica* **4** (4).

Brown, K. (1962). "Termite control research in Uganda." Government Printer, Entebbe.

Brown, L. H. (1963). Wild animals, agriculture and animal industry. *Publ. I.U.C.N. N.S.* No. 1.

Brown, L. H. (1965). "Africa. A natural history." Hamish Hamilton, London.

Buechner, H. K. and Dawkins, H. C. (1961). Vegetation change induced by elephants and fire in the Murchison Falls National Park, Uganda. *Ecology* **42**.

Burtt, B. D. R. (1929). Fruit and seed dispersal by mammals in Singida. *J. Ecol.* **17** (2).

Burtt, B. D. (1942). Some East African vegetation communities. *J. Ecol.* **30**.

Carter, G. S. (1956). "The papyrus swamps of Uganda." Heffer, Cambridge.

Coe, M. J. (1964). Colonization of the nival zone of Mt Kenya. *Proc. E. Afr. Acad.*

Dale, I. R. (1952). Is East Africa drying up?. *E. Afr. Agric. J.* **17**.

Dale, I. R. (1954). Forest spread and climatic change in Uganda during the Christian era. *Emp. For.-Rev.* **33**.

Dale, I. R. and Greenway, P. J. (1961). "Kenya trees and shrubs." Buchanans Estates, Nairobi.

Darling, F. F. (1960). "Wildlife in an African territory." Oxford University Press, London.

Darling, F. F. (1960). An ecological reconnaissance of the Mara plains in Kenya. *Wildl. Monogr.* **5**.

Davies, W. and Skidmore, C. L. (1966). "Tropical pastures." Longmans, London.

Edwards, D. C. (1940). Vegetation map of Kenya with particular reference to the grassland types. *J. Ecol.* **28**.

Edwards, D. C. and Bogdan, A. V. (1951). "Important grassland plants of Kenya." Government Printer, Nairobi.

Eggeling, W. J. (1934). Notes on the flora and fauna of a Uganda swamp. *Uganda J.* **1**.

Eggeling, W. J. (1947). Observations on the ecology of the Budongo rain forest, Uganda. *J. Ecol.* **34**.

Eggeling, W. J. (1948). A review of some vegetation studies in Uganda. *Uganda J.* **12**.

Eggeling, W. J. and Dale, I. R. (1951). "The indigenous trees of the Uganda Protectorate." Government Printer, Entebbe.

Elliot, H. F. L. (ed.) (1964). The ecology of man in the tropical environment. *I.U.C.N. Publ.* No. 4.

Gillman, C. (1949). A vegetation types map of Tanganyika territory. Amer. Geogr. Soc., New York.

Glover, J. (1963). The elephant problem at Tsavo. *E. Afr. Wildl. J.* **1**.

Glover, P. E. (1964). Termitaria and vegetation patterns on the Loita plains of Kenya, *J. Ecol.* **52**.

Glover, P. E. (1968). Fire and other influences on savanna. *E. Afr. Wildl. J.* **6**.

Greenway, P. J. (1943). Second draft report on vegetation classification. *E. Afr. Pasture Research Conf., Nairobi*.

Grimwood, I. R. (1963). "The fauna and flora of East Africa." *I.U.C.N.N.S. Publ.* No. 1.

Harker, K. W. (1961). "An illustrated guide to the grasses of Uganda." Government Printer, Entebbe.

Harrison, H. (1936). The Shinyanga game experiment, a few of the early observations. *J. Anim. Ecol.* **5**.

Harrison, J. L. (1962). The distribution of feeding habits among animals in a tropical rain forest. *J. Anim. Ecol.* **31**.

Harrison, W. V. (1961). "Termites: their recognition and control." Longmans, London.

Harroy, J. P. (1949). "Afrique: terre qui meurt." M. Hayez, Brussels.

Hedburg, O. (1951). Vegetation belts of the East African mountains. *Svensk. Bot. Tidskr.* **45**.

Hedburg, O. (1961). The phytogeographical position of the afroalpine flora. *In* "Recent advances in botany." University of Toronto Press, Toronto.

Hedburg, O. (1964). Features of Afroalpine plant ecology. *Acta Phytogeographica Suecica, Upsala*.

Holloway, C. W. (1962). "The impact of big game on forest policy on Mt Kenya and N.E. Aberdares forest reserve." Government Printer, Nairobi.

Hopkins, B. (1965). "Forest and savanna." Heinemann, Ibadan and London.

Hubbard, C. E. (1926—27). "East African pasture plants—grasses." Crown Agents, London.

Hubbard, C. E. and Milne-Redhead, E. (eds.) (1959). "Flora of East Tropical Africa." Crown Agents, London.

Huxley, J. (1962). East Africa: the ecological base. *Endeavour* **21**.

Karani, P. K. (1969). Flowering and fruiting in Uganda forests. *Occasional Paper. Uganda Forest Dept., Entebbe*.

Keay, R. W. J. (1959). "Explanatory notes on vegetation map of Africa south of the Tropic of Cancer." Published for AETFAT by Oxford University Press, London.

Kirkpatrick, T. W. (1957). "Insect life in the tropics." Longmans, London.

Lake Manyara Conference (1961). On land management problems in areas containing game. *E. Afr. Agr. For. J.* **27**.

Langdale Brown, I., Osmaston, H. H. and Wilson, J. G. (1964). "The vegetation of Uganda and its bearing on land-use." Government Printer, Entebbe.

Langlands, B. W. (1962). "Burning in East and Central Africa: an ecological discussion." Conference Paper, Makerere College, Uganda.

Lamprey, H. F. (1962). East African wildlife as a natural resource. "The natural resources of East Africa." (Russell, E. W., ed.) East African Literature Bureau.

Lamprey, H. F. (1963a). Ecological separation of the large mammal species in the Tarangire Game Reserve, Tanganyika. *E. Afr. Wildl. J.* **1**.

Lamprey, H. F. (1963b). The survey and assessment of wild animals and their habitat in Tanganyika. *I.U.C.N. N.S. Publ.* No. 1.

Ledger, H. P. (1964). The role of wildlife in African agriculture. *E. Afr. Agric. For. J.* **30**.

Lind, E. M. (1956). Studies in Uganda swamps. *Uganda J.* **20**.

Lind, E. M. and Tallantire, A. C. (1962). "Some common flowering plants of Uganda." Oxford University Press, London.

Maeterlinck, M. (1939). "The life of the white ant." Dodd Mead, New York.

Marais, E. N. (1937). "The soul of the white ant." Dodd Mead, New York.

Mitchell, B. L. (1961). Ecological effects of game control measures in African wilderness and forested areas. *Kirkia* **1**.

Morison, C. G. T., Hoyle, A. C. and Hope Simpson, H. F. (1948). Tropical soil-vegetation catenas and mosaics. *J. Ecol.* **36**.

Osmaston, H. A. (1962). "The vegetation of the national parks, Uganda." National Parks Handbook.

Osmaston, H. A. (1964). Pollen and seed dispersal of *Chlorophora excelsa* and *Parkia filicoidea* with special reference to fruit bats (*Eidolon helvum*). *In* "Vegetation of Uganda" (Langdale Browne *et al.*). Entebbe.

Pearsall, W. A. (1957). Ecological survey of Serengeti. *Oryx* **4**.

Petrides, G. A. (1956). Big game densities and range carrying capacity in East Africa. *Trans. N. Amer. Wildl. Conf.* **21**.

Petrides, G. A. (1963). Ecological research as a basis for wildlife management in Africa. *I.U.C.N. NS Publ.* No. 1.

Petrides, G. A. and Swank, W. G. (1963). Range carrying capacities and population densities for large mammals in Queen Elizabeth National Park, Uganda. *Proc. Symp. Afr. Mammals Zool. Soc. S. Afr.*

Phillips, J. F. V. (1930). Fire: its influence on biotic communities and physical factors in S. and E. Africa. *S. Afric. J. Sci.* **27**.

Phillips, J. (1931). A sketch of the floral regions of Tanganyika territory. *Trans. Roy. Soc. S. Afr.* **19**.

Phillips, J. (1959). "Agriculture and ecology in Africa." Faber and Faber, London.

Phillips, J. F. V. (1965). Fire—as master or servant, its influence on the bio-climatic regions of trans-Saharan Africa. *Proc. 4th Ann. Tall. Timbers Fire Ecology Conf., Florida*.

Rattray, J. M. (1960). "The grass cover of Africa." F.A.O.

Richards, P. W. (1952). "The tropical rain forest. An ecological study." Cambridge University Press, London.

Riney, T. (1963). The impact of introductions of large herbivores on the tropical environment. *Proc. Techn. Meet. I.U.C.N., Nairobi.*

Robyns, W. (1937). "Aspects de végétation des parcs nationaux du Congo Belge." Inst. Parcs Nat. du Congo Belge, Brussels.

Russell, E. W. (ed.) (1962). "The natural resources of East Africa." East African Literature Bureau.

Salt, G. (1954). A contribution to the ecology of upper Kilimanjaro. *J. Ecol.* **42**.

Schnell, R. (1950). "La fôret dense." Lechevalier, Paris.

Shantz, H. L. and Marbut, C. F. (1923). The vegetation and soils of Africa. *Amer. Geogr. Soc. Res. Series* No. 13.

Sharfe, S. H. (1954). "Dwellers in darkness: an introduction to the study of termites." Longmans, London.

Simon, N. M. (1962). "Between the sunlight and the thunder; the wildlife of Kenya." Collins, London.

Swank, G. W. and Petrides, G. A. (1957). "A study to determine the effect of vegetation and other factors on game populations in Queen Elizabeth Park, Uganda." Uganda Game and Fisheries Dept., Entebbe.

Swynnerton, C. F. M. (1936). The tsetse flies of East Africa. *Trans. Ent. Soc., London* **84**.

Talbot, L. M. (1963). Comparison of the efficiency of wild animals and domestic livestock in utilization of East African rangelands.

Talbot, M. and Talbot, M. H. (1963). The high biomass of wild ungulates on East African savanna. *Trans. N. Amer. Wildl. Conf.* **28**.

Thomas, A. S. (1942). Note on the distribution of *Chlorophora excelsa. Emp. For. J.* **21**.

Thomas, A. S. (1945). The vegetation of some hillsides in Uganda. *J. Ecol.* **33**.

Trapnell, C. G. and Langdale Brown, I. (1962). The natural vegetation of East Africa. *In* "The natural resources of East Africa" (Russell, E. W., ed.). East African Literature Bureau.

UNESCO (1963). A review of the natural resources of the African continent. UNESCO, Paris.

Verheyen, R. (1951). "Contribution a l'étude éthologique des mammifères du parc national de l'Upemba." Inst. des Parcs Nat. du Congo Belge, Brussels.

Verschuren, J. (1958). "Ecologie et biologie des grands mammifères. Exploration du parc national de la Garamba." Inst. des Natr Parcs du Congo Belge, Brussels.

Vesey-Fitzgerald, D. F. (1955). "The vegetation of the outbreak areas of the red locust in Tanganyika and N. Rhodesia." Anti-locust research centre, London.

Vesey-Fitzgerald, D. F. (1955). The topi herd. *Oryx* **3**.

Vesey-Fitzgerald, D. F. (1960). Grazing succession among East African game animals. *J. Mammal.* **41**.

Vesey-Fitzgerald, D. F. (1963). Central African grasslands. *J. Ecol.* **51**.

Vesey-Fitzgerald, D. F. (1964). Mammals of the Rukwa Valley. *Tanganyika Notes and Records* **62**.

Vesey-Fitzgerald, D. F. (1965). The utilization of natural pastures. *E. Afr. Wildl. J.*

Wayland, E. J. (1940). Desert versus forest in East Africa. *Geogr. J.* Vol. **96**.

Welch, J. R. (1960). Observations on deciduous woodlands in the eastern province of Tanganyika. *J. Ecol.* Vol. **48**.

Wild, H. (1952). The vegetation of Southern Rhodesia termitaria. *Rhod. Agric. J.* **49**.

Woosnam, R. B. (1907). Ruwenzori and its life zones. *Geogr. J.* **30**.

Chapter IV: Time Perspectives

Andrews, C. W. (1906). "A descriptive catalogue of the Tertiary vertebrata of the Fayum, Egypt." British Museum, London.

Aubreville, A. (1949). Climats, fôrets et désertification de l'Afrique tropicale. *Soc. Edit. Geogr. Marit. Colon., Paris.*

Bakker, E. M. Van Zinderen (1962). Botanical evidence for quarternary climates in Africa. *Ann. Cape Prov. Mus.* **2**.

Bakker, E. M. Van Zinderen (1964). Pollen analysis and its contribution to the palaeoceology of the Pleistocene in Southern Africa. *In* "Ecological Studies in Southern Africa." (Davis, D. H. S., ed.) W. Junk, The Hague.

Bishop, W. W. (1958). Miocene mammals from the Napak Volcano, Karamoja, Uganda. *Nature* **182**.

Bishop, W. W. (1963). The later Tertiary and Pleistocene in eastern equatorial Africa. *In* "African ecology and human evolution" (Howell, F. C. and Bourliére, F., eds). Viking Fund Publications in Anthropology, No. 36, New York.

Bishop, W. W. and Whyte, F. (1962). Tertiary mammalian fauna and sediments in Karamoja and Kavirondo. E.A. *Nature* **196**.

Bishop, W. W. and Clark, J. D. (eds.) (1967). "Background to evolution in Africa." University Chicago Press, Chicago.

Bond, G. (1963). Pleistocene environments in southern Africa. *In* "African ecology and human evolution" (Howell, F. C. and Bourliére, F., eds). Viking Fund Publications in Anthropology, No. 36, New York.

Bone, E. L. and Singer, R. (1965). Hipparion from Langebaanweg, Cape Province, and a revision of the genus in Africa. *Ann. South Afr. Mus.* **48** (16).

Booth, A. H. (1958). The zoogeography of W. African primates: a review. *Bull. Inst. Fr. Afr. Noire* **20**.

Brown, L. H. (1965). "Africa. A natural history." Hamish Hamilton, London.

Carcasson, R. (1964). A preliminary survey of the zoogeography of African butterflies. *E. Afr. Wildlife J.* **2**.

Chapin, J. P. (1932—54). The birds of the Belgian Congo. *Bull. Amer. Mus. Nat. Hist.* **65** (75, 75A, 75B).

Chesters, K. I. M. (1957). Miocene flora of Rusinga Island. *Plaeontographica* B. **161**.

Clark, J. D. (1959). "Prehistory of South Africa." London.

Clark, Sir W. E. Le Gros and Leakey, L. S. B. (1951). The Miocene Hominoidea of East Africa. "Fossils, mammals of Africa", No. 1. British Museum, London.

Clark, Sir W. E. Le Gros (1960). "The antecedents of man." Harper and Row.

Clark, Sir W. E. Le Gros (1964). "The fossil evidence for human evolution." University of Chicago Press.

Coetzee, J. A. (1964). Evidence for a considerable depression of the vegetation belts during the Upper Pleistocene on the East African mountains. *Nature, London* **204**

Cole, S. "Prehistory in East Africa." Weidenfeld and Nicholson, London.

Cooke, H. B. S. (1958). Observations relating to Quarternary environments in East and South Africa. *Trans. Geol. Soc. S. Africa* **61** (Annexure).

Cooke, H. B. S. (1963). Pleistocene mammal faunas of Africa with particular reference to southern Africa. *In* "African ecology and human evolution" (Howell, F. C. and Bourliére, F., eds). Viking Fund Publications in Anthropology, No. 36. New York.

Cooke, H. B. S. (1967). The Pleistocene sequence in South Africa and problems of correlation. *In* "Background to evolution in Africa" (Bishop, W. W. and Clark, J. D., eds). University of Chicago Press, Chicago.

Cooke, H. B. S. (1968). The fossil mammal fauna of Africa. *Quart. Rev. Biol.* **43** (3).

Corbet, J. B. and Hanks, J. (1968). A revision of the elephant shrew family *Macroscelidae*. *Bull. Brit. Mus. Nat. Hist.* Vol. **16**, No. 2.

Corner, E. J. H. (1954). The evolution of tropical forest. "Evolution as a process" (Huxley, J., Hardy, A. C. and Ford, E. B., eds). Allen and Unwin, London.

Darlington, P. J. (1957). "Zoogeography." Wiley, New York.

Darwin, C. "The origin of species and the descent of man."

Davis, D. H. S. (1962). Distribution patterns of southern African *Muridae* with notes on some of their fossil antecedents. *Ann. Cape Prov. Mus.* **2**.

De Beer, G. R. (1958). "Embryos and ancestors." Oxford University Press, London.

Dekeyser, P. L. and Villiers, A. (1954). Essai sur le peuplement zoologique terrestre de l'ouest Africain. *Bull. Inst. Fr. Afr. Noire* **16** (3).

Dobzhansky, T. (1959). Evolution in the tropics. *Ann. Sci.* **38**.

Eisentraut, M. (1963). "Die Wirbeltiere des Kamerungebirges." Paul Parey, Hamburg and Berlin.

Flint, R. F. (1959). Pleistocene climates in E. and S. Africa. *Bull. Geol. Soc. Am.* **70**.

Flower, W. H. and Lydekker, R. (1891). "An introduction to the study of mammals living and extinct." London.

Goodman, M. (1963). Man's place in the phylogeny of the primates as reflected in serum proteins. *In* "Classification and human evolution" (Washburn, S. L., ed.). Viking Fund Publications in Anthropology, No. 37, New York.

Gentry, A. W. (1968). Historical zoogeography of antelopes. *Nature, London* **217**, No. 5131.

Goldschmidt, R. (1940). "The material basis of evolution." Yale University Press, New Haven.

Hall, B. P. (1960). The faunistic importance of the scarp of Angola. *Ibis* **102**.

Hanney, P. and Morris, B. (1962). Some observations on the pouched rat in Nyasaland, *J. Maum.* **43**, No. 2.

Heim De Balzac, H. (1967). Faits nouveaux concernant les myosorex (soricidae) de l'Afrique Orientale. *Mammalia*, **31**, 4.

Howell, F. C. and Bourliére, F. (eds.) (1963). "African ecology and human evolution." Viking Fund Publications in Anthropology, No. 36, New York.

Huxley, J. S. (1941). "The uniqueness of man." London.

Huxley, J. S. (1942). "Evolution, the modern synthesis." London.

Huxley, J. S. (1954). The evolutionary process. *In* "Evolution as a process," (Huxley, J., Hardy, A. C. and Ford, E. B., eds). Allen and Unwin, London.

Hopwood, A. T. (1954). Notes on the recent and fossil mammalian faunas of Africa. *Proc. Linn. Soc., London* **165**.

Keast, A. (1968). The southern continents as backgrounds for mammalian evolution. *Quart. Rev. Biol.* **43** (3).

Klinger, H. P., Hamerton, J. L., Mutton, D. and Lang, E. M. (1963). The chromosomes of Hominoidea. *In* "Classification and human evolution" (Washburn, S. L., ed.). Viking Fund Publications in Anthropology, No. 37, New York.

Lavocat, R. (1959). Origine et affinites des rongeurs de la sous-famille de *Dendromurinae. Compt. Rend. Acad. Sci., Paris* **248**.

Leakey, L. S. B. (1951). "Olduvai Gorge." Cambridge University Press, London.

Leakey, L. S. B. (ed.) (1952). "Proceedings of the 1947 Pan-African congress on prehistory Nairobi." Oxford University Press, London.

Leakey, L. S. B. (1965). "Olduvai Gorge, 1951—61", Vol. 1. "Fauna and background." Cambridge University Press, London.

Livingstone, D. A. (1962). Age of deglaciation in the Ruwenzori range, Uganda. *Nature* **194**.

Lonnberg, E. (1929). The development and distribution of the African fauna in connection with and depending upon climatic changes. *Archiv. Zoolog.* **21A**, No. 4.

Matthew, W. D. (1915). Climate and evolution. *Ann. N.Y. Acad. Sci.* **24**.

Mayr, E. (1942). "Systematics and the origins of species." Columbia University Press, New York.

Mayr, E. (1954). Genetic environment and evolution. *In* "Evolution as a process" (Huxley, J., Hardy, A. C. and Ford, E. B., eds). Allen and Unwin, London.

Mayr, E. (1957). Species problem. *Amer. Ass. Adv. Sci. Publ.* No. 50.

Mayr, E. (1963). Evaluation of fossil hominids. *In* "Classification and human evolution" (Washburn, S. L., ed.). Viking Fund Publications in Anthropology, No. 37. New York.

McCall, A., Baker, R. and Walsh, H. (1967). Stratigraphy at Olduvai. *In* "Background to evolution in Africa" (Bishop, W. W. and Clark, D. J., eds). University of Chicago Press, Chicago.

Misonne, X. (1963). "Les rongeurs du Ruwenzori et des régions voisines." Inst. des Parcs Nat. du Congo et du Rwanda, Brussels.

Monod, T. (1963). The late Tertiary and Pleistocene in the Sahara. *In* "African ecology and human evolution" (Howell, F. C. and Bourliére, F., eds). Viking Fund Publications in Anthropology, No. 36. New York.

Moreau, R. E. (1935). Some eco-climatic data for closed evergreen forest in tropical Africa. *J. Linn. Soc. Zool.* **39**.

Moreau, R. E. and Pakenham, R. H. W. (1941). The land vertebrates of Pemba, Zanzibar and Mafia. *Proc. Zool. Soc., London* A **110**.

Moreau, R. E. (1944). Kilimanjaro and Mt Kenya: some comparisons with special reference to the mammals and birds. *Tanganyika Notes and Records* **18**.

Moreau, R. E. (1945). Mt Kenya, a contribution to the biology and bibliography. *J. E. Afr. Nat. His. Soc.* **18**.

Moreau, R. E. (1952). Africa since the Mesozoic with particular reference to certain biological problems. *Proc. Zool. Soc., London* **121**.

Moreau, R. E. (1958). The *Malimbus* spp. as an evolutionary problem. *Rev. Zool. Bot. Afr.* **57**.

Moreau, R. E. (1963). Vicissitudes of the African biomes in the late Pleistocene. *Proc. Zool. Soc., London* **141**.

Moreau, R. E. (1966). "The bird faunas of Africa and its islands." Academic Press, London.

Moreau, R. E. (1969). Climatic change and the distribution of forest vertebrates in West Africa. *J. Zool., London*.

Morrison, M. E. S. (1961). Pollen analysis in Uganda. *Nature* **190**.

Mortelmans, G. and Monteyne, R. (1962). Le quarternaire du Congo occidental et sa chronologie. *Ann. Mus. Congo Belge*, **40**.

Mortelmans, G. and Nenquin, J. (eds). "Actes du IVe congrès Pan-Africain de prehistoire." Terruren, 1962 (Leopoldville, 1959).

Nilsson, E. (1952). Pleistocene climatic changes in East Africa. Proceedings of 1947 Pan-African congress on prehistory, Nairobi (Leakey, L. S. B., ed.), Oxford.

Rahm, U. (1966). Les mammifères de la fôret equatoriale de l'est du Congo. *Ann. Mus. Roy. Afr. Centrale Tervuren Zool.* **149**.

Rensch, B. (1959). "Evolution above the species level." Methuen, London.

Romer, A. S. (1945). "Vertebrate paleontology." University of Chicago Press, Chicago.

Savage, R. J. G. (1967). "Aspects of Tethyan biogeography." (Adams, C. G. and Ager, D. V., eds). The Systematics Association Publication No. 7, London.

Simons, E. C. and Wood, J. (1967). Early Cenozoic mammalian faunas Fayum. *Peabody Mus. N. H. Yale Bull.* No. 28.

Simpson, G. G. (1940). Mammals and landbridges. *J. Washington Acad. Sci.*

Simpson, G. G. (1953). "The major features of evolution." Columbia University Press, New York.

Simpson, G. G. (1953). "Evolution and geography." University of Oregon Press, Corvallis.

Simpson, G. G. (1965). Note on Fort Ternan beds of Kenya. *Amer. J. Sci.*

Verheyen, W. N. (1961). Contribution à la craniologie comparée des primates. *Mus. Roy. Afr. Centrale Tervuren, Belgique Ser. Sci. Zool.* **105**.

Washburn, S. L. (ed.) (1963). "Classification and human evolution." Viking Fund Publications in Anthropology, No. 37. New York.

Wayland (1940). Desert versus forest in East Africa. *Geogr. J.* **96**.

Willett, H. C. (1953). Atmospheric and oceanic circulation as factors in glacial—interglacial climates. *In* "Climatic change: evidence, causes and effects" (Shapley, H., ed.). Harvard University Press, Cambridge, Mass.

Wood (1959). Eocene radiation and phylogeny of the rodents. *Evolution* **13**.

Zeuner, F. E. (1962). "Dating the past." Methuen, London.

Appendix II: Anatomy of Mammals

Clark, W. E. Le Gros and Medawar, P. B. (eds) (1945). "Essays on growth and form." Clarendon Press, Oxford.

Cunningham, D. J. (1941). "Cunningham's textbook of anatomy." Oxford University Press, London.

Cuvier, G. and Laurrillard, M. (1849). "Anatomie comparée." Paris.

Ellenberger, W., Baum, H. and Dittrich, H. (1901). "Handbook of anatomy of animals for artists," 5 vols. London.

Grasse, P. (1955). "Traité de zoologie)", Vol. 17 "Mammifères systématique Masson et Cie", Paris.

Gray, H. (1958). "Gray's anatomy," 32nd ed. London.

Gray, J. (1953). "How animals move." Cambridge University Press, London.

Huxley, T. H. (1871). "Manual of the anatomy of vertebrated animals." London.

Muybridge, E. (1899). "Animals in motion." London.

Owen, R. (1866—8). "On the anatomy of vertebrates," 3 vols. London.

Romer, A. S. (1949). "The vertebrate body." W. B. Saunders, Philadelphia and London.

Straus, Durckheim H. (1845). "Anatomie descriptive et comparative de chat." Paris.

Stubbs, G. (1776) (reprinted 1938). "The anatomy of the horse." London.

Thompson, D'Arcy W. (1917). "On growth and form." Cambridge University Press, London.

Young, J. Z. (1950). "The life of vertebrates." Clarendon Press, Oxford.

Young, J. Z. (1957). "The life of mammals." Clarendon Press, Oxford.

Primates

Andrew, R. J. (1964). Displays of the primates. *In* "Evolutionary and genetic biology of primates," Vol. II. Academic Press, New York and London.

Ansell, W. F. H. (1960). "Mammals of Northern Rhodesia." Government Printer, Lusaka.

Armstrong, L. (1964). Life Magazine, New York.

Asdell, S. A. (1946). "Patterns of mammalian reproduction." Constable, London.

Azuma, S. and Toyoshima, A. (1962). Progress report of the survey of chimpanzee in their natural habitat. *Primates* **3** (2).

Bates, G. L. (1905). Notes on the mammals of the S. Cameroons and the Benito. *Proc. Zool. Soc., London*, 65—85.

Baudenon, P. (1949). Contribution à la connaissance du Poto de Bosman dans le Togo-Sud. *Mammalia* **13**.

Baumgartel, W. (1959). The last British gorillas. *The Geographical Magazine* **32**, 1.

Beatty, E. H. (1951). A note on the behaviour of the chimpanzee. *J. Mammal.* **32** (1).

Bingham, H. C. (1932). "Gorillas in a native habitat." Carnegie Inst. Washington Publ. 426.

Blackwell, K. (1969). Rearing and breeding of Demidoff's galagos. *Int. Zoo Yearbook* **9**.

Blackwell, K. and Menzies, G. (1968). Observations of pottos. *Int. Zoo Yearbook* **8**.

Blake, P. A. (1967). Some aspects of blue monkey social organization. *Proc. E. Afr. Acad.*

Blake, P. A. (1968). A free-ranging hybrid, *Cercopithecus mitis stuhlmanni-ascanius schmidti. Folia Primat.* **9**.

Bolwig, N. (1959). A study of the behaviour of the chacma baboon. *Behaviour* **14**.

Bolwig, N. (1962). Behaviour of patas monkey, *Erythrocebus patas. Behaviour* **21**.

Booth, A. H. (1954). A note on the colobus monkeys of the Gold and Ivory Coast. *Ann. Mag. Nat. Hist.* **7** (12).

Booth, A. H. (1955). Speciation in the Mona monkeys. *J. Mammal.* **36**.

Booth, A. H. (1956a). The *Cercopithecidae* of the Gold and Ivory Coast. *Ann. Mag. Nat. Hist.* **9** (12).

Booth, A. H. (1956b). The distribution of primates in the Gold Coast. *J. W. Afr. Sci. Ass.* (2).

Booth, A. H. (1958). The zoogeography of W. African primates. *Bull. Inst. Fr. Afr. Noire* **102**.

Bosman, Van W. (1704). "Beschrijving van de Guineze Goudkust." Utrecht.

Brain, C. K. (1965). Observations on the behaviour of the vervet monkey (*C. aethiops*). *Zoologica Africana* (1).

Bueltner, Janusch J. and Andrew, R. J. (1962). The use of incisors by primates in grooming. *Ann. J. Phys. Anthrop.* **20**.

Bueltner, Janusch J. (1963). An introduction to the primates. *In* "Evolutionary and genetic biology of the primates," Vol. I. Academic Press, New York and London.

Bueltner, Janusch J. (1964). The breeding of galagos in captivity and some notes on their behaviour. *Folia Primat.* **2**.

Bugher, J. C. (1951). The mammalian host in yellow fever. *In* "Yellow Fever" (Strode, G. K., ed.). McGraw Hill, New York.

Butler, H. (1960). Some notes on the breeding cycle of the Senegal galago (*G. senegalensis*) in the Sudan. *Proc. Zool. Soc., London* **135**

Butler, H. (1967). Seasonal breeding of the Senegal galago (*G. senegalensis*) in the Nuba Mts., Sudan. *Folia Primat.* **5**.

Buxton, A. P. (1951). Observations on the night resting habits of monkeys in a small area on the edge of the Semliki Forest, Uganda. *J. Anim. Ecol.* **20**.

Cansdale, G. S. (1944). *Galago demidovii. J. Soc. Preserv. Fauna Emp.* **50**.

Carpenter, R. (1942). Societies of monkeys and apes. *Biol. Symp. New York*.

Chalmers, N. (1968). The social behaviour of free-living mangabeys in Uganda. *Folia Primat.* **8**.

Chalmers, N. (1968). Group composition, ecology and daily activities of free-living mangabeys in Uganda. *Folia Primat.* **8**.

Chalmers, N. (1968). The visual and vocal communication of free-living mangabeys in Uganda. *Folia Primat.* **9**.

Chapin, J. P. (1925). The crowned eagle, ogre of Africa's monkeys. *Nat. Hist. N.Y.* **25**.

Clarke, W. E. Le Gros (1959). "The antecedents of man." Edinburgh University Press.

Cowgill, U. M. (1964). Visiting in *Perodicticus. Science* **146**.

Cowgill, U. M. (1968). Displacement behaviour in *Perodicticus. Folia Primat.* **8**.

Dandelot, P. (1959). Note sur la classification des Cercopitheques du groupe *aethiops. Mammalia* **23**.

Dandelot (1962). Confrontation de Cercopithecus avec son habitat. *C. R. Soc. Biogeogr.* **343**.

Dart, R. A. (1963). Carnivorous propensities of baboons. *In* "The primates." Zoological Society Symposium, No. 10. Academic Press, London and New York.

Darwin, C. (1872). "The expression of the emotions in man and animals." London.

Dekeyser, P. L. (1955). "Les mammifères de l'Afrique noire francaise." Inst. Fr. Afr. Noire.

De Vore, I. (1963). "Baboon ecology and human evolution." *In* "Classification and human evolution" (Washburn, S. L., ed.), Viking Publications in Anthropology, No. 37. New York.

Dobroruka, J. (1966). Zur Artbildung der Mangaben Gattung. *Rev. Zool. Bot. Afr., Brussels.*

Dobroruka, J. (1966). Zur Farbung der Brazza. *Saugetier kundliche mitteilungen.*

Dominique, P. C. (1966). Notes sur le potto *Perodicticus potto. Biologia Gabonica* **2**, **4**.

Donisthorpe, J. (1958). A pilot study of the mountain gorilla, *G.g. beringei* in Southwest Uganda. *S. Afr. J. Sci.* (54).

Doyle, G. A., Pelletier, A. and Bekker, T. Courtship, mating and parturition in the lesser bushbaby (*G. senegalensis*). *Folia Primat.* **7**.

Durrell, G. (1954). "The Bafut Beagles." London.

Eimerl, S. and De Vore, I. (1966). "The primates." Time-Life, New York.

Elliot, D. G. (1913). "A review of the primates." Amer. Mus. Nat. Hist. N.Y.

Fraser, A. F. (1968). "Reproductive behaviour in Ungulates." Academic Press, London and New York.

Freedman, L. (1962). Growth of muzzle length relative to calvaria length in *Papio. Growth* **26**.

Geidion, S. (1962). "The eternal present." Oxford University Press, London.

Gerard, P. (1929). Contribution a l'étude de la placentation chez les lemuriens propos d'une anomalie de la placentation chez *Galago demidoffi. Arch. Anat. microsp.* **25**.

Gerard, P. (1932). Etudes sur l'ovogenese et l'ontogenese chez les lemuriens du genre *Galago. Arch. Biol. (Belg.)* **43**.

Goodall, G. (1965). Chimpanzees of the Gombe Stream Reserve. "Primate behaviour" (De Vore, I., ed.), Holt Rinehart and Winston, New York.

Goodall, J. (1963). Feeding behaviour of wild chimpanzees. Zoological Society Symposium No. 10.

Goodman, M. (1963). Man's place in the phylogeny of the primates as reflected in serum proteins. *In* "Classification and human evolution" (Washburn, S. L., ed.). Viking Publications in Anthropology, No. 37, New York.

Grassé, P. P. (ed.) (1955). "Traité de zoologie", Vol. 17. "Mammifères systématique." Masson et Cie, Paris.

Haddow, A. J. (1952). Field and laboratory studies on an African monkey (*Cercopithecus ascanius*) *schmidti. Proc. Zool. Soc., London* **122**.

Haddow, A. J. (1956). Blue monkey group in Uganda. *Uganda Wildl. and Sport* (1).

Haddow, A. J., Smithburn, K. C., Mahaffy, A. F. and Bugher, S. C. (1947). Monkeys in relation to epidemiology of yellow fever in Bwamba County, Uganda. *Trans. Roy. Soc. Trop. Med. Hyg.* (40).

Haddow, A. J. and Ellice, J. M. (1964). Studies on bushbabies (*Galago* spp.) with special reference to the epidemiology of yellow fever. *Trans. Roy. Soc. Trop. Med. Hyg.* **58**.

Hall, K. R. L. and Gartlan, J. S. (1965). Ecology and behaviour of the vervet monkey (*C. aethiops*) Lolui Island, Lake Victoria. *Proc. Zool. Soc., London* (145).

Hall, K. R. L. (1966). Behaviour and ecology of the wild patas monkey (*Erythrocebus patas*) in Uganda. *J. Zool., London* **148**.

Hall Craggs, E. C. B. (1965). An osteometric study of the hindlimb of the Galagidae. *J. Anat., London* **99**.

Hall Craggs, E. C. B. (1965). An analysis of the jump of the lesser galago (*G. senegalensis*). *J. Zool.* **147**.

Hartley, E. G. (1966). "B" virus; Herpes virus simiae. *Lancet* **i**.

Hayman, R. W. (1937). A note on *Galago senegalensis inustus. Ann. Mag. Nat. Hist.* **20**.

Hayman, R. W. (1950). Guerezas or Colobus monkeys. *Zoo Life* No. **5**, 3.

Hill, J. E. and Carter, T. D. (1941). The mammals of Angola, Africa. *Bull. Ann. Mus. Nat. Hist.* **78**.

Hill, W. C. Osman (1953—1966). Primates. *In* "Comparative anatomy and taxonomy." Edinburgh University Press.

Hofer, H. (1957). Uber die Bewegungen des Potto. *Natur. Volk.* **87**.

Hollister, N. (1924). East African Mammals in the U.S. Natural History Museum. *Bull. U.S. Nat. Hist. Mus.* **99**.

Honigmann, H. (1936). On the speed of digestion in some mammals. *Proc. Zool. Soc., London.*

Hoof, J. A. R. Van (1962). Facial expressions in higher primates. *In* "Evolutionary aspects of animal communication and imprinting and early learning." Zoological Society Symposium No. 8. Academic Press, London and New York.

Huxley, J. S. (1932). "Problems of relative growth." Methuen, London.

International Zoo Yearbook (1959—63) (Jarvis, C., ed.). Vols. I—V. Zoo Soc., London.

I.U.C.N. (1964). Animals and plants threatened with extinction. Survival Commission of I.U.C.N.

Itani, J. (1966). Social organization of chimpanzees. *Shizen* **21** (8).

Itani, J. and Suzuki, A. (1967). The social unit of chimpanzees. *Primates* **8** (4).

Izawa and Itani, J. (1966). Chimpanzees in Kasakati basin, Tanganyika. (1) Ecological study in the rainy season. *Kyoto Univ. Afr. Studies* **1**.

Jolly, C. J. (1963). A suggested case of evolution by sexual selection in primates. *Man* **221**.

Jolly, C. J. (1965). The origins and specializations of the long-faced *Cercopithcoidea*. University of London Ph.D. Thesis.

Jolly, C. J. (1966). Introduction to the *Cercopithecoidea* with notes on their use as laboratory animals. *In* "Some recent developments in comparative medicine." Zoological Society Symposium, No. 17. Academic Press, London and New York.

Jolly, C. J. (1966). The fossil history of the baboons. 2nd Int. Symp. on baboon, San Antonio.

Jones, C. and Sabater, Pij. (1968). Comparative ecology of *Cercocebus albigena* and *Cercocebus torquatus*. *Folia Primat.* **9**.

Kawabe, M. (1966). One observed case of hunting behaviour among wild chimpanzees living in the savanna woodland of Western Tanzania. *Primates* **7** (3).

Kawai, M. and Mizahura, H. (1959). An ecological study of the wild mountain gorilla (*G. g. berengei*). *Primates* **2**.

Keith (1899). On the chimpanzees and their relationship to the gorilla. *Proc. Zool. Soc., London.*

Kellog, W. N. (1933). "The ape and the child." McGraw-Hill, New York and London.

Klinger, H. P., Hamerton, J. L., Mutton, D. and Lang, E. M. (1963). The chromosomes of the hominoidea. *In* "Classification and human evolution" (Washburn, S. L. ed.). Viking Publications in Anthropology, No. 37. New York.

Kohler, W. (1925). "The mentality of apes." Harcourt Brace and World Inc., New York.

Kolar, K. (1965). Einige Mitteilungen uber Fortpflanzung und Jugendentwicklung von *Galago senegalensis*. *Zool. Gart.* **31**.

Kortland, A. (1962). Chimpanzees in the wild. *Scient. Amer.* **206**.

Kortland, A. (1967). Experimentation with chimpanzees in the wild. *In* "Progress in primatology" (Starck, D., ed.). Stuttgard.

Lancaste, J. B. and Lee, R. B. (1965). The annual reproductive cycle in monkeys and apes. *In* "Primate behaviour: field studies of monkeys and apes" 486—511 (De Vore, I., ed.). Holt Rinehart and Winston, New York.

Lawrence, B. and Washburn, S. L. (1936). On a new race of *Galago demidovii*. *Occ. paper. Boston Soc. Nat. Hist.* VIII.

Linn, I. (in press). A Longevity record for *Galago demidovii*.

Loveridge, A. (1951). "Tomorrow's a holiday." R. Hale, London.

Lowther, F. De L. (1939). The feeding and grooming habits of the Galago. *Zoologica* **24**.

Lowther, F. De L. (1940). A study of the activities of a pair of *G. senegalensis moholi* in captivity. *Zoologica* **25**.

Lydekker, R. (1905). Colour evolution in guereza monkeys. *Proc. Zool. Soc., London.*

Malbrandt, R. and Maclatchy, A. (1949). Faune de l'équateur Africain francais. "Mammifères" Vol. II. Lechevalier, Paris.

Marler, P. (1965). Communication in monkeys and apes. *In* "Primate behaviour" (De Vore, I., ed.). Holt Rinehart and Winston, New York.

Matschie, P. (1893). *Galago zanizbaricus. S.B. Ges. naturf. Fr. Berl.* **III**.

Mayr, E. (1963). "Animal species and evolution." Oxford University Press, London.

Merfield, F. G. and Miller, H. (1956). "Gorilla hunter." Ferrer and Strauss, New York.

Miller, G. S. (1931). Human hair and primate patterning. *Smithsonian Misc. Coll.* **85**, 10.

Montagna, W. and Ellis (1959). Observations on the skin of the Potto. *Am. J. Phys. Anthrop.* **17**.

Montagna, W. (1962). The skin of lemurs. *Ann. N.Y. Acad. Sci.* **102**.

Montagna, W. and Yun, I. S. (1962). Further observations on the skin of *Perodicticus potto. Ann. J. Phys. Anthrop.* **20**.

Morris, D. (1962). "Biology of art." Methuen, London.

Morris, D. (ed.) (1967). "Primate ethology." Weidenfeld and Nicholson, London.

Morris, D. (1968). "The naked ape." London University Press.

Morris, R. and Morris, D. (1966). "Men and apes." Hutchinson and Co., London.

Møller, H. (1969). Unpublished summary of field-observations on *C. l'hoesti*.

Napier, J. R. (1964). Evolution of bipedal walking in the hominids. *Arch. Biol.* **75**.

Napier, J. R. and Napier, P. H. (1967). "A handbook of living primates." Academic Press, London, New York.

Neumann, O. (1896). Uber die geografische Verbreitung der Colobusaffen in Ostafrica and deren Lebensweise Sitzungsber. *Ges. Naturfr., Berlin*.

Nishida, T. (1967). Savanna living chimpanzees. *Shizen* **22** (8).

Nissen, H. W. (1931). A field study of the chimpanzee. *Comp. Psych. Monograph*.

Osborn, R. M. (1963). Behaviour of the mountain gorilla. *In* "The Primates." Zoological Society Symposia, No. 10, Academic Press, London and New York.

Pitman, C. R. S. (1935). The gorillas of the Kayonza Region. *Proc. Zool. Soc., London*.

Pitman, C. R. S. (1954). The influence of the Belgian Congo on the distribution of Uganda's primates and some of their characteristics. *Ann. Mus. Roy. Congo Belge Zool.* **1**. Miscellanea Zoologica, Tervuren.

Pilbeam, D. and Walker, A. (1968). A new cercopithecoid tooth from Napak, Uganda. *Nature* **220**.

Pocock, R. I. (1907). A monographic revision of the monkeys of the genus *Carcopithecus. Proc. Zool. Soc., London* **677**.

Primate Records. E.A. Inst. Virus Research, Entebbe.

Rahm, U. (1960). Quelques notes sur le Potto de Bosman. *Bull. Inst. Fr. Afr. Noire* **22**.

Rahm, U. and Christiaensen, A. (1963). Les mammifères de la règion occidentale du lac Kivu. *Mus. Roy. Afr. Centr., Tervuren.*

Raven, H. C. (1950). The anatomy of the gorilla. Columbia University Press, New York.

Read, H. (1936). "Art and society." Faber and Faber, London.

Reynolds, V. and F. (1965). Chimpanzees of the Budongo forest. *In* "Primate behaviour" (De Vore, I., ed.). Holt, Rinehart and Winston, New York.

Rode, P. (1936). Sur quelques caractères differentiels de la tête osseuse des Cercopitheques et des Cercocebes. *Mammalia* **1**.

Rode, P. (1937). "Les primates de l'Afrique." Librarie Larose, Paris.

Rode, P. (1938). Considerations sur la systématique des simiens africains. *Mammalia* **2**.

Ross, R. W. (1953). Studies on bushbabies (*Galago* spp.) experimentally infected with yellow fever virus. *Rep. Virus Res. Inst., Entebbe* No. 3.

Rowell, T. E. (1966). Hierarchy in the organization of a captive baboon group. *Anim. Behaviour* **14**; 430—433.

Rowell, T. E. (1966). Forest living baboons in Uganda. *J. Zool., London* **149**.

Rowell, T. E. (1967). Variability in the social organization of primates. *In* "Primate ethology" (Morris, D., ed.). Weidenfeld and Nicholson, London.

Sahlins and Elman, R. (1960). "Evolution and culture." University of Michigan Press, Ann Arbor.

Sanderson, I. T. (1940). The mammals of the North Cameroons forest area. *Trans. Zool. Soc., London* **24**.

Sanderson, I. T. (1957). "The monkey kingdom." Hamish Hamilton, London.

Sauer, E. G. F. and Sauer, E. M. (1963). The S.W. African bushbaby of the *G. senegalensis* group. *J. S.W. Afr. Scient. Soc.* **16**.

Schaller, G. B. (1963). The mountain gorilla. "Ecology and behaviour." University of Chicago Press, Chicago.

Schenkel, R. and L. Schenkel (in press). On sociology of free-ranging Colobus (*C. guereza caudatus*).

Schouteden, H. (1947). De Zoogdieren van Belgish-Congo. *Ann. Mus. Roy. Congo Belge* **3**, Ser. 2.

Schultz, A. H. (1963). Age changes, sex differences and variability as factors in the classification of primates. *In* "Classification and human evolution" (Washburn, S. L., ed.). Viking Publications in Anthropology, No. 37. New York.

Schwartz, E. (1927). Erythrism in monkeys of the genus *Cercopithecus*. *Ann. Mag. Nat. Hist.* **9** (19).

Schwartz, E. (1928a). The species of the genus *Cercocebus*. *Ann. Mag. Nat. Hist.* **10**, 1.

Schwartz, E. (1928b). Notes on the classifications of the African monkeys of the genus *Cercopithecus*. *Ann. Mag. Nat. Hist.* **10**, 1.

Schwartz, E. (1929). On the local races and distribution of the Black and White Colobus monkeys. *Proc. Zool. Soc., London.*

Schwartz, E. (1930). *Galago senegalensis inustus. Rev. Zool. Bot. Afr.* **19**.

Schwartz, E. (1931a). On the African long-tailed lemurs or galagos. *Ann. Mag. Nat. Hist.* **10**, 7.

Scott, J. (1963). Factors determining skull form in primates. *In* "The Primates." Zoological Society Symposium No. 10. Academic Press, London and New York.

Simonds, P. E. (1965). The bonnet macaque in S. India. "Primate behaviour" (De Vore, I., ed.). Holt, Rinehart and Winston, New York.

Simons, E. L. (1965). New fossil apes from Egypt. *Nature* **205**.

Simons, E. C. (1967). Early Cenozoic mammalian faunas Fayum. *Peabody Mus. N.H., Yale. Bull.* No. 38.

Simons, E. I. and Pilbeam, D. R. (1965). Preliminary revision of the *Dryopithecinae. Folia Primat.* **3**.

Simpson, D. I. H. (1965). Resistance of *Galago demidovii thomasi* to infection with yellow fever virus. *Ann. Trop. Med. Parasit.* **59**, No. 4.

Smithburn, K. C. and Haddow, A. J. (1949). The susceptibility of African wild animals to yellow fever. I. Monkeys. *Amer. J. Trop. Med.* **29**.

Snow, C. C. and Vice, T. (1965). Organ weight allometry and sexual dimorphism in the olive baboon (*Papio anubis*). *In* "The baboon in medical research" (Vagtborg, H., ed.). Austin University of Texas Press.

Starck, D. and Frick, H. (1958). Beobachtungen am athiop Primaten. *Zool. Jahrb. Abt. f. Systemat. Okol. Geogr.* **86**.

Stott, K. (1960). A note on *Cercopithecus mitis. J. Mammal.* **41**.

Straus, W. L. and Wislocki, G. B. (1932). On certain similarities of the sloth and slow lemur. *Bull. Mus. Comp. Zool., Harvard* **74**.

Struhsaker, T. T. (1966). Population dynamics of vervet monkey (*C. aethiops*). *In* "Social communications among primates." University of Chicago Press, Chicago.

Struhsaker, T. T. (1966). Vocal communication by vervet monkeys. *In* "Social communication among primates." University of Chicago Press, Chicago.

Struhsaker, T. T. (1967). Behaviour of the vervet monkey (*C. aethiops*). *Univ. Calif. Publ. Zool.* **82**.

Suckling, J. A., Suckling, E. E. and Walker, A. (1969). Suggested function of the vascular bundles in the limbs of *Perodicticus potto. Nature* **221**, No. 5178.

Sugiyama, Y. (1967). Forest-living chimpanzees. *Shizen* **22** (8).

Sugiyama, Y. (1968). Social organization of chimpanzees in the Budongo Forest, Uganda. *Primates* **9**.

Sugiyama, Y. (1969). Social behaviour of chimpanzees in the Budongo Forest, Uganda. *Primates* **10**.

Suzuki, A. (1969). An ecological study of chimpanzees living in savanna woodland. *Primates* **10** (2).

Tappen, N. C. (1960). Distribution of African monkeys. *Curr. Anthropology* **1**.

Tappen, N. C. (1963). Genetics and systematics in the study of primate evolution. Zoological Society Symposium, No. 10.

Thompson, D'Arcy W. (1942). "On growth and form." Cambridge University Press, London.

Uganda Game Dept. Annual Reports. Government Printer, Entebbe.

Ullrich, Von W. (1956). Das Verhalten des Schimpanzen, *Pan troglodytes* beimsprung. *Saugetierkundl. Mitt.* **2**.

427

Ullrich, Von W. (1961). Zur Biologie and Soziologie der Colobus Affen (*Colobus guereza caudatus*). *Der Zool. Gart.* Band **25** Heft 6.

Vagtborg, H. (1965). "The baboon in medical research." Aust. Univ. of Texas.

Vallois, H. V. (1955). Ordre des primates. *In* "Traité de zoologie," tome XVII (Grassé, P., ed.). Mammifères systématique, Masson, Paris.

Van Lawick-Goodall, J. (1967). "Mother-offspring relationships in free-ranging chimpanzees" (Morris, D., ed.). Weidenfeld and Nicholson, London.

Verheyen, W. N. (1959). Summary of the results of a craniological study of the African primate genera *Colobus* and *Cercopithecus*. *Rev. Zool. Bot. Afr.*

Verheyen, W. N. (1962). Contribution à la craniologie comparée des primates. *Mus. Roy. Afr. Centr., Tervuren. Ser. 8. Sci. Zool.* **105**.

Wade, P. (1958). Breeding seasons among mammals in the lowland rain forest of north Borneo. *J. Mammal.* **39**.

Washburn, S. L. (ed.) (1963). "Classification and human evolution." Viking Fund Publs. in Anthrop. No. 37. New York.

Washburn, S. L. and De Vore, I. (1961). The social life of baboons. *Scient. Amer.* **204**.

Washburn, S. L., Jay, P. C. and Lancaster, J. B. (1965). Field studies of old world monkeys and apes. *Science, N.Y.* **150**.

Wickler, W. (1967). Socio-sexual signals and their intra-specific imitation among primates. *In* "Primate ethology" (Morris, D., ed.). Weidenfeld and Nicholson, London.

Yasuda, J., Aoki, A. and Montagna, W. (1961). Observations on the skin of galagos. *Amer. J. Phys. Anthrop.* No. 19.

Yerkes, R. M. and Yerkes, A. W. (1929). "The great apes." Yale University Press.

Zuckerman, S. (1926). Growth changes in the skull of the baboon. *Proc. Zool. Soc.* **3**.

Zuckerman, S. (1932). "The social life of monkeys and apes." London.

Hyrax

Brauer, A. (1913). Zur Kenntnis des Gebisses von *Procavia*. Sonder-Abdruck aus den Sitzungsberichten der Gesellschaft Naturforschender Freunde. Berlin. No. 2.

Bothma, J. Du P. (1966). *Hyracoidea. In* "Preliminary identification manual for African mammals." Smithsonian Institution, Washington.

Coe, M. J. (1962). Notes on the habits of the Mt Kenya hyrax, *Procavia johnstonii mackinderi*. *Proc. Zool. Soc., London* **138**.

Coetzee, C. G. (1966). The relative position of the penis in S.A. dassies as a character of taxonomic importance. *Zoologica Africana*.

Fox, H. (1933). Some notes upon the nature, health and maintenance of the Hyrax. *Proc. Amer. Phil. Soc.* **72**.

Gray, J. E. (1933). Revision of the species of hyrax founded on the specimens in the British Museum. *Ann. Mag. Nat. Hist.* **4**, 1.

Hahn, H. (1934). Die Familie der *Procaviidae. Zeitschr. f. Saugetierk* **9**.

Hanse (1962). Preliminary study of *Procavia* in relation to farming. Rep. Deb. Nat. Conserv., Cape Town.

Hatt, R. T. (1933). An annotated catalogue of the *Hyracoidea* in the American Museum of Natural History with a description of a new species from the Lower Congo. *Amer. Mus. Novitates* **594**. 1.

Hatt, R. T. (1936). The hyraxes collected by the Amer. Mus. Congo Expedition. *Bull. Amer. Mus. Nat. Hist.* **72**.

O'Donohue, P. N. (1963). Reproduction in the female hyrax, *Dendrohyrax arboreus ruwenzorii. Proc. Zool. Soc., London* **141**.

Rahm, U. (1957). Der Baum oder, *Dendrohyrax dorsalis. Zool. Garten Lpz.*

Richard, P. B. (1964). Notes sur la biologie dudaman des arbres *Dendrohyrax dorsalis. Biologia Gabonica* **I**, 1.

Sale, J. B. (1965). The *Hyracoidea* : A review of the systematic position and biology of the hyrax. *J. E. Afr. Nat. Hist. Soc.* **23**. No. 5.

Sale, J. B. (1965). The feeding behaviour of rock hyraxes in Kenya. *E. Afr. Wildl. J.* **3**.

Sale, J. B. (1965). Observations on parturition and related phenomena in the hyrax. *Acta Tropica* **22** (1).

Sale, J. B. (1965). Hyrax feeding on a poisonous plant. *E. Afr. Wildl. J.* **3**.

Sale, J. B. (1966). Habitat of rock hyrax. *J. E. Afr. Nat. Hist. Soc.* **25** (3).

Sale, J. B. (1966). Daily food consumption and mode of ingestion of hyrax. *J. E. Afr. Nat. Hist. Soc.* **25** (3).

Sale, J. B. (1969). Breeding in rock hyrax. *In* "Biology of reproduction in mammals." Blackwell, Oxford.

Thomas, O. (1892). On the species of the *Hyracoidea. Proc. Zool. Soc., London.*

True, F. W. (1890). Description of two new species of mammals from Mt Kilimanjaro, E.A. *Proc. U.S. Nat. Mus.* **13**.

Turner, M. I. M. and Watson, R. M. (1965). An introductory study on the ecology of hyrax, *Dendrohyrax brucei* and *Procavia johnstonii*, in the Serengeti National Park. *E. Afr. Wildl. J.* No. 3.

Van Der Horst, C. J. (1941). On the size of the litter and the gestation period of *Procavia capensis. Science, N.Y.* **93**.

Wislocki, G. B. and Van Der Westhuysen, O. P. (1940). The placentation of *Procavia capensis* with a discussion of the placental affinities of the *Hyracoidea. Contr. Embryol.* **28** (171).

Orycteropus

Anthony, R. L. F. (1934). Donnée nouvelle sur l'évolution de la morphologie dentaire et cranienne des Tubulidentata (Orycteropes). *Bull. Soc. Zool.* **59**. Paris.

Bequaert, J. (1922). The predaceous enemies of ants. *Bull. Amer. Mus. Nat. Hist.* **65**.

Bigourdan, J. (1950). Sur quelques caractères et habitudes de l'orycterope. *Premiere conf. Int. Ouest* **I**.

Bridges (1958). The aardvark. *In* "Animal Kingdom." New York Zool. Soc. (Feb.).

Broom, R. (1909). On the milk dentition of Orycteropus. *Ann. S. Afr. Mus., Cape Town,* **V.**

Colbert, E. H. (1941). A study of *Orycteropus gaudreyi* from the island of Samos. *Bull. Am. Mus. Nat. Hist.*

Derscheid, J. M. (1925). Un Orycterope de l'Aruwimi. *Rev. Zool. Afr.* **13**; *Bull. Cercle Zool. Congolais* **2** (1).

Foran, I. (1961). The aardvark. *Wildlife.* Sept.—Dec. issue, Nairobi.

Frechkop, S. (1937). Sur les extrémités de l'orycterope. *Bull. Mus. Roy. Hist. Nat. Belg.* **13.**

Fossati, L. (1937). Abitudini dell'Oritteropo d'Eritrea. Notizie e varieta. *Natura* **28** (1).

Gray, J. E. (1865). Revision of the genera and species of entomophagous Edentata founded on the examination of the specimens in the British Museum. *Proc. Zool. Soc., London.*

Hatt, R. T. (1932). "The aardvark of the Haut-Uele." American Museum Novitates No. 535.

Hatt, R. T. (1934). The pangolins and aardvarks collected by the American Museum Congo Expedition. *Bull. Am. Mus. Nat. Hist.* **66.**

Hediger (1951). Observations sur la psychologie animale dans les Parcs Nationaux du Congo Belge. Inst. des Parcs Nat. du Congo Belge.

Jaeger, H. F. (1913). "Anatomische Untersuchungen des Orycteropus." Stuttgart.

Mitchell, T. (1965). *Puku Occ. Papers* **3.**

Pocock, R. I. (1924). External characters of *Orycteropus afer. Proc. Zool. Soc., London* **697.**

Sampsell, R. (1969). Handrearing an aardvark. *Int. Zoo Yearbook* **9.**

Sonntag, C. F., Woollard, H. H. and Clark, W. E. le Gros (1925—1926). A monograph of *Orycteropus afer. Proc. Zool. Soc., London.*

Thomas, O. (1890). A milk dentition in Orycteropus. *Proc. Roy. Soc., London* **67.**

Verheyen, B. (1951). Contribution a l'étude éthologique des mammifères du parc national de l'Upemba. Inst. des Parcs Nat. du Congo Belge.

Verschuren, J. (1958). Ecologie et Biologie des grands mammifères. *Explorat. Parc. Natn. Garamba.*

Pangolin

Albrecht, P. (1883). Note sur le pelvisternum des Edentates. *Bull. Acad. Roy. Belg. VI* **16,** 5.

Anthony, R. (1919). Anatomie de la queue des pangolins. *Bull. Mus. Nat. Hist., Paris.*

Beebe, C. W. (1914). The pangolin or scaly ant-eater. *Zool. Soc. Bull., New York* **17.**

Bequaert, J. (1922). The predaceous enemies of ants. *Bull. Amer. Mus. Nat. Hist.* **45.**

Cansdale, G. S. (1947). West African tree pangolins. *Zoo Life, London* **4.**

Dekeyser, P. L. (1953). Les Pangolins. *Notes Afr.* **57.**

Ehlers, E. (1894). Der Processus Xiphoideus und seine Muskulatur von *Manis* spp. *Zool. Misc. Gottingen.* **34.**

Eisentraut, M. (1957). Weisbauch-schuppentier *Manis tricuspis. Zool. Garten.* **23**.

Frechkop, S. La locomotion et la structure des Tatous et des Pangolins. *Ann. Soc. Zool. Belg.* **80**.

Hatt, R. T. (1934). The pangolins and aardvarks collected by the Amer. Mus. Congo Expedition. *Bull. Amer. Mus. Nat. Hist.* **66**.

Jouffroy, F. K. (1966). Musculature de l'avant-bras et de la main, de la jambe et du pied chez *Manis gigantea. Biologia Gabonica* **2**, 3.

Krieg, H. and Rahm, U. (1961). Das Verhalten der Schuppentiere (*Pholidota*). *Handb. Zool.* **8**.

Malbrant, R. and Maclatchy, A. (1949). "Faune de l'équateur Africain francais." "Mammifères" Vol. 2. Lechavalier, Paris.

Mohr, E. (1961). Schuppentiere. Die Neue Brehm Bucherei.

Pages, E. (1965). Les pangolins du Gabon. *Biologia Gabonica* **1**, 3.

Pocock, R. I. (1924). The external characters of the pangolins (*Manidae*). *Proc. Zool. Soc., London.*

Rahm, U. (1955). Beobachtungen an den Schuppentieren. *Rev. Suisse Zool.* **62**.

Rahm, U. (1957). Notes on the pangolins of the Ivory Coast. *J. Mammal* **37**.

Rahm, U. (1960). The pangolins of West and Central Africa. *African Wildl.* **14**.

Schouteden, H. (1930). Les Pangolins. *Rev. Zool. Bot. Africaine* **18** (4). *Bull. Cercle Zool. Congolais* (6) **3**.

Sikes, S. K. (1962). Pangolins. *African Wildlife* **16**.

Sweeney, R. C. H. (1956). Notes on *Manis temmincki. Ann. Mag. Nat. Hist., London.*

Tims, H. W. M. (1908). Tooth vestiges and associated mouth parts in the Manidae. *J. Anat. Physiol.* **42**, 375—387.

Vincent, F. (1964). Quelques observations sur les pangolins (*Pholidota*). *Mammalia* **28** (4).

Weber, M. (1891). "Beitrag zur Anatomie und Entwickelung des Genus Manis Zoologische Egebnisse eine Reise in Niederlandish Ost Indien." E. Brill. Leiden.

Wright, A. C. A. (1954). The magical importance of Pangolins among the Basukuma. *Tanganyika Notes and Records* No. 36.

Dugong

Annandale, N. (1906). The habits of the dugong. *J. Asiat. Soc., Calcutta,* series 2—1.

Bertram, C. K. R. and Bertram, G. C. L. (1968). Sirenia as aquatic meat-producing herbivores. *In* "Comparative nutrition of wild animals." Zoological Society Symposium, No. 21, pp. 385—391. Academic Press, London.

Bertram, G. C. L. and Bertram, R. C. K. (1966). The dugong. *Nature, Lond.* **209**, 938—939.

Dexler, H. and Eger, O. (1911). Beitrage zur Anatomie des Saugerruckenmarkes I. Halicore Dugong. *Morphol. Jahrb.* **43**.

Dollman, G. (1933). Observations on dugong skulls from Mafia Island. *Proc. Linn. Soc.* (Jan.) and *Nat. Hist. Mag. B.M.* **4**.

Gallus (1961). "Fabulous beasts." Harrap and Co., London.

Gohar, H. A. F. (1957). The Red Sea dugong. *Public. Marine Biol. Stud. Ehardaga* **9**.

Gudernatsch, J. F. (1908). Zur Anatomie und Histologie der Verdanungstraktes von Halicore Dugong. *Morphol. Jahrb.* **37**.

Harry, R. R. (1956). "Eugenie" the dugong mermaid. *Pacific Discovery* **9**, 1.

Harrison, R. J. and King, J. E. (1965). "Marine mammals." Hutchinson, London.

Hill, W. C. O. (1945). Notes on the dissection of two dugongs. *J. Mammal.* **26**.

Hirasaka, K. (1934). The distribution of sirenians in the Pacific. *Proc. 5th Pacific Sco. Congress, Canada, 1933.*

Isaac, F. M. (1968). Marine botany of the Kenya coast: 4 Angiosperms. *J. E. Afr. Nat. Hist. Soc. and Nat. Mus.* **27**, No. 1 (116).

Jarman, P. J. (1961). Status of the dugong in Kenya. *E. Afr. Wildl. J.*

Jones, S. (1959). On a pair of captive dugongs. *J. Mar. Biol. Ass., India* **1** (2).

Jones, S. (1968). Notes on dugong. *Int. Zoo Yearbook.*

MacInnes, I. G. (1951). Australian Fisheries Handbook. Sydney.

Mann, T. (undated). Notes on the dugong. M/S unpublished.

McMillan, L. (1955). The dugong. *Australian Geog. Walkabout Mag.* **21**, 2.

Murie, J. (1874). On the form and structure of the Manatee. *Trans. Zool. Soc., London* **8**.

Owen, R. (1839). On the anatomy of the dugong. *Proc. Zool. Soc., London* **28**.

Petit, G. (1927). Contribution a l'étude de la morphologie externe des sireniens. *Bull. Mus. Nat. Hist.* No. 5, Paris.

Savory, B. (1958). A note on the dugong. *Tanganyika Notes and Records* No. 51.

Schevill and Watkins (1965). Underwater calls of *Trichechus*. *Nature* **205** (4969).

Simpson, G. G. (1932). Fossil sirenians. *Bull. Amer. Mus. Nat. Hist.* **59**.

Simpson, G. G. (1932). Fossil sirenia of Florida and the evolution of the sirenia. *Bull. Amer. Mus. Nat. Hist.* **59**.

Sithu, H. L. A. Aung (1967). A brief note on dugongs. *Int. Zoo Yearbook* **7**.

Steller, G. W. (1751). De bestiis marinis. *Novi comm. Acad. Sci. Imp. Petropolitanae.*

Wendt, H. (1956). "Out of Noahs Ark." Wiedenfeld and Nicholson, London.

Williams, J. H. (1960). "In quest of a mermaid." Rupert Hart Davis, London.

Ziemzen, A. (1957). "Sirenien oder Seekuhe." Verlag Wittenberg.

Systematic Index

forest, 210

H

Haemoschus, 68
  aquaticus, 28, 45
Hare, 54, 105
  rock, 37
  spring, 10, 41, 63, 86, 94, 95
Hartebeeste, 33, 40, 79, 93
  lelwel, 93
Halitherium, 388
Hedgehog, 20, 37, 38, 42, 53, 57
Heliophobius, 33, 93
  argenteocinereus, 81
Heliosciurus, vi
  gambianus, 72
  rufobrachium, 72
  ruwenzorii, 68
Helogale, 38
Herbivores, 40, 41, 55, 60
Herpestes javanicus, 11
  sanguineus, 78
Hesperosiren, 388
Heterocephalus, 38, 62
  glaber, 19
Heterohyrax, 77, 330, 332, 349
  antinae, 331
  brucei, 18, 330, 332, 340 et seq., 345
    bakeri, 341
    diesneri (nyansae), 341
    hindei (albipes), 341
    kempi, 332, 341
    lademanni, 332, 341
    manningi, 332
    munzneri, 341
    prittwitzi, 341
    songeae (frommi), 341
Hipparionid, 59
Hippopotamus, 42
Hippopotamus, 40, 42, 53, 59
  pygmy, 67
Hipposideros, 92
  caffer, 18
  commersoni gigas, 93
Hippotragus niger, 33, 81
Hog, forest, 27
Hominidae, 99
Hominoid, 102, 111
Hominoidea, 99, 101, 102, 140
Homo, 102
  erectus, 107
  sapiens, 99, 107
Horse, 53

hyaena, 61
Hyaena, 41, 44, 53, 59, 92, 105
  brown (*see Hyaena brunnea*), 61
  striped (*see Hyaena hyaena*), 61
Hyaenodont, 54, 55
Hybomys univittatus, 73, 74
Hylarnus harrisoni, 45
Hylobates, 116
Hylochoerus, 27
  meinertzhageni, 72
Hylomyscus denniae, 16, 26
  stella, 72
Hyracoid, 54, 55
Hyracoidea, 37, 329 et seq., 388
Hyrax, 18, 26, 28, 37, 53, 68, 74
  Bruce's (*see Heterohyrax brucei*), 340 et seq.
  rock (*see Procavia*, 334 et seq.
  tree (*see Dendrohyrax* sp.), 12, 72, 73, 77, 344 et seq.
  yellow spotted rock (*see Heterohyrax brucei*), 340 et seq.
Hystrix, vi, 105
  africae-australis, 81
  cristata, 81

I

Ichneumia albicauda, 43
Idiurus zenkeri, 45
Impala, 35, 36
Insectivore, 1, 37, 54

J

Jackal, 105
  black-backed (*see Canis mesomelas*), 41, 61

K

Kenyapithecus, 102
Klipspringer, 37, 77, 94, 95
Kongoni, 40
Kudu (*see Strepsiceros strepsiceros*), 19, 36, 61

L

Langur, 57, 149
Lemuroidea, 329
Leopard (*see Felis pardus*), 26, 78, 105, 106, 122, 184, 187, 349
Lepus, 105
  capensis, 16

438

Subject Index

Colouring in cercopithecoids, 203, 206
    et seq., 255
Cultivation mosaic, 50
Cultivation, pressure of, 42, 48, 109, 120

D

Defence, in,
    potto, 285
    tree pangolin, 375
Dentition, of,
    aardvark, 379
    dugong, 391
    hominids, 105
    hyraxes, 329
    monkeys, 142 et seq.
    prosimians, 273
Desert, 38
    deserts of the past, 63
    semi-, 37
Disease, influence on primate distr.
    108, 109, 136, 178, 225, 245
Dispersal routes, 70
Display in gorilla, 122
Dissection, of,
    aardvark, 382
    baboon, 181
    chimpanzee, 130, 131
    dugong, 392, 393
    dugong (head), 396
    giant pangolin, 360, 363
    greater galago, 295
    hyrax, 330, 331
    lesser galago, 320
    mangabey, 191
    neglectus monkey, 200
    patas monkey, 266
    pied colobus, 167
    potto, 279
    red colobus, 158
    tree pangolin, 373
Distribution, of,
    aardvark, 380, 381
    apes, 116
    baboon, 178
    black mangabey, 193
    black and white colobus, 164
    Brazza's monkey, 253
    Bruce's hyrax, 341
    chimpanzee, 127, 129
    crested mangabey, 199
    duikers, 68, 71
    dugong (world), 391

elephant shrew, 66
giant pangolin, 359
gorilla, 120
greater galago, 291, 293
ground pangolin, 365
hartebeeste, 80
l'Hoest's monkey, 229
lesser galago, 317
man, 109
mitis monkey, 235, 250
mona monkey, 250
needle-clawed galago, 325
neglectus monkey, 74
oryx, 63
patas monkey, 263, 265
pied colobus, 152, 164
Potamogoretacea, 391
pottos, 281
pygmy galago, 301
red colobus, 150, 156
redtail monkey, 221, 228
rock hyrax, 335
tree hyrax, 345, 346
tree pangolin, 371
vervet monkey, 211, 213, 214
Zanzibar galago, 309, 313
Distribution, of,
    Cercopithecidae, 53, 207, 310
        in Bwamba, 47
    Manidae (world), 353

E

East Africa,
    climate, 12
East African "overlap" region, v, vi, 8
    topography, 8
Endemic species, 8, 81
Environment, East African, 8
Evolution, of,
    cercopithecidae, 140 et seq.
    cercopithecus, 201 et seq.
    hyrax, 329 et seq.
    man, 105 et seq.
Evolution,
    climatic change, 62
    continental zoogeography, 52
    deserts of the past, 63
    forests of the past, 63
    mountains and refuges, 67
    northern and southern forests, 69

F

Fayum, 55